Risk, Reliability, Uncertainty, and Robustness of Water Resources System

Risk, Reliability, Uncertainty, and Robustness of Water Resources Systems is based on the Third George Kovacs Colloquium organized by the International Hydrological Programme (UNESCO) and the International Association of Hydrological Sciences. Thirty-five leading scientists with international reputations provide state-of-the-art reviews of topical areas of research on water resources systems, including aspects of extreme hydrological events, floods and droughts, water quantity and quality, dams, reservoirs and hydraulic structures, evaluation of sustainability and climate change impacts. In addition to discussing essential challenges and research directions, the book will assist in applying theoretical methods to the solution of practical problems in water resources. The authors represent several disciplines such as hydrology; geography; civil, environmental, and agricultural engineering; forestry; systems sciences; operations research; mathematics; physics and geophysics; ecology; and atmospheric sciences.

This review volume will be valuable for all those with an interest in risk, reliability, uncertainty, and robustness of water resources systems, including graduate students, scientists, consultants, administrators, and practicing hydrologists and water managers.

Janos J. Bogardi is Chief of the Section on Sustainable Water Resources Development and Management, Division of Water Sciences, UNESCO, Paris, France. He was the Chair Professor for Hydraulics, Hydrology and Quantitative Water Resources Management at Wageningen Agricultural University, the Netherlands. He has published more than 140 articles and other publications and has edited three books.

Zbigniew W. Kundzewicz is a Professor and Head of Laboratory of Climate and Water Resources at the Research Centre of Agricultural and Forest Environment, Polish Academy of Sciences, Poznań, Poland. He is also Head of the Water Resources Research Unit at the Potsdam Institute for Climate Impact Research in Potsdam, Germany. He is the author of more than 130 scientific publications, two books (in Polish), and editor or co-editor of six books in English. He is Editor of the International Association of Hydrological Sciences and Editor-in-Chief of *Hydrological Sciences Journal.*

The **International Hydrological Programme** (IHP) was established by the United Nations Educational, Scientific and Cultural Organisation (UNESCO) in 1975 as the successor to the International Hydrological Decade. The long-term goal of the IHP is to advance our understanding of processes occurring in the water cycle and to integrate this knowledge into water resources management. The IHP is the only UN science and educational programme in the field of water resources, and one of its outputs has been a steady stream of technical and information documents aimed at water specialists and decision makers.

The **International Hydrology Series** has been developed by the IHP in collaboration with Cambridge University Press as a major collection of research monographs, synthesis volumes, and graduate texts on the subject of water. Authoritative and international in scope, the various books within the series all contribute to the aims of the IHP in improving scientific and technical knowledge of fresh water processes, in providing research know-how, and in stimulating the responsible management of water resources.

TITLES IN PRINT IN THE SERIES:

M. Bonell, M. M. Hufschmidt, and J. S. Gladwell *Hydrology and Water Management in the Humid Tropics: Hydrological Research Issues and Strategies for Water Management*
Z. W. Kundzewicz *New Uncertainty Concepts in Hydrology*
R. A. Feddes *Spaces and Time Scale Variability and Interdependencies in the Various Hydrological Processes*
J. Gibert, J. Mathieu, and F. Fournier *Groundwater and Surface Water Ecotones: Biological and Hydrological Interactions and Management Options*
G. Dagan and S. Neuman *Subsurface Flow and Transport: A Stochastic Approach*
D. P. Loucks and J. S. Gladwell *Sustainability Criteria for Water Resource Systems*
Jan C. van Dam *Impacts of Climate Change and Climate Variability on Hydrological Regimes*
J. J. Bogardi and Z. W. Kundzewicz *Risk, Reliability, Uncertainty, and Robustness of Water Resources Systems*

INTERNATIONAL HYDROLOGY SERIES

Risk, Reliability, Uncertainty, and Robustness of Water Resources Systems

Edited by

Janos J. Bogardi
*Division of Water Sciences,
UNESCO, Paris, France*

Zbigniew W. Kundzewicz
*Research Centre of Agricultural
and Forest Environment,
Polish Academy of Sciences, Poznań, Poland
and Potsdam Institute for Climate
Impact Research, Potsdam, Germany*

CAMBRIDGE UNIVERSITY PRESS
Cambridge, New York, Melbourne, Madrid, Cape Town, Singapore, São Paulo

Cambridge University Press
The Edinburgh Building, Cambridge CB2 2RU, UK

Published in the United States of America by Cambridge University Press, New York

www.cambridge.org
Information on this title: www.cambridge.org/9780521800365

First published 2002
This digitally printed first paperback version 2005

A catalogue record for this publication is available from the British Library

Library of Congress Cataloguing in Publication data
Risk, reliability, uncertainty, and robustness of water resources systems / edited by
Janos J. Bogardi, Zbigniew W. Kundzewicz.
p. cm. – (International hydrology series)
Includes bibliographical references.
ISBN 0-521-80036-6
1. Hydrology – Statistical methods – Congresses. 2. Uncertainty (Information
theory) – Congresses. I. Bogardi, Janos. II. Kundzewicz, Zbigniew. III. Series.
GB656.2.S7 R57 2001
551.48′072 – dc21 00-065073

ISBN-13 978-0-521-80036-5 hardback
ISBN-10 0-521-80036-6 hardback

ISBN-13 978-0-521-02041-1 paperback
ISBN-10 0-521-02041-7 paperback

In memoriam
Barbara Bogardi, MD
1948–1997

Contents

Contributors

PROFESSOR A. BÁRDOSSY
Institute for Hydraulic Engineering
University of Stuttgart
D-70550 Stuttgart
Germany

DR. B. C. BATES
CSIRO
Land and Water
Private Bag PO
Wembley, Western Australia 6014
Australia

DR. C. J. BELL
Hughes STX Corporation
Hydrological Sciences Branch
NASA Goddard Space Flight Center
Greenbelt, MD 20771
USA

DR. M. J. BENDER
Lyonnaise South East Asia (ASTRAN)
903, 203 Jalan Bukit Bintang
55100 Kuala Lumpur
Malaysia

PROFESSOR JANOS J. BOGARDI
UNESCO, Division of Water Sciences
75732 Paris
France

PROFESSOR L. DUCKSTEIN
System and Industrial Engineering
The University of Arizona
Tucson, AZ 85721
USA

DR. T. FAHMY
Anjou Recherche
92982 La Défense
France

DR. N. M. FENNESSEY
Dept. of Civil and Environmental Engineering
University of Massachusetts
North Dartmouth, MA 02747
USA

DR. D. GATEL
Compagnie Générale des Eaux
92982 La Défense
France

PROFESSOR ADRIAN V. GHEORGHE
Federal Institute of Technology
GI-ETH, Winterthurerstr 190
8057 Zurich
Switzerland

DR. O. GILARD
Division Hydrologie – Hydraulique, Cemagref
3 bis quai Chauveau
69009 Lyon
France

PROFESSOR W. KINZELBACH
Institute of Hydromechanics and Water Resources Engineering
Federal Institute of Technology
ETH Hoenggerbert, CH-8093 Zurich
Switzerland

DR. V. KLEMEŠ
3460 Fulton Road
Victoria, BC V9C 3N2
Canada

PROFESSOR ROMAN KRZYSZTOFOWICZ
Department of Systems Engineering
University of Virginia, Thornton Hall
Charlottesville, VA 22903
USA

PROFESSOR Z. W. KUNDZEWICZ
ZBSRiL PAN
60-809 Poznan
Poland
and PIK, Telegratenberg D14412 Potsdam, Germany

DR. B. J. LENCE
Dept. of Civil Engineering
University of British Columbia
2324 Main Mall
Vancouver, BC V6T 1Z4
Canada

DR. G.-M. LI
Institute of Hydromechanics and Water
Resources Engineering
Federal Institute of Technology
ETH Hoenggerbert, CH-8093 Zurich
Switzerland

PROFESSOR DANIEL P. LOUCKS
Civil and Environmental Engineering
Cornell University
Ithaca, NY 14853
USA

DR. DARKO MILUTIN
AHT International GmbH
Huyssenallee 66–68
D-45001 Essen
Germany

DR. R. J. MOORE
Institute of Hydrology
Wallingford OX10 8BB
UK

PROFESSOR H. P. NACHTNEBEL
Department for Water Management
Hydrology and Hydraulic Engineering
University for Agricultural Sciences (BOKU)
Vienna
Austria

PROFESSOR N. OKADA
Disaster Prevention Research Institute
Kyoto University
Uji Kyoto, 611
Japan

DR. E. PARENT
ENGREF
75014 Paris
France

PROFESSOR ERICH J. PLATE
University of Karlsruhe
Kaiser Str. 12
D-76128 Karlsruhe
Germany

MR. D. E. RUPP
Ph.D. Candidate
Oregon State University
Corvalis, OR
USA

DR. ANDRZES RUSZCZYŃSKI
Department of Management Science and Information
Systems, Rutgers University
94 Rockefeller Rd.
Piscataway, NJ 08854
USA

PROFESSOR URI SHAMIR
Water Research Institute,
Technion Israel Institute of Technology
Haifa Israel

DR. BIJAYA P. SHRESTHA
CH2M Hill
2485 Natomas Park Drive, Suite 600
Sacramento, CA 95833
USA

Dr. R. R. Suresh
Dept. of Civil and Environmental Engineering
Tufts University
Medford, MA 02155
USA

Professor Kuniyoshi Takeuchi
Dept. of Civil and Environmental Engineering
Yamanashi University
Kofu 400
Japan

Dr. A. Tecle
School of Forestry
Northern Arizona University
Flagstaff, AZ 86011-5018
USA

Mr. J. F. Thomas
Resource Economic Unit, Perth
35 Union Street
Subiaco, Western Australia 6008
Australia

Professor. S. Vassolo
Institute of Hydromechanics and Water
Resources Engineering
Federal Institute of Technology
ETH Hoenggerbert, CH-8093 Zurich
Switzerland

Professor R. M. Vogel
Dept. of Civil and Environmental Engineering
Tufts University
Medford, MA 02155
USA

Professor B. C. Yen
University of Illinois at Urbana-Champaign
205 N. Mathews Ave.
Urbana, IL 61801
USA

1 Introduction

JANOS J. BOGARDI* AND ZBIGNIEW W. KUNDZEWICZ**

We are pleased to offer the reader a volume consisting of contributions of the Third George Kovacs Colloquium held in UNESCO, Paris from September 19 to 21, 1996. It is a continuation of a series of biannual international scientific meetings organized jointly under the auspices of the International Hydrological Programme (IHP) of UNESCO and the International Association of Hydrological Sciences (IAHS) in the challenging fields of water resources research. These meetings commemorate the late Professor George Kovacs, established authority in hydrology, who paid valuable service to both organizations convening this Colloquium. Professor Kovacs was Chairman of the Intergovernmental Council of the IHP of UNESCO and President of the IAHS.

The theme of the Colloquium, "Risk, Reliability, Uncertainty, and Robustness of Water Resources Systems," denotes- an essential recent growth area of research into water resources, with challenges and difficulties galore. The two-and-a-half-day Colloquium included twenty-four oral presentations covering a broad range of scientific issues. It dealt with different facets of uncertainty in hydrology and water resources and with several aspects of risk, reliability, and robustness.

The contributions to the Colloquium concentrated on the state-of-the-art approaches to the inherent problems. They also outline the possible future, identify challenging prospects for further research and applications. Presentations included both theoretical and applied studies, while several papers dealt with regional problems. Methodological contributions focused on underlying concepts and theories.

The presentations at the Colloquium, based on invitation, were delivered in three categories: keynote lectures, invited lectures, and young scientists' communications. Contributions belonging to all three categories are included in this volume.

Uncertainty means absence or scarcity of information on prior probabilities, that is, a situation when very little is known for sure. The term "risk" is usually used in a situation when it is possible to evaluate the probability of outcomes. Uncertainty in water resources may have different sources. It may result from the natural complexity and variability of hydrological systems and processes and from our inability to understand them. On the other hand, human behavior itself has a strong uncertainty component. Priorities, preferences, and judgment of consequences of future societies are largely uncertain: unexpected shifts and unpredictable changes cannot be ruled out.

Among the plethora of uncertainties in hydrology and water resources research, one thing is certain: water-related problems have grown and will continue to grow in a world of high population growth, with consequences such as increased need for food and justified aspirations of nations and individuals of better living conditions. The present convenors feel that, in most cases, contributions to the Kovacs Colloquium presented research that was problem-driven rather than method-driven.

The problems tackled at the Colloquium vary considerably as to the degree of complexity, from high degrees of sophistication to down-to-earth approaches. These latter call for explanatory data analysis in the sense of meticulous work: trying to unveil patterns existing in raw data and decipher the message that the raw data may carry. While studying uncertainty, one often starts from the most essential points and disregards the rest. In order to build up a more general, rough image, it is necessary to forget things of lesser importance, to neglect small impacts. Further, it is prerequisite to identify the bottleneck, that is, the weakest link in the system. Improvements of better parts may not help much if the weakest link in the chain still exists.

Von Neumann once said that the incremental value of information is highest if the existing information is at its lowest level. The real problems encountered are so complex and burning at the same time that a rough, approximate solution is very welcome. The time for refining detail may follow.

* Division of Water Sciences, UNESCO, Paris, France.
** Research Centre of Agricultural and Forest Environment, Polish Academy of Sciences, Poznań, Poland and Potsdam Institute for Climate Impact Research, Potsdam, Germany.

In a number of contributions to the Colloquium, common sense rules were reiterated. Even if calls to be careful with extrapolations, to avoid abusing methods beyond their legitimate applicability, to check whether the simplified assumptions hold, may seem trivial, experience shows that many offenses against these rules still do occur.

The idea of re-initialization was considered by the convenors several times during this Colloquium. This is a notion from the realm of automatic optimization, where a minimum (or maximum) of a complicated, multidimensional functional is being sought. Taking a gradient as the direction of search is fine, but there may exist better directions auguring a faster convergence. Such directions determined by possibly sophisticated algorithms work well at the beginning but may gradually degenerate and deteriorate their performance. The notion of re-initialization means that after some number of iterations, one should go back to common sense, back to basics. Abandoning a complicated way of determining direction of search, one goes back to the good (even if not the best) and safe gradient direction.

There is a strong need for a holistic, cradle-to-grave, perspective and an interdisciplinary approach to solve complex problems. Participants of the Colloquium reached far beyond classical hydrology and water resources research. They considered pre- and posthydrological components in the interdisciplinary chain. Examples of prehydrological components include meteorology and climatology, while the posthydrological ones are social, psychological, institutional implications.

A statement has been issued that availability of a perfect long-term drought forecast in several areas would not solve all the problems since many countries or regions are not ready to use the information and to prepare effectively for a drought. The infrastructure and institutions are inadequate, and widespread illiteracy is a significant barrier.

This book is organized around seven thematic clusters. Following this introduction are two contributions devoted to underlying concepts and notions. Gheorghe's integrated regional risk assessment and safety management refer to a much broader, environmental perspective than water context alone. In his philosophical contribution, Klemes warns mathematical modelers that sometimes their attempts may be described as "the unbearable cleverness of bluffing."

The largest theme, comprising five contributions, is related to floods and droughts. Many facets of risk are inherent in floods. There is a social risk: How to build the flood defense system? How to place the compromise between the will of providing adequate protection and limited resources? To what flood frequencies should the defenses be sufficient? There is also a personal risk: How to behave in case of floods? How to react to forecasts?

Moore presents aspects of the existing flood protection system in the United Kingdom that relate to the theme of the Colloquium. Krzysztofowicz advocates the advantages of probabilistic hydrometeorological forecasting giving the clients significantly more information than lumped yet unreliable figures. Gillard presents a French concept of flood risk management being a combination of two elements for every place of concern: one related to land use and the other related to flood frequency. Thomas and Bates review policy responses to increasing climate variability in Australia, with particular reference to droughts. Okada's chapter on a community's disaster risk awareness raised considerable discussion at the Colloquium. One of the discussers expressed the opinion that the water profession should encourage journalists to write and publish articles rather than to build new reservoirs. This statement illustrates the power of the media and the potential of demand management as a possible activity on the demand, rather than the traditional supply-side approach (planning new reservoirs to meet every foreseen supply). However, it is not unlikely that drought forecasts and vast media coverage may also cause a natural human reaction that will not lead to water savings: to catch as much water as possible, to fill all available storages, bath tubs and buckets, to water gardens and lawns, to intensify washing.

Three papers were presented on quantity and quality aspects of hydrological sciences. The chapter by Kinzelbach, Vassolo, and Li deals with groundwater systems, studying capture zones of wells; Fahmy, Parent, and Gatel present the results of studies of uncertainty in modeling quality of drinking water, using the Bayesian approach. Tecle and Rupp report on stochastic rainfall-runoff modeling for a semi-arid forested environment.

Another area of water resources research with a very high uncertainty component is that of climate change impacts on water resources. Within this theme, two contributions were presented at the Colloquium and reproduced here. Vogel and coauthors consider the issue of reliability, resilience, and vulnerability of water supply systems, while Bárdossy and Duckstein report on their work on hydrological risk under nonstationary conditions. In this latter work, the perspective embraces both prehydrological systems (circulation of the atmosphere) and the posthydrological ones (ecology and health).

The general area of water resources systems is represented in five chapters, demonstrating methodology and case studies. Two contributions present methodology and application of fuzzy compromise programming to water resources systems planning (Bender), and a new variant of stochastic branch-and-bound method for water quality management (Lence and Ruszczyński).

Yen's chapter presents methodology of analysis of system and component uncertainties, while Shrestha studies uncertainty and risk of water resources system in changing climates. Shamir offers an overview of problems in theory and practice.

Two contributions (Loucks and Nachtnebel) undertake a very ambitious aim to measure sustainability. These methodological works do not end up with real-world applications, yet are of much interest and possibly set the stage for further attempts.

Finally, among the three contributions on reservoirs and hydrological structures, Takeuchi presents his opinion on the future of reservoirs and criteria of their development and management. Milutin and Bogardi report on the use of performance criteria for multi-unit reservoir operation and water allocation problems. Plate's chapter discusses risk management for hydraulic structures.

A range of interpretation of risk, reliability, vulnerability, and robustness has been noted in the present proceedings volume. The verbal and mathematical terminologies clash, as we still explore the sociopsychological implications of risk and related terms (concepts) while the quest of a natural scientist and engineer is clearly targeting the mathematical definitions and quantification of these performance indices.

The debate is left open. The reader will be confronted with the multiple uses of these terminologies. Is it a failure of the editors not to provide consistency? We claim that it is not. A scientific area as volatile as risk, reliability, uncertainty, and robustness considerations in water resources management cannot and should not be regulated during this phase of rapid development. The fascinating fact of new development implies the lack of a "guided tour," but it also implies a chance for discoveries. Readers may find their own definitions on the nucleus of the idea to be developed further.

Therefore, we are proud to be associated with this endeavor as convenors of the Colloquium and as editors. We take responsibility for shortcomings of the book and calculated risks, while giving credit to the authors, whose enthusiastic participation in the Third George Kovacs Colloquium laid the solid foundation for this book. It was also source of inspiration for us and a very rewarding experience of scientific cooperation.

The word "kovacs" in Hungarian (similarly, kowal, kovar in Slavic languages) means blacksmith, and this name is appropriate to the situation. Kovacs Colloquium indeed helps in forging progress in hydrological sciences.

We are confident that the readers will agree that the Colloquium was an excellent tribute to the late Professor George Kovacs, to his scientific, managerial, and human virtues, and to his broad smile that many of us remember so well.

Janos J. Bogardi and Zbigniew W. Kundzewicz
Convenors and Editors
Paris – Poznań, February 1998

2 Integrated regional risk assessment and safety management: Challenge from Agenda 21

ADRIAN V. GHEORGHE*

Motto: Sustainable is what people agree is sustainable.

ABSTRACT
This chapter introduces the field of integrated regional risk assessment and safety management for energy and other complex industrial systems. The international initiative includes compilation of methods and guidelines, and development of various models and decision support systems to assist implementation of various tasks of risk assessment at the regional level. The merit of GIS methodology is highlighted.

2.1 INTRODUCTION

Almost ten years after the UNCED (United Nations Conference on Environment and Development), Rio de Janeiro, Brazil, 1992, some progress has been achieved in relation to the protection of the environment, development policies, and strategical future topics. A number of issues were addressed by UNCED – Agenda 21 that were connected with the topic of this chapter.

Issue 1. Achieving sustainable development, environmental protection shall constitute an integral part of the development process.

Issue 2. Environmental issues are best handled with the participation of all concerned citizens.

Issue 3. National authorities should endeavor to promote the internalization of environmental costs.

Issue 4. Information for decision making would involve:
- bridging the data gap;
- improving availability of information.

Issue 5. Emergency planning and preparedness are integral parts of a coherent sustainable development.

Regional risk assessment and safety planning is a coordinated strategy for risk reduction and safety management in a spatially defined region, across a broad range of hazard sources. It deals equally with normal operation of plants as well as with accidental situations, including synergetic effects.

Regional safety planning requires:

- a framework approach, including a consistent and state-of-the-art methodology;
- legal conditions;
- political will.

Integrated Regional Risk Assessment and Safety Management (IRRASM) has as an overall goal to design, analyze, and conduct practical risk assessment and safety management activities at the regional level for minimizing risks to people and the environment.

Methods and models to be used for IRRASM studies must be specific to the level of details and application as presented in Figure 2.1. In the integration process, a number of models for each individual level, such as engineering, management, politics are available for use and therefore their application is tailored to the area of interest in IRRASM.

Risk, in the content of IRRASM, indicates "the possibility, with a certain degree of probability, of damage to health, environment and goods, in combination with the nature and magnitude of the damage."

A number of indicators are designated to highlight a measure of risk, namely:

- annual fatality rates;
- mean fatalities per year;
- individual and societal risk criteria;
- F–N curves, etc.

Targets at risk, when developing scenarios for risk analysis at the regional level, are:

- people;
- ecological systems;
- water systems;

* Swiss Federal Institute of Technology, Zurich, Switzerland.

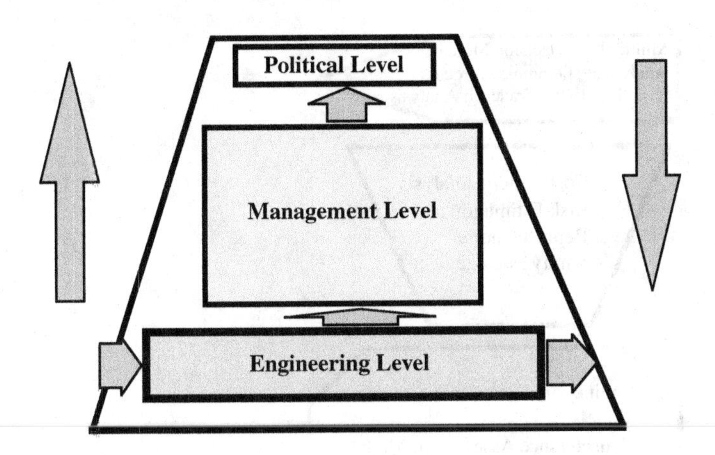

Figure 2.1. A hierarchical approach to problem-solving issues for IRRASM.

- economic resources and other associated infrastructures, etc.

A methodological framework for dealing with the complex tasks of regional safety planning includes:

- development of guidelines;
- adoption of validated models for calculating either probabilities or various types of consequences;
- databases which must include information on a variety of data related to the use of models (e.g., reliability data, emission factors, severe accidents information, etc.);
- knowledge bases that should incorporate expert judgment and non-quantitative information on various aspects related to the regional safety planning;
- adequate tools (e.g., Decision Support Systems – DSS) to assist calculation and representation of various risks which might (co)-exist at the regional level. Specialized DSSs should be addressed to emergency planning and preparedness in relation to safety management;
- GIS (Geographical Information Systems) would be part of the tools available for representing and managing risks at the regional level;
- training and adequate case studies would be a necessary activity within IRRASM.

2.2 REGIONAL SAFETY PLANNING

Definition 2.1

IRRASM is a multidisciplinary process: engineers, computer scientists, and model builders play a central role in the risk assessment stage. Social scientists can contribute with practical advice to the embedding process concerning hazard sources and help communal organizations to deal with such problems, taking into account local economic conditions and political reality. IRRASM involves a complex set of actions for risk reduction and safety management in a defined region across a large number of hazard sources (during normal operation and accidental situations) that includes synergetic effects.

In the process of analyzing risk at the regional level, specific models are available. Integration of risk is achieved in the decision-making process and for this, access to various models, databases, other modern representation environments, e.g., GIS (Geographic-Information Systems), is necessary. In Figure 2.2 a representation of the access of specific models to various levels of use is given. The process of achieving this is rather complex, and involves knowledge of operation research techniques, decision analysis and engineering-economic systems, physical models for pollutant dispersion, etc., databases and knowledge bases.

In Figure 2.2, a detailed representation is achieved in order to portray the hierarchical arrangements in problem solving of risk assessment and safety management when dealing with a large variety of hazardous sources, activities, and decision makers.

At Level I, specific use is made of multicriteria decision models, analytical hierarchical techniques, and instruments to deal with the risk sensitivity phenomenon or the trade-off analysis. At Level II, models and instruments of work have to be adequately tailored to decisions regarding management, and in this case cost benefit analysis, risk estimation and representation, and safety management models are appropriate and instrumental in solving practical problems of risk management. Level III type models involve engineering-economic and simulation models as well as consequence assessment models and tools. Approaches close to the concept of LCA (Life Cycle Analysis) are significant when dealing with various types of impacts and risks that could come within the regional risk assessment.

When working for IRRASM, it is also of relevant importance and use to have access to specialized databases and knowledge retrieval systems. The integration of results from applying these tools and techniques is necessary at all levels of the decision-making process.

2.2.1 Defining a region

The appropriate basis for area selection depends on particular circumstances of each case. Suggested factors when defining a region are:

- The area should be selected for its physical, industrial, and economic characteristics and not necessarily on administrative boundaries.

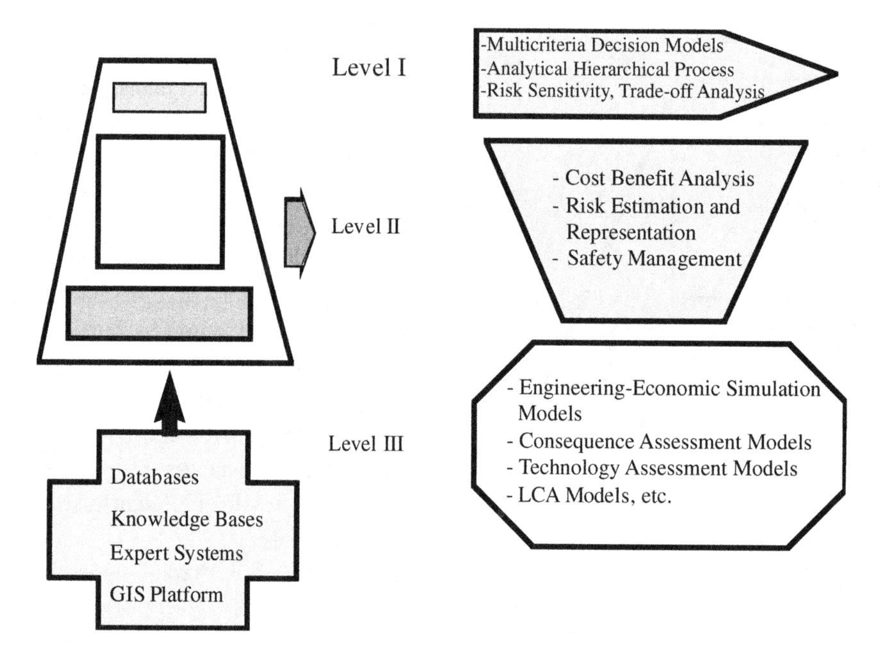

Figure 2.2. Models and knowledge infrastructures and their integration for IRRASM.

- It should be defined on the basis of the facilities and systems of concern and the potential areas that can be directly affected.
- Hard boundaries should not be drawn before an initial hazard analysis and prioritization.
- It is important that as many as possible of the authorities with risk management roles or relevant information become involved.
- Some risk sources will have potential for effects well beyond the immediate area.

2.2.2 Objectives and scope for an IRRASM study

One or more of the following major *objectives* could be considered:

- prioritize hazards in a region;
- evaluate and verify individual/societal risk criteria;
- identify sources of continuous emissions and estimate risks to various targets in the region;
- perform accident and consequence assessment;
- integrate various types of risk in the region;
- design emergency response plans.

The *scope* of these studies includes:

- sources of continuous emissions and accidental releases;
- scenario for accidental states;
- risk assessment for environment and public;
- safety management actions;
- risk management actions.

2.2.3 Hazard identification

A first step in the methodology of regional safety planning is that of hazard identification. The specific aspects within the hazard identification phase are:

- potential hazard sources estimation;
- continuous emissions and their risk to health and environment;
- major accidents from fixed installations and storage;
- transportation of dangerous goods;
- wastes and their associated technology.

2.2.4 A need for prioritization of risks at the regional level

Large and complex industrial areas include various risk sources and activities (e.g., operating process plants, storage terminals, transport activities). The process also goes to the level of an individual plant. A cumulative assessment of such risks should include a detailed hazard analysis and quantified risk assessment for all industrial facilities and associated activities.

Van den Brand methodology

There are a number of methods dedicated to risk prioritization. The van den Brand methodology is based on a step-by-step approach.

Step 1. The user must decide if there are any relevant industrial activities.

Step 2. By using basic information on a given kind of activity and of substances handled, one can determine the kind of average effect that can occur in the case of an accident.

Step 3. By comparing the possible affected area with actual or planned populations living in that area, it is possible to estimate the likely consequences.

Step 4. Assess the probability of such an event. This can be calculated by using probability numbers for different kinds of industrial activity and "correcting" these numbers by using correction factors based on the specific circumstances.

Step 5. Both consequences and probability numbers are visualized as a level of risk in a graphic form (risk matrix).

It is possible to agree on risk criteria for decision making. One limitation of using such a risk graph is that it evaluates fatalities as the only indicator for consequences.

Fuzzy sets approach (ETH Zurich)

As was highlighted above, quantitative methods in IRRASM normally use Boolean logic and classical set theory. In the overall risk estimation process, problems arise when fitting knowledge and experiences (which are not "crisp"). In the classical set theory a set A is the combination of well-distinguished objects x in the universe X. A fuzzy set A doesn't distinguish objects in this way and x shows qualities of other objects. The other main difference to classical set theory is the usage of linguistic variables, given by the quintuple $\{C, T(A), U, G, M\}$. An example to describe "incidental loss" is given:

1. Expression $C = \{\text{consequences}\}$
2. Term $T(C)$ fixes the range of C:
   $T(C) = \{\textit{negligible, marginal, critical, catastrophic}\}$
3. All values in $T(C)$ are represented as a fuzzy set in the universe U
4. G gives the description for C, e.g., G: = "catastrophic" means 75% up to 100%
5. Membership functions for every value of C is associated.

An interface with the prioritization scheme of van den Brand follows. Classifications are widely used in IRRASM studies in order to assess consequences or probabilities for risk calculations. In practice it is often difficult to establish and separate such classes from each other. The documentation of membership grades or functions to characterize various issues of risk makes an IRRASM transparent to the risk assessment process. The fuzzy set theory allows flexibility in handling risk at the regional level. A large number of consequence indicators – up to nine – were integrated into a comprehensive methodology.

Calculation and representation of risk involves a significant number of subjective factors. The fuzzy set tool used for IRRASM has the ability to formulate and document all steps in the risk assessment process which, by definition, has a large degree of subjectiveness. The risk evaluation process can be considered in a more flexible manner, by using normal language, jointly with scientific and engineering descriptions and calculations.

2.3 ON SOME ORGANIZATIONAL ASPECTS

The following procedural steps are suggested in order to address and implement IRRASM studies:

- The organization that intends to undertake the study should formulate the study objectives and draft a project proposal, including the timetable, the manpower, and the financial and other resource requirement.
- The initiating organization should ensure that all the relevant organizations, industry and institutions are involved, on the basis of the draft project proposals. These organizations should decide on the conditions under which they wish to participate and on whether the proposed objectives and the draft study proposal require any modifications to fit their needs. They should also decide on the practical forms in which they are prepared to participate, be it manpower, information sources, or funds. Should any adjustments applicable to the objectives of the study be made, joint agreement must be reached by all the participating organizations. They may also establish a joint coordinating committee. Industry participation in these studies is considered essential and every attempt should be made to ensure the cooperation of industry from the outset.
- A steering committee for the project should be established by the participating organizations, specifying its responsibilities and terms of reference. For complex and sensitive projects, a supervisory steering committee (with political representatives) may be formed, again specifying its duties and responsibilities.

- The steering committee should establish working groups. The steering committee should formulate the project proposal into a detailed plan and establish working groups to carry out various analyses. If external consultants are necessary, the steering committee should make tenders for the work and choose the best person for the job. The working groups should undertake the various analyses associated with the project.
- The steering committee should accept, if necessary after some modification, the final report of the working groups and prepare its own covering report, including conclusions and recommendations.
- The participating organizations should receive the reports and decide on: the final conclusions and recommendations, the policy changes to be implemented, and which of the proposed actions should be carried out, including final prioritization and action plans for implementation.

The participating organizations should put their decisions into effect, ensuring that the responsibilities and procedures are properly arranged to monitor and evaluate the implementation process. They should evaluate, together or separately, the results of their risk management policy, implemented on the basis of the results of the study.

2.4 TECHNIQUES FOR INTERACTIVE DECISION PROCESSES IN IRRASM

New techniques are available today in order to assist the integration of models, citizens, and the potential decision makers. One of these techniques is known as *the cooperative discourse model*. A formal representation of the use of this technique is given in Figure 2.3. Various actors are involved in the decision-making process. Specific steps one can consider include concerns and criteria, assessment of options, evaluation of options. Potential products in the cooperative discourse model are value tree representation, performance profile, priority options.

The actors involved in the process of IRRASM work and implementation are stakeholder group, experts, citizens, sponsors, research team.

Step 1 (Concerns and criteria) involves the following tasks: elicitation of value trees for each group, additions to concern list and the generation of options, additions and modifications of concern list, input to concern list (generation of options), transformation of concerns into indicators.

Step 2 (Assessment of options) involves the following tasks: suggestions for experts (group specific assessments), Group Delphi (collection of expert judgments), transformation of expert judgments into group utilities, incorporation of institutional knowledge, verification of expert judgments (literature search and independent review).

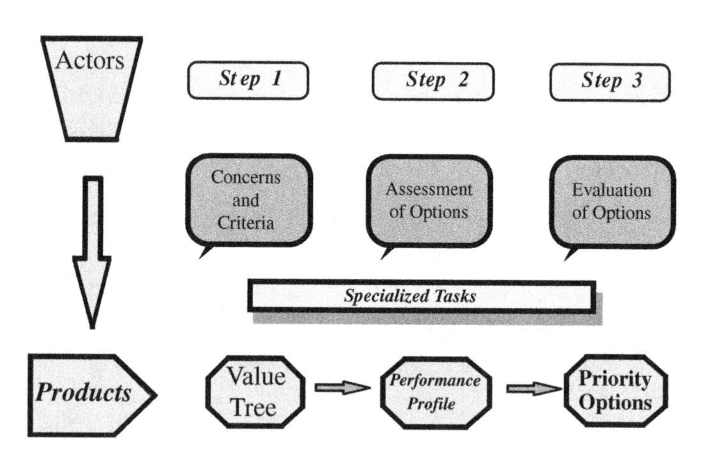

Figure 2.3. The cooperative discourse model described in a number of major steps.

Step 3 (Evaluation of options) involves the following tasks: witnesses to citizen panels, participation as discusser or videotaped presenter, option evaluation and recommendation, compilation of citizen report.

Figure 2.4 is a representation of the cooperative discourse model. It is argued that "because no single mode of discourse can fulfil all of the needs of competence and service all the various actors in an efficient manner, co-operative discourse is a hybrid of different discourse settings. Each discourse setting is oriented toward facilitating a discussion about a primary type of *validity claim*. The differentiation is based in a conceptualisation of four different types of actors: sponsor and research team, experts, stakeholders, and citizens; and four different types of validity claims: communicative, cognitive, normative, and expressive" (Webler 1994).

In the context of this chapter, a *validity claim* is defined as a type of statement that makes an appeal to acceptance and it is fundamentally different in the sense that the appeal must be validated by the group according to a unique set of criteria.

2.5 THE USE OF DSS FOR INTEGRATED RISK ASSESSMENT STUDIES

It is evident that the information technology has large capabilities in assisting various stages of the IRRASM work and dissemination of results. An integrated approach is adequate; in Figure 2.5 it is considered the overall integration to be achieved within the technical, economical, environmental, and legal

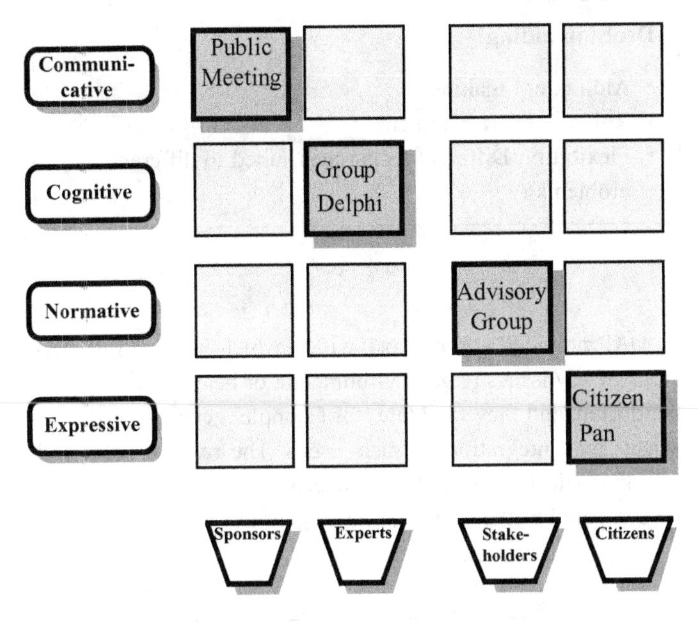

Figure 2.4. Structural approach to cooperative discourse model.

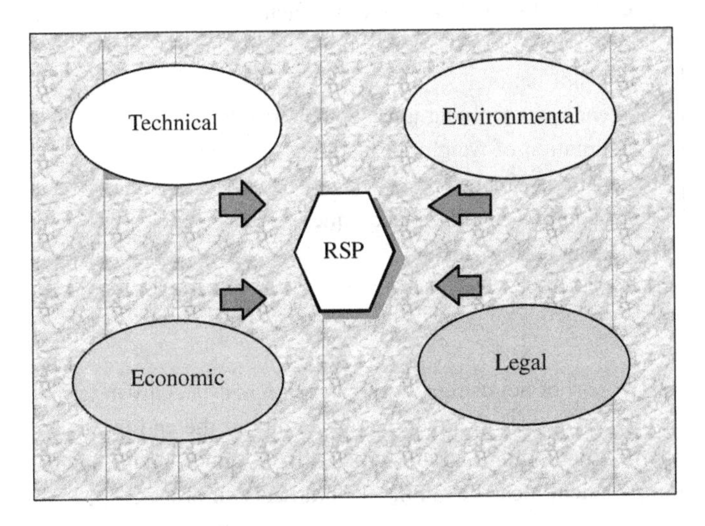

Figure 2.5. Integration of technic, economic, environmental, and legal aspects for IRRASM.

framework. The use of CD-ROM technology is ready to assist various stages in the practical process of IRRASM. Next, some basic definitions and further extensions related to DSSs and decision analysis framework are given.

Definition 2.2

A decision support system (DSS) is defined as a computer-based system that supports technological and managerial decision making by assisting in the organization of knowledge about ill-structured, semistructured, or unstructured issues.

A general taxonomy for DSSs is given next:

- foundation/management theory
- group DSS
- routing DSS
- database management systems
- multiple criteria DSS
- marketing DSS
- multiple criteria decision making
- management science.

Recent experiences with DSS for energy, risk,[1] and environmental management[2] indicate the use of the following methodologies: ad hoc design, expert systems, operational research, genetic algorithms, neural networks, fuzzy logic.

2.5.1 Decision process and the role of models and tools in IRRASM

It is acknowledged that the process of initiating, promoting risk analysis, and implementing safety management studies for large industrial complexes involves complex decisions as well as the participation of many actors. The process of integrating various aspects of risk such as environment, health, performance of hazardous installations, safety culture, management, involves decision-aiding techniques known as decision analysis, which are close to the field of management science.

There are, in general, positive and negative aspects associated with decision making. Indeed, many decisions are made intuitively by experts and do not use structured processes or techniques. For many decision problems related to energy, the solutions and then advantages and disadvantages may not be immediately apparent because of the complexity of the issues involved. There is a need for systematic processes to be followed that help structural thinking and analysis, and allow different viewpoints to be taken into consideration. Structuring helps avoid inappropriate ad hoc decisions and allows the process of reaching a decision to be more open and the decision itself to be more readily defensible (decisions made today very often have long-term effects). In the end, the use of various decision-aiding techniques and the overall process and technology of decision analysis allows the integration of various risks at regional and area levels. The integration of various risks within the decision-making process is the appropriate mechanism that allows displaying various risks and choosing the most appropriate resilient solutions.

[1] See Beroggi and Wallace 1995.
[2] See Paruccini 1993.

There are many inputs, influences, and constraints that a decision maker will consider when deciding which actions to initiate regarding risk reduction or safety management to a particular situation.

Comment

Decision-aiding techniques (DAT) are tools for decision makers; they are decision-aiding techniques, but not decision-making techniques. A large number of tools are available to assist in solving and structuring decisions of such complexity.

The main stages of DAT with reference to IRRASM are:

Step 1. Define and describe the problem (e.g., selecting an appropriate regional risk scenario).
Step 2. Consider and define appropriate quality assurance requirements.
Step 3. Formalize the descriptive model of the problem (e.g., options for alternative production technologies, electricity generation, and system constraints).
Step 4. Obtain the necessary information for modeling.
Step 5. Analyze, in order to determine the set of alternatives and criteria.
Step 6. Ensure selection of the proper method to make the decision regarding the proper integration of various criteria and their optimization.
Step 7. Establish a clear record of the process and any decisions taken as a result of the integration process of various types of criteria and constraints.

To make a decision means to select a method of action, out of a set of possible alternatives. The decision process is complex and sometimes iterative; the set of alternatives or criteria may vary from iteration to iteration.

Decision making involves three major elements:

a. alternatives, among which the "best" one will be chosen
b. criteria for judgment
c. methods for selecting one alternative from the whole set.

Decision-aiding process (DAP) is needed because it helps to generate a degree of shared understanding among interested parties who are concerned with issues on energy mix planning. There are a few issues to be highlighted: in this domain one has to involve complexity, uncertainty, or even fuzziness, multiple criteria with the objectives in conflict, and group interests. DAP can provide models that integrate these features into an adequate methodology.

Decision aiding

- Aiding, not making
- Role of expert judgment
- Flexibility: Different techniques suited to different problems

DAP provides a framework within which informed discussions of key issues (e.g., environmental or health risks) can be conducted, and they facilitate, for example, generation of consensus and integration of such issues. The relative costs and benefits to the user of applying a decision-aiding process should be compared to the costs and benefits of not using one.

Overview of DAP

- Choice of options and relevant factors
- Assessment of different options according to each relevant factor
- Techniques yielding unacceptable results
- Estimation of weighting factors
- Sensitivity analysis
- Probability-encoding techniques
- Presentation of results

The cost of not using a DAP can be high, especially in terms of inefficient use of human resources and in the end the incapacity of integrating various types of risks and selective safety improvement measures in the overall decision process.

Techniques yielding unacceptable results. It should be noted that there are techniques in use for the purpose of making decisions that may not yield acceptable results, even though they appear to provide the user with a very simple methodology. Examples are the "break-even" technique and the "successive goal" technique. The first of these is based on the law of diminishing returns according to which, beyond a certain point, further expenditure is not accompanied by a comparable reduction in the associated risk. In this respect, the performance of each option can be plotted on a graph in order to determine the break point that separates the efficient from the inefficient options. However, because the relative importance of each of the different factors is not taken explicitly into account, this method cannot be recommended, for certain applications, such as the type of safety decisions associated with the design and operation of nuclear or chemical facilities available in the analyzed region. The second technique is based on the successive elimination of those options that do not achieve the previously identified goal. Although it is a very simple form of multi-criteria analysis, its drawback is that it does not actually compare the real advantages and disadvantages of each particular option. When the best option has been selected, one has to check the robustness of this choice in relation to the variety of criteria and indicators, and a sound cost-effectiveness strategy. Various uncertainties or value judgments have to be tested in a sensitivity analysis phase. At the end of this selection process, results have to be presented as simply and as clearly as possible.

The following steps have to be considered when involving DAP:

1. Choice of options and relevant technical, environmental, health, and economic factors.
2. Assessment of options according to each relevant factor and the relative importance of each factor in the political or technical decision environment.
3. Choice of the most appropriate and relevant decision-aiding techniques.
4. Estimation of weighting factors for the criteria involved.
5. Use of sensitivity analysis.
6. Presentation and interpretation of results.

The wide practice of the use of decision analysis shows that no decision-aiding technique is appropriate for all. Examples of the types of problems in the field of risk assessment and safety management with reference to IRRASM, where DAT can be helpful, include the following:

- Siting of new electricity generation industries, landfills, or other hazardous installations, taking into account factors such as alternative land uses, environmental effects, and population density.
- Development of strategies for emergency planning (e.g., sheltering, evacuation, relocation), taking into account the social implications of moving people in all types of weather, as well as financial and logistical considerations.
- Deriving levels of pollution contamination in water and soil at which emergency countermeasures may be introduced following an accident.
- Determining requirements for additional safety equipment.
- Determining the appropriate degree of redundancy of safety devices and plant equipment, taking into account possible safety effects (e.g., domino effect) on other plant equipment.
- Optimizing the protection and other aspects of maintenance and repair operations, choosing between major alternatives such as replacement or repair of large components, reliability-centered maintenance programs, adequate inspection programs, etc.
- Selection of optimum plant equipment and procedures for normal and abnormal circumstances.

Decisions related to energy mix selection involve negotiations between actors with different viewpoints or value judgments. One objective of the negotiations is to determine the "choice factors" that are representative of the different viewpoints.

The range of possible decisions in any investigated area is generally limited by certain constraints that may be imposed, for example, by a regulatory authority at the federal or cantonal level, by operating policies, or by generally accepted good practice. Other constraints may be less tangible, such as acceptance by the public or by the industry practice, limited perceived threats to environmental quality and health risks, and political conditions. New techniques are now available (decision conferencing[3]) which enable the integration of various aspects into a comprehensive decision policy.

2.5.2 Decision-aiding techniques in use for safety management

A number of decision-aiding techniques are currently available. The application of Cost-Effectiveness Analysis (CEA), Cost-Benefit Analysis (CBA), Multi-Attribute Utility Technique (MAUT), and Multi-Criteria Outranking Technique (MCOT) are considered in greater detail (see footnote below).[4]

[3] One has to make a clear distinction between dialogue process and decision conferencing (they are aiming at reaching a decision).

[4] *Cost-Effectiveness Analysis (CEA)*: This technique can be applied when two related factors are considered, one of which is a cost. The relative interest of each option is assessed through a "cost-effectiveness ratio" illustrating the amount of effort (cost) devoted to a decrease of one unit of risk. This technique is relatively easy to apply and can be recommended when only two relevant factors are envisaged because in this case the sensitivity analysis is very easy to make. *Cost-Benefit Analysis (CBA)* originates from the economic theory of welfare; it compares the benefits and harm associated with different options. The principal characteristic of this technique is that all relevant factors have to be expressed in monetary terms. When all these factors are expressed in the same unit it is very easy to aggregate all these different costs in a total cost. The best option is then the one presenting the minimum total cost. *Multi-Attribute Utility Technique (MAUT)*: The essence of this technique is to define a scoring scheme (or a multi-attribute utility function), measured on a scale between 0 and 1, with the property that if the score (or utility) is the same for two options there is no preference for one or the other. If however the utility for option i exceeds that for option j, then option i is preferred to

Multi-attribute techniques: main drawbacks

- Preferences are clearly subjective; conclusions open to question; difficult to "sell" to physical scientists
- Difficult to determine weights and probabilities
- Strictly, conclusions are only applicable for exact problem studied
- Limited generic application; provide attribute lists; provide scoping results

2.5.3 Overview of decision analysis for IRRASM activities

Definition 2.3

Decision analysis results from combining the fields of systems analysis[5] and statistical decision theory.[6] The methodology of decision analysis assists logical decisions in complex, dynamic, and uncertain situations.

Some important aspects related to the above field are:

- Decision analysis specifies alternatives, information, and preferences of the decision maker and then finds the logically implied decision.
- Decision making requires choosing between alternatives, mutually exclusive resource allocations that will produce outcomes of different desirabilities with different likelihoods.
- Decision making requires specifying the amount of uncertainty that exists given available information.[7]

option *j*. Such utility functions are established for each relevant factor (or attribute) and then aggregated in a total utility function representing the global interest for each option. The MAUT is an interesting decision analysis model for high-level strategy problems involving different factors (quantitatively or qualitatively). Recent experiments show that this technique can be implemented in practice when dealing with different potential actors involved in a complex decision-making environment. Fuzzy sets theory can also be adopted to this technique in order to accommodate for imprecision in evaluating the factors. *Multi-Criteria Outranking Technique (MCOT)*: The aggregation techniques described in the previous sections combine all the evaluations according to the different relevant factors into a single figure of merit expressing the global interest of each option. Instead of expressing the performances of each option in terms of a single overall figure of merit, outranking techniques compare each option *i* to every option *j* in order to evaluate whether option *i* outranks (or is preferable to) option *j*.

[5] Systems analysis captures the interactions and dynamic behavior of complex situations.

[6] Statistical decision theory is concerned with logical decisions in simple, uncertain situations.

[7] Decision analysis treats uncertainty by encoding informed judgment in the form of probability assignments to events and variables.

- Decision analysis determines the decision maker's trade-offs between monetary and nonmonetary outcomes and also establishes his preferences for outcomes that are risky or distributed over time.
- Decision analysis provides a formal, unequivocal language for communication among the people included in the decision-making process.

Definition 2.4

Decision basis, as part of the decision analysis, consists of quantitative specification of the three elements of the basis: the alternatives, the information, and the preferences of the decision maker.

- There are two essential steps in any decision analysis: the development[8] and the evaluation[9] of the decision basis;
- Practice of decision analysis has been extended recently to the level of corporation management, policy decision makers, technology selection.

2.5.4 Decision Support Systems and the IRRASM process

DSS of potential interest within IRRASM and an associated "*dialogue process*" are:

- Single user or group DSS
- Single or multiple criteria DSS[10]
- Multiple criteria decision making[11]
- Application domain[12]
- Operational[13] or strategic[14]

Remark 2.1

For a multicriteria DSS in use for IRRASM there is a close interconnection between alternatives, criteria,[15] and indicators.[16]

[8] The decision maker must elicit each of the three elements from the decision maker.

[9] This is achieved by using sensitivity analysis and value of information calculations.

[10] A multicriteria DSS needs capabilities delivered by a model management system (structuring the alternatives), an information management system (databases, knowledge bases, GIS, user interfaces, scenario construction or alternative generation module).

[11] For example, as decision-aid or decision-making tools.

[12] For example, risk and environmental management, emergency management and preparedness, etc.

[13] Maintainability of energy systems, reliability-centered maintenance for nuclear power plants.

[14] Strategic electricity/energy system development.

[15] For an energy environment multicriteria DSS, criteria used are, e.g., consequences to humans, consequences to the environment, economic consequences, social consequences.

[16] Suitable indicators can often be found with user participation.

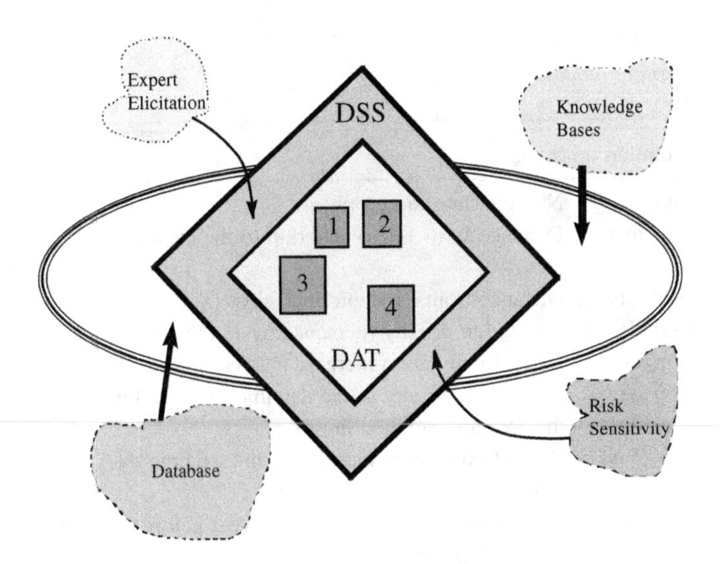

Figure 2.6. DSS, DAT (1 – cost engineering models, 2 – decision analysis models, 3 – operation research models, 4 – advanced AI tools).

Decision analysis methodology and appropriate models are often integrated into the structure of a given DSS. A DSS could have a wider applicability range than decision analysis tools (DAT) in the overall decision-aiding process[17] (see also Figure 2.6).

Advanced DSS for IRRASM have a more complex structure. Figure 2.7 represents such an architecture, taking into account the following:

- the core of DSS which would include:
- the comparative assessment layer which is designed to compare various types of risks either environmental or health and further display the results;
- the application area, e.g., transportation stage, production, storage, etc.;
- the decision-aiding technique layer which would include a variety of tools and methods in order to assist the multi-criteria decision process;
- risk as a common denominator aspects which would be considered across the various stages of the multilayer DSS.

Regarding emergency preparedness and planning at regional levels (see Figure 2.8), one has to consider that the probability of such an event could be equal to one; consequences have to be calculated and emergency rules and procedures have to be associated to the overall framework of DSSs. Figure 2.8 indicates how such a procedural approach has to be followed.

[17] Decision analysis tools are often considered as specialized, and in-depth instruments which are aiming to assist in the decision-making process.

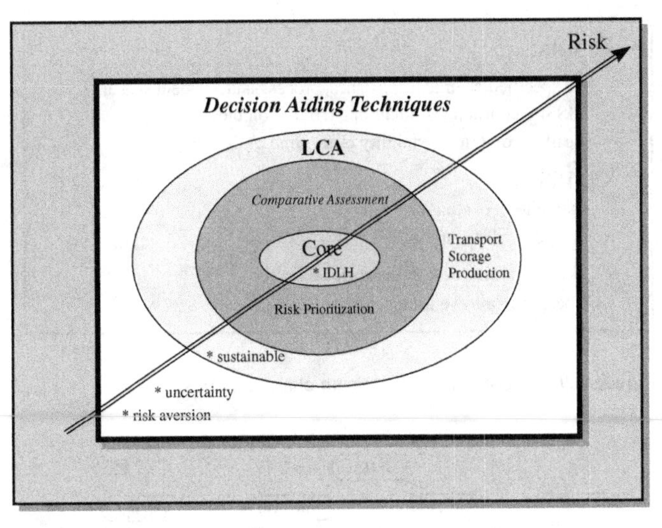

Figure 2.7. A complex approach to DSS for IRRASM problem solving.

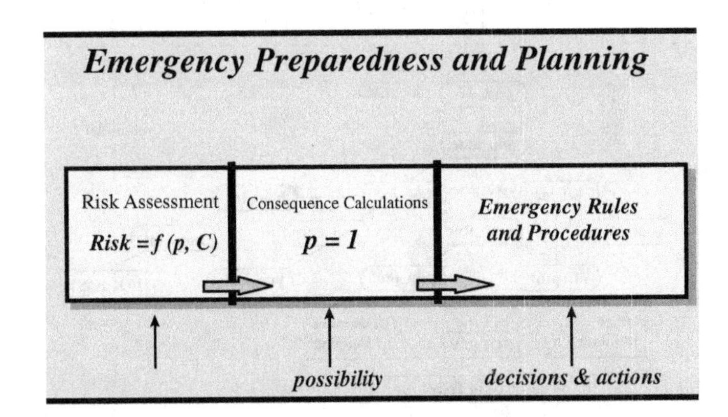

Figure 2.8. A framework for IRRASM emergency preparedness and planning activities.

An example of a DSS dedicated to emergency planning and preparedness (ETH-CHEMRISK) for chemical accidents is given in Figure 2.9.

2.6 THE USE OF GIS TECHNOLOGY FOR IRRASM

The use of GIS technology can be tracked down along the two broad categories of risk assessment tools: (i) preventive risk assessment and (ii) corrective risk assessment. In the preventive category a working example may be "The Estimation of the Environmental Impacts of Tokyo Bay Development Programs by Modeling Air Pollution Concentration." The project was introduced as a case illustrating the implementation of the so-called intelligent decision support systems by

- Inputs:
  - the specified chemical substance, process and accident scenario
  - GIS site coordinates, including terrain complexity
  - weather conditions, stability class, time of the accident
- Outputs:
  - source term evaluation
  - toxic dose distribution
  - likely fatalities; risk at the regional level
  - emergency intervention zones

Figure 2.9. ETH-CHEMRISK main characteristics.

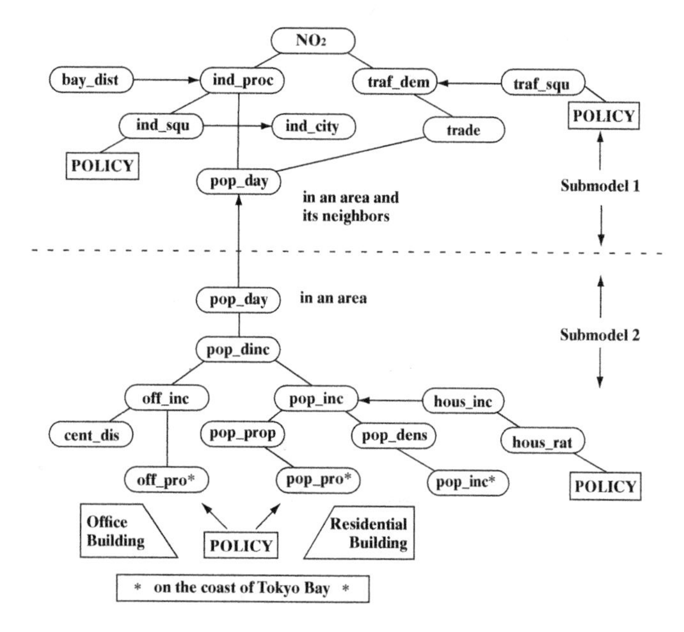

Figure 2.10. Hierarchical model structure in the Tokyo Bay Project (compiled from Kainuma, Nakamori, and Morita 1991).

advanced methods, including interactive modeling, fuzzy modeling and simulation, and linguistic fuzzy modeling. It sets into operation a hierarchical model structure (Figure 2.10) aimed at determining the influence of building new offices and houses along the coast of the Tokyo Bay, in Japan.

It is interesting to note that, though the GIS notion is not explicitly emphasized in the project presentation, it impregnates the entire structure and functioning of the inferring device. This fact would be clearly transparent through the list of main variables with which the models operate (Table 2.1).

There are two ways the outputs of the multicriteria assessment performed are presented. The one in Figure 2.11, plotting NO_2 concentrations against model's variables, is obviously for the expert's eye only. Deriving a planning decision from such diagrams is a near-impossible job.

Table 2.1. *List of main variables in the Tokyo Bay Project models*

Notation	meaning
NO_2	NO_2 concentration (ppb)
bay_dist	Distance from Tokyo Bay (km to the center of the area)
cent_dis	Distance from the center of Tokyo (km)
pop_dens	Population density (persons/km^2 in the area)
pop_inc	Rate of population increase (%/year in the area)
pop_day	Population density in the daytime (persons/km^2 in the area and neighborhood)
pop_dinc	Rate of population increase in the daytime (%/year in the area)
pop_inc*	Rate of population increase the year before (%/year in the area)
pop_pro*	Rate of population increase the year before in the neighborhood affecting the area (%/year in the area)
ind_proc	Industrial shipment density (104 yen/km^2 in the area and neighborhood)
ind_city	Urban industrial shipment density (104 yen/km^2 in the area and neighborhood)
trade	Density of the whole sale and retail sales (104 yen/km^2 in the area and neighborhood)
traf_de	Weighted traffic density (104/(km^2·h))
ind_squ	Land use for industry (% in the area and neighborhood)
traf_squ	Land use for traffic (% in the area and neighborhood)
hous_rat	Land use for housing (% in the area)
hous_inc	Rate of increase of housing area (%/year in the area)
off_inc	Rate of increase of offices (%/year in the area)
off_pro*	Rate of increase of offices the year before in the neighborhood affecting the area (%/year in the area)

Source: Compiled from Kainuma, Nakamori, and Morita 1991.

To put results in an operational perspective, the authors felt the need for the alternative way depicted in Figure 2.12, and had the results overlaid in GIS fashion over the base map of the area of interest.

Most risk assessment tools falling into the second, corrective category would explicitly recognize the need for a GIS format. Figure 2.13 summarizes the generic structure of a nuclear accident consequence assessment software. The left-hand side shows a variety of base- and input maps (layers). A variety of models, given in the central column, would operate in a complex fashion with the input layers to generate the output maps (layers) depicted in the right-hand side of the chart.

This may be a typical example of a nonlinear I/O GIS structure, linking inputs to outputs via complex models involving

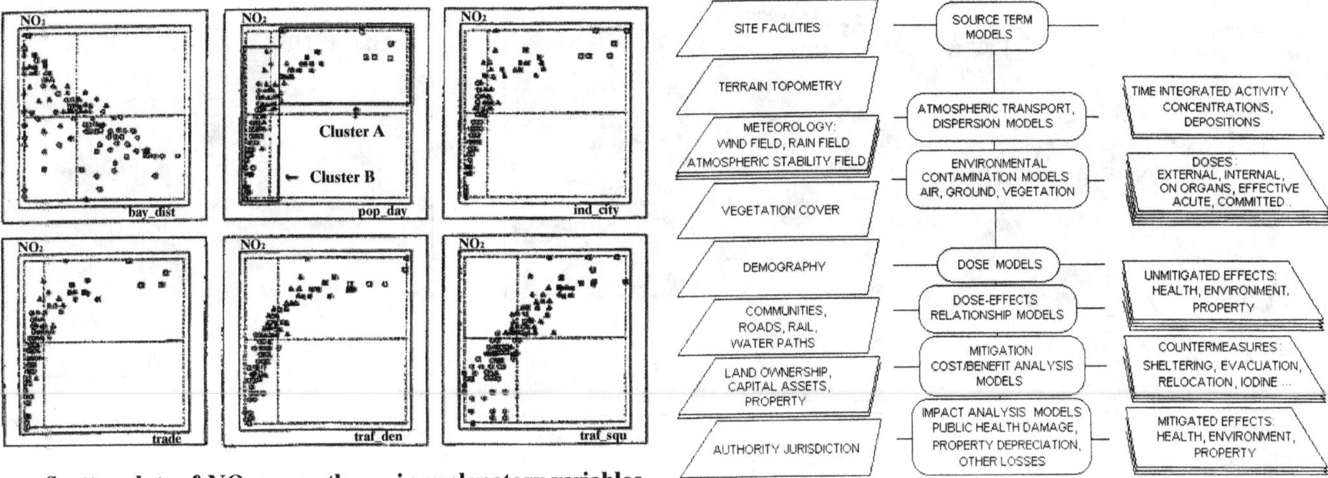

Figure 2.11. As explained.

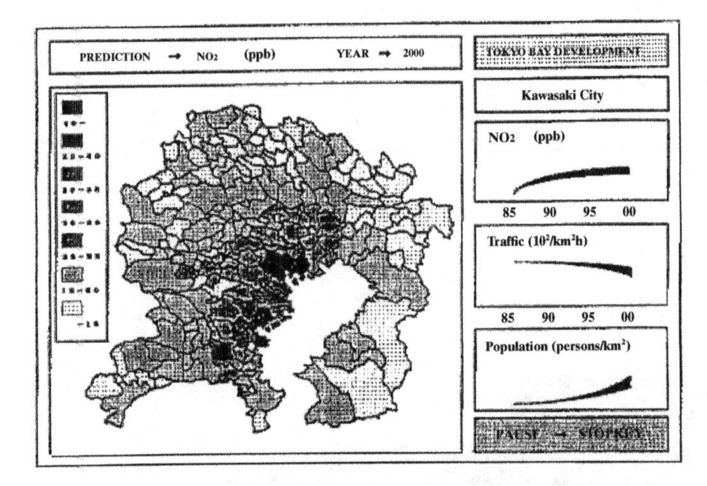

Figure 2.12. GIS-wise output of a dynamic fuzzy simulation with the Tokyo Bay Model (Kainuma et al. 1991).

analytical functions, logical constructs, and rules. An instance of GIS-formatted DSS for nuclear accident consequence assessment is ETH-NUKERISK, developed within the framework of Polyproject on "Risk and Safety Technical Systems" at ETH-Zurich. Figures 2.14–2.16 illustrate the research product.

Though not represented in Figure 2.16, a comprehensive database has to support the risk assessment-dedicated GIS, in most of its input component layers, as well as the models' workings. The same applies to chemical accident consequence assessment. The concept of a merger between a GIS and a database is illustrated in Figure 2.17, the software package ETH-CHEMRISK.

Figure 2.13. Generic chart of a GIS-formatted software for nuclear accident consequence assessment.

Figures 2.18 and 2.19 illustrate the GIS participation in the code's workings, emphasizing the important role of the topometric layer in a complex terrain like the one prevailing in Switzerland. Indeed, in an essentially hilly/mountainous topography the mere recourse to "roughness corrections" to conventional Gaussian models seems to be insufficient, and hydrodynamic flow models are in order, to account for the actual complexities encountered in the atmospheric transport and dispersion of pollutants.

The experience gained with the development and operation of ETH-NUKERISK and ETH-CHEMRISK is currently called upon in developing a comprehensive and, it is hoped, more performant framework for the needs of Project Integrated Risk Assessment and Free Market, by KOVERS.[18]

2.7 THE KOVERS APPROACH

To substantiate the terms of reference of the Project Integrated Risk Assessment and Free Market, KOVERS has commissioned a number of research contracts on a variety of topics including risk assessment software development. One such endeavor has taken the following strategy of approach:

Phase 1. Develop the tools that make the prerequisite in a state-of-the-art risk assessment: the databases; and the GIS;

Phase 2. Assimilate/develop an appropriate class of analytical models, both in the preventive and the corrective category;

[18] Centre of Excellence "Risk and Safety Sciences," Swiss Federal Institute of Technology, Zurich.

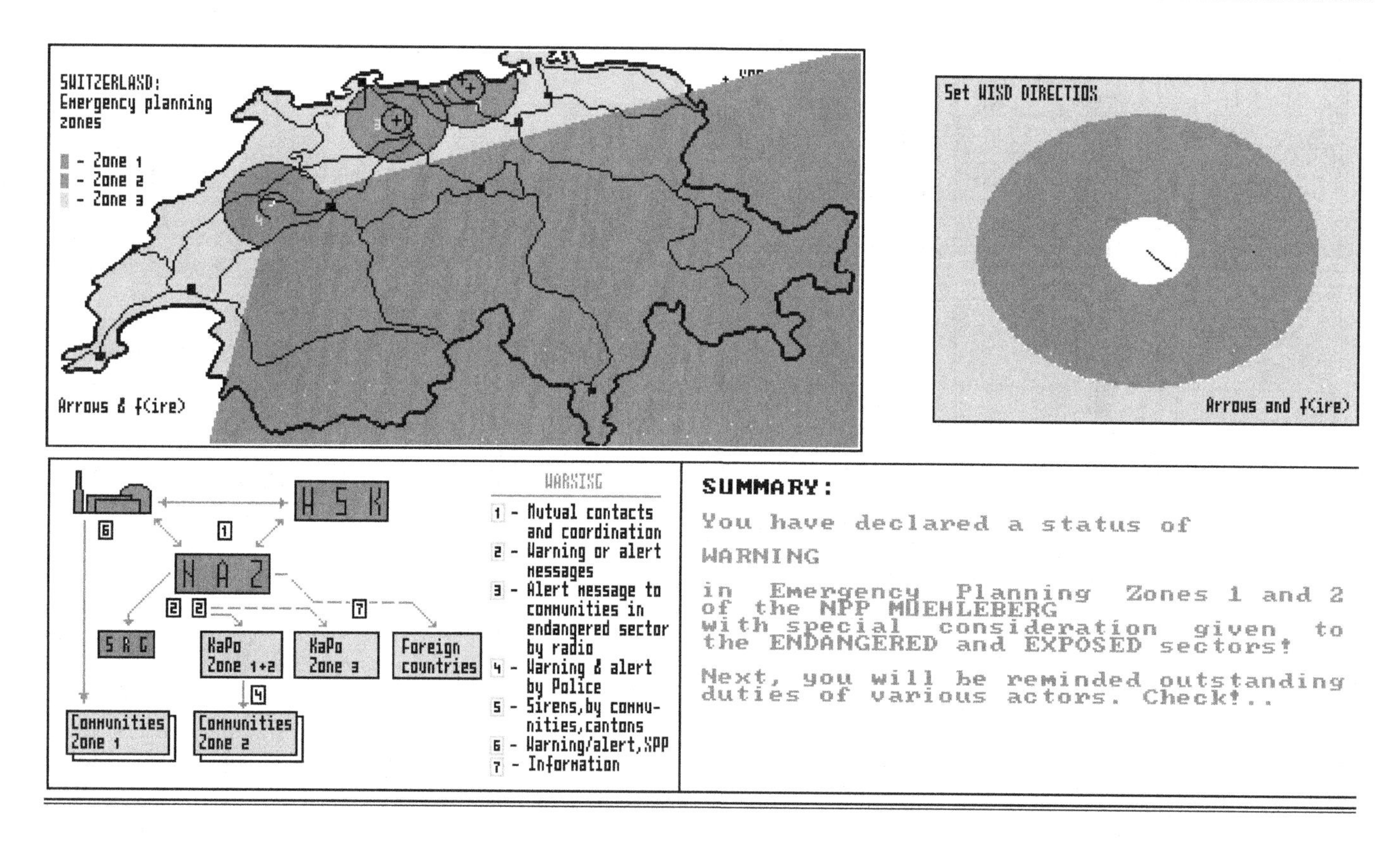

Figure 2.14. An ETH-NUKERISK working screen.

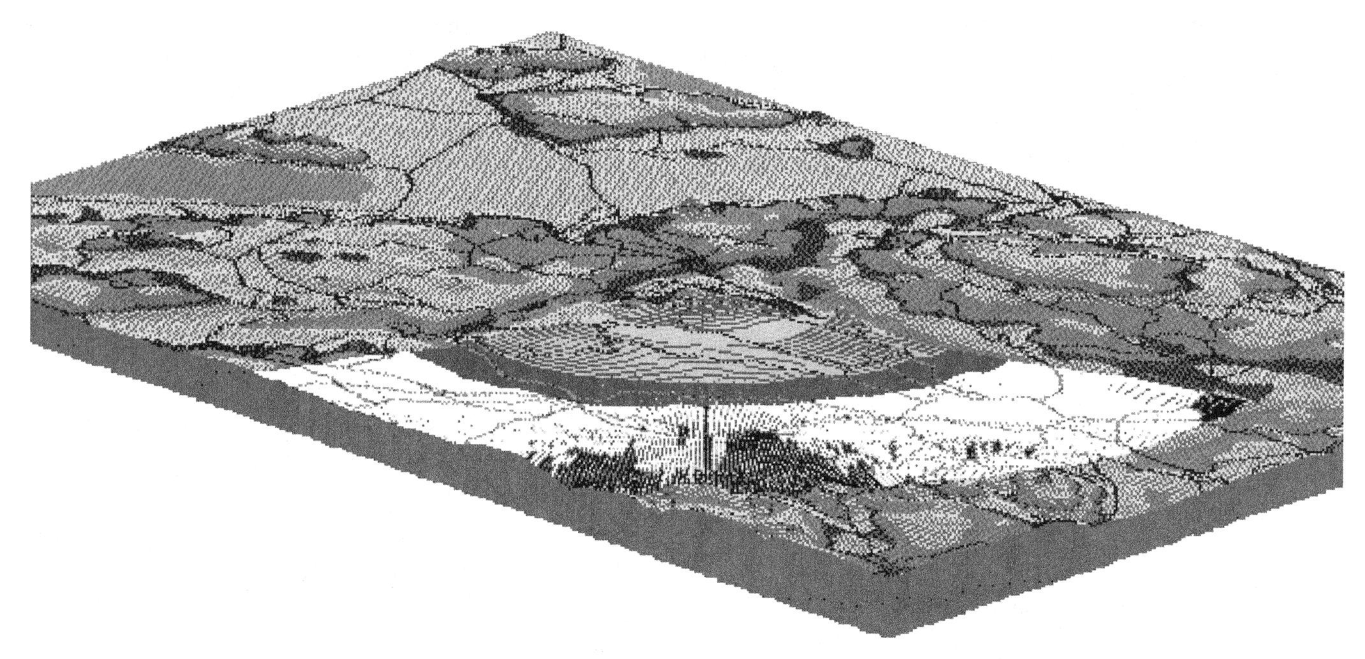

Figure 2.15. ETH-NUKERISK GIS-formatted output overlaying expected health effects of environmental contamination to a base map (Muehleberg area topometry) and an input map (major roads).

Figure 2.16. Typical output map in the GIS framework of ETH-NUKERISK: the expected intervention zones in case of a major accidental radioactive release at Muehleberg Nuclear Power Station.

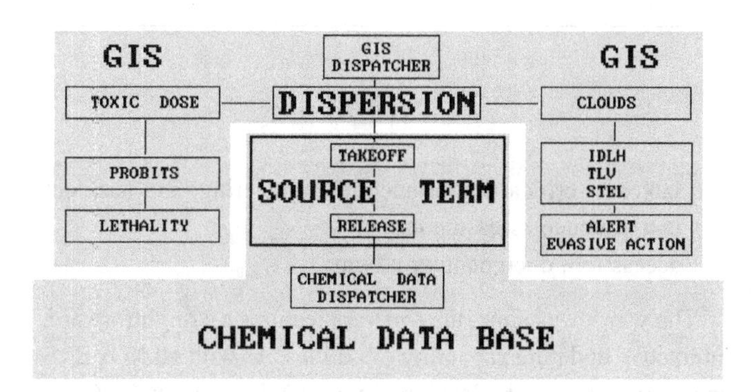

Figure 2.17. GIS, database, and model interaction in ETH-CHEMRISK.

Phase 3. Articulate the tools above in the framework of higher level hierarchical, mostly logical and rule-based, models in order to achieve the standards of a multicriterial, lifetime-oriented, and business-relevant risk assessment.

The primary objective of interest was chosen to be the lifetime risk of dangerous substances, of which a primary target has been selected: the epichlorohydrin – an essential intermediate product of many chemical industries in Switzerland and elsewhere.

As of the date of this briefing, three components have been developed:

KOVERS Monographs of Dangerous Substances;
KOVERS Chemicals Database; and
SWISS GIS by KOVERS.

2.7.1 KOVERS monographs of dangerous substances

The package KOVERS MONOGRAPHS OF DANGEROUS SUBSTANCES aims to eventually cover several dozen substances of primary relevance in chemical risk assessment. Currently it is represented by one pilot module: EPICHLOROHYDRIN. The following box, which actually is an excerpt of the code's Help files, briefly explains the concept:

> KOVERS MONOGRAPHS OF DANGEROUS SUBSTANCES is a pilot software product done in compliance with the guidelines established in the background document.

The complete List of Index Items comprises:

1. About . . .
2. Introduction
3. General Presentation
4. Chemical and Physical Properties
5. Analytical Methods
6. Industrial Production
7. Uses
8. Disposal of Wastes
9. Environmental Transport and Distribution
10. Occurrence
11. Occupational Exposure
12. General Population Exposure
13. Absorption
14. Body Organs Distribution
15. Metabolic Transformation and Excretion
16. Acute Aquatic Toxicity
17. Short-Term Exposures, Animals
18. Oral Exposure, Animals
19. Subcutaneous Exposure, Animals
20. Inhalation Exposure, Animals
21. Effects on the Eyes and Skin, Animals
22. Carcinogenicity via Short-Term Oral Exposure, Animals
23. Carcinogenicity via Prolonged Oral Exposure, Animals
24. Carcinogenicity via Inhalation Exposure, Animals
25. Carcinogenicity via Subcutaneous Exposure, Animals
26. Carcinogenicity via Intraperitoneal Exposure, Animals
27. Carcinogenicity via Dermal Exposure, Animals
28. Mutagenicity, Animals
29. Mutagenic Tests with Positive Results, Plants, Animals
30. Effects on Reproduction, Animals
31. Teratogenicity, Animals
32. Controlled Studies on Effects, Man
33. Accidental Exposures, Man
34. Sensitization, Man
35. Carcinogenic Effects, Man
36. Mutagenic Effects, Man
37. Effects on Reproduction, Man
38. Evaluation of Health Risks for Man
39. Regulations/Guidelines on Occupational Exposure
40. Regulations/Guidelines on Ambient Air Levels
41. Regulations/Guidelines on Surface Water Levels
42. Regulations/Guidelines on Levels in Food
43. Regulations/Guidelines on Occupational Exposure
44. Standards on Labeling and Packaging
45. Standards on Storage and Transport
46. References

The codes in the series are intended to offer a fast and versatile access to data and knowledge about a variety of substances the production, transportation, processing, distribution, and disposal of which may pose risks to life, the public health, the quality of the environment, and business.

With codes such as EPICHLOROHYDRIN users may:

- get an authorized (WHO/ILO endorsed) summary account on the targeted substance;
- access topically oriented information via a subject index;
- access a comprehensive reference list;
- cross-reference any word in the displayed files via a convenient search-and-deliver mechanism;
- access the targeted substance file in KOVERS CHEMICAL DATABASE – another KOVERS project covering more than 700 chemicals;
- access a computational facility providing formulae and tables of critical importance in engineering, and accident consequence assessments;
- access to a documentary album.

The way it was done, the software features a straightforward, interactive updating capability of all files. As with all KOVERS codes, the product is developed under Microsoft WINDOWS 3.1(1) operating system, thus offering a standard, user-friendly interface.

The computerized monograph largely works in a hypertext fashion. The computational section of the monograph is also believed to be of assistance.

2.7.2 KOVERS chemical database

The code KOVERS CHEMICAL DATABASE covers a series of features of a few hundred substances that would fall under the label "dangerous." An explanatory excerpt of the database's Help follows:

KOVERS CHEMICAL DATABASE is a pilot software product done in compliance with the guidelines established in the background document: The code is intended to offer a fast and versatile access to data and knowledge about a variety of substances the production, transportation, processing, distribution, and disposal of which may pose risks to life, the public health, the quality of the environment, and business.

The code consists of two modules:

the DATABASE; and
the MANAGER.

With the DATABASE users may:

get files consisting of 27 different features of 701 chemicals;
get one selected feature, out of 27, for all 701 chemicals;
amend/update the existing chemical files;
enter new chemicals according to the database format;
lock out one or several chemicals, so that the database be restricted to only a limited, relevant variety;
unlock for use any previously locked out chemical;
at any time, fully restore the database to its original contents;
access a comprehensive knowledge base;
access an (expandable) image album;
use the facilities above in conjunction with all standard WINDOWS facilities and other applications that access the WINDOWS Clipboard.

With the MANAGER users may:

sort chemicals in the decremental or incremental order by any of the quantitative features available, thus obtaining interesting rankings;
sort chemicals by contents, either elemental or in regard to a specific radical;
sort chemicals by any of the 54 varieties of fire hazards they may be prone to;
sort out the chemicals that would normally be in gaseous state at a user-specified temperature;
sort out the chemicals that would normally be in liquid state at a user-specified temperature;
sort out the chemicals that would flash, at a user-specified temperature;
sort out the chemicals that would normally be less dense than air at 0 degrees Celsius;
sort out the chemicals that would normally be denser than air at 0 degrees Celsius;
sort out the chemicals that would normally be lighter than water at 4 degrees Celsius;
sort out the chemicals that would normally be heavier than water at 4 degrees Celsius;
sort out the chemicals that are soluble, or miscible, or decompose;
sort out the chemicals that are insoluble;
again, access a comprehensive knowledge base;
access an (expandable) image album;
use the facilities above in conjunction with standard WINDOWS facilities and other applications that access the WINDOWS Clipboard.

The CHEMICAL DATABASE is designed to work on-line with a series of computer applications envisaged within the KOVERS projects.

Equipped with a cross-reading device allowing a sorting-out of chemicals-by-features, and of features-by-chemicals; ranking and categorizing devices; a knowledge base given an analytic description of a variety of models of direct relevance in risk assessment; and an open image base, the product is likely to meet most of the terms of reference discussed in this briefing.

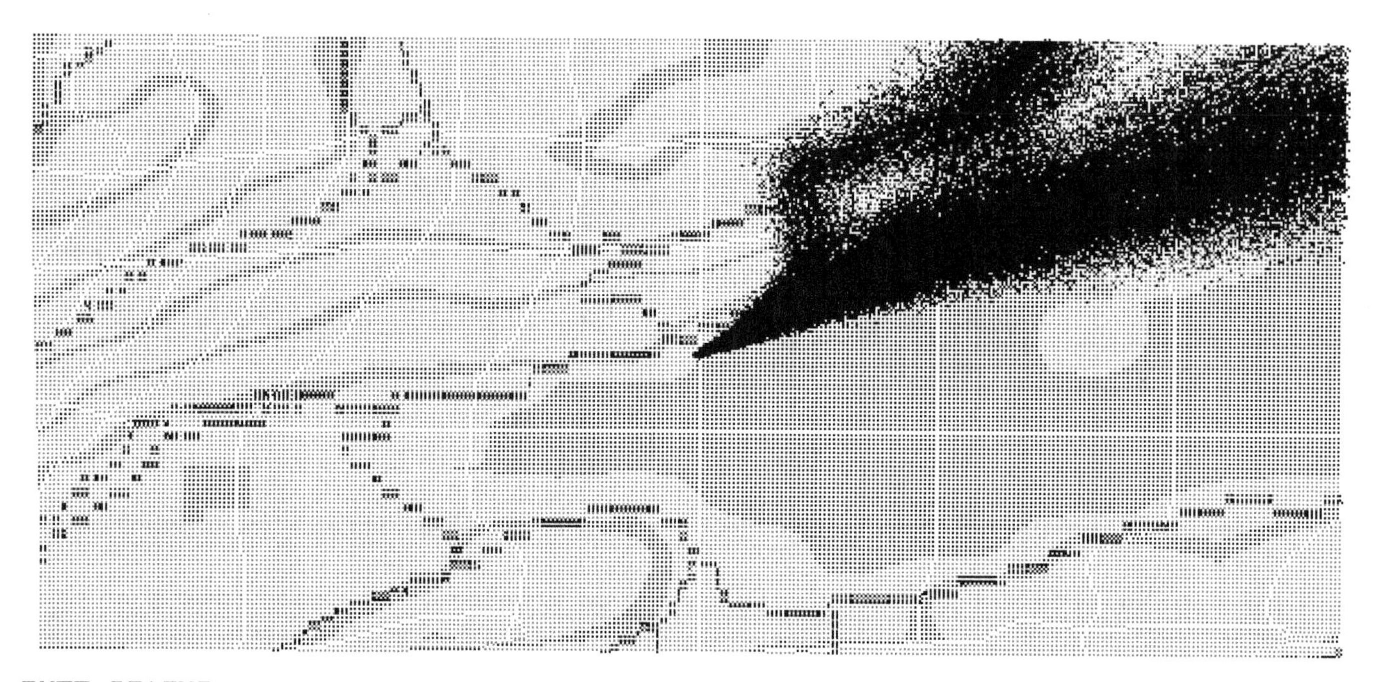

Figure 2.18. Intermediate output map in ETH-CHEMRISK, overlaying a time series of complex puff trajectories from a buoyant gas release near the shores of Bieler See, distorted by the heated slopes of the Juras plus an inversion layer, to a base map (local topometry) plus an input map (roads).

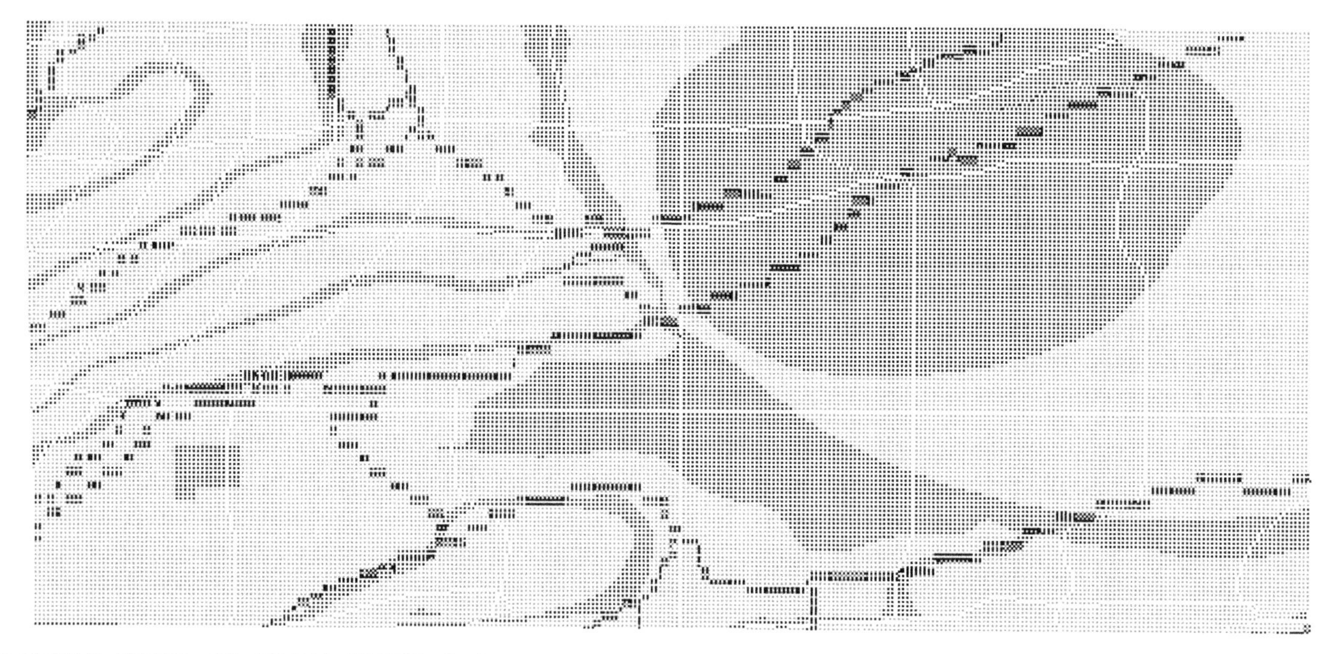

Figure 2.19. Output map in ETH-CHEMRISK: isopleths of likely risk of fatalities inferred with probit functions from time-integrated, terrain-distributed concentrations consecutive to the accident simulated in Figure 2.18.

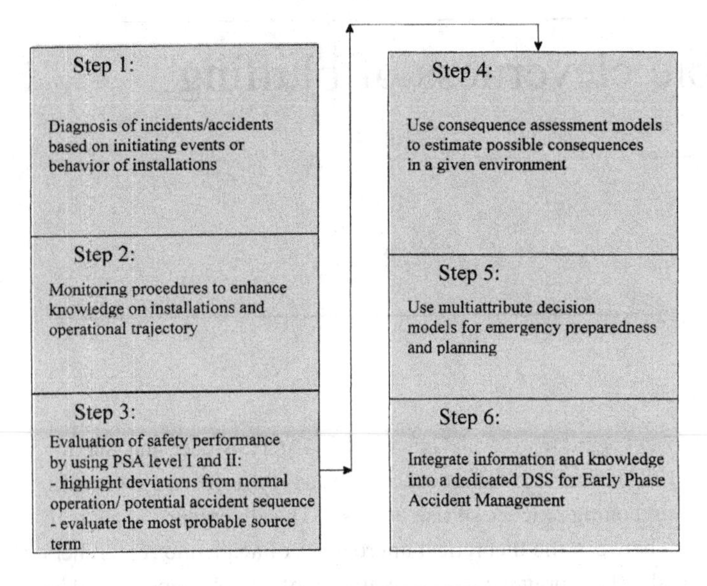

Figure 2.20. A stepwise approach to IRRASM.

A stepwise approach to enacting risk analysis and safety management at the regional level is given in Figure 2.20. This figure includes, logically, the potential use of PSA[19] Level I and II approaches (or their associates QRA[20]) in order to identify solutions for the early phase accidental situations.

A number of lessons are important to remember when dealing with regional safety planning:

- risk is more than just a number;
- not all risks are equal;
- risk integration is possible up to some degree;
- one has to compare risks from various industries and activities at the regional level;
- balance risk-benefit in order to achieve high economic development and efficiency in a sustainable manner;
- introduce results from comparative risk assessment and IRRASM into the decision-making process by accepting levels of tolerable risk at the societal level.

[19] Probabilistic Safety Assessment.
[20] Quantitative Risk Assessment.

2.8 CONCLUSIONS

Regional risk assessment involves an interdisciplinary activity. Results of such studies are integrated into development programs, safety-related works, environmental impact assessment, and the like. The use of modern instruments such as DSS, GIS, etc., are becoming essential for a broad approach to risk issues and adequate management decisions.

REFERENCES

Beroggi, G. and Wallace, W. A., eds. (1995) *Computer Supported Risk Management.* Kluwer Academic, Dordrecht.

Gheorghe, A. V., ed. (1996) Integrated regional health and environmental risk assessment and safety management. Special Issue, *International Journal Environment and Pollution*, 6: 4–6.

Gheorghe, A. V., Krause, J.-P., and Mock, R. (1996) *Integration von Fuzzy-Logic fur eine regionale Umweltrisiko-Analyse*, vdf Publishing House, Zurich, Polyproject Documents no. 12.

Gheorghe, A. V. and Nicolet-Monnier, M. (1995) *Integrated Regional Risk Assessment*, Vols. I and II. Kluwer Academic, Dordrecht.

Gheorghe, A. V. and Vamanu, D. (1996) *Geographical Information Systems (GIS) in the Risk Assessment Business: A Briefing.* KOVERS, ETH Zurich, Working Paper no. 2.

Paruccini, M., ed. (1993) *Applying Multiple Criteria Aid for Decision to Environmental Management.* Kluwer Academic, Dordrecht.

Renn, O. and Goble, R. (1996) A regional concept of qualitative growth and sustainability – support for a case study in the German state of Baden-Wurttemberg, *Int. J. Sustain. Dev. World Ecol.* 3.

Webler, T. (1994) *Experimenting with a New Democratic Instrument in Switzerland*, Report of the Polyproject on "Risk and Safety of Technical Systems," ETH Zurich, Switzerland, June.

3 Risk analysis: The unbearable cleverness of bluffing

V. KLEMEŠ*

ABSTRACT
A critical contemplation is offered of what is known and what is not known about the hydrological, economic, and societal uncertainties that are eagerly and routinely subjected to all kinds of mathematical prestidigitation in the process of risk analysis of water resources systems. An illustration of the issues is provided by a reality check on some real-life situations, mostly based on the author's own experience from the past forty years. Three simple recommendations are offered that may bring risk analysis down to earth.

3.1 INTRODUCTION

Spurred by the newly available computing technology, the influx of mathematical statistics and probability theory into the analysis of uncertainty in hydrology and water resources in the 1960s was powerful and unprecedented. Also unprecedented has been its side effect – an outburst of "applications" of various spurious theories suddenly made so easy by the computer. By the mid-1970s the danger of this malignant growth was already evident and solitary warning voices could be heard. One of the clearest and most passionate was that of the late Myron Fiering of Harvard University: "Fascination with automatic computation has encouraged a new set of mathematical formalisms simply because they now can be computed; we have not often enough asked ourselves whether they ought to be computed or whether they make any difference . . . we build models to serve models to serve models to serve models, and with all the computation, accumulated truncation, roundoff, sloppy thinking, and sources of intellectual slippage, there is some question as to how reliable are the final results" (Fiering 1976). A partial answer to this question was provided a few years later by Fiering's colleague Peter Rogers who, in a memorable review of a book describing models for assessment of water-quality-based environmental hazards (Rogers 1983), opined that the models themselves were "hazardous to use . . . for practical applications or policy decisions."

* Victoria, BC, Canada.

With regard to risk analysis for dam safety, Moser and Stakhiv (1987) observed: "There seems to be a growing divergence in the burgeoning practice of risk analysis. This divergence appears to be between the theoretical approaches of academic researchers and the pragmatic orientation of practitioners . . . The dichotomy stems from the pursuit of traditional and more refined, but narrowly focussed, analytical techniques that allow decision makers to become increasingly precise about very restricted and constrained formulations of complex social problems."

In the intervening years, mathematical models for risk analysis, decisions under uncertainty, etc., have been further advanced, their theory refined, and their divergence from reality has often reached what seems to be a point of no return: They have become an end in themselves, intellectual parlor games played behind a façade of practical-looking jargon. It comes as no surprise that, in his highly acclaimed book on the history of risk, Peter L. Bernstein (1996) suspects that risk management has run out of control because people are putting blind faith in formulae. Indeed, paraphrasing Rogers, one is tempted to say: "Reliance on the so-called rigorous mathematical approaches to risk analysis may pose the greatest risk to the planning and design of water resources systems," a statement meant to be factual rather than flippant.

This claim is based on these facts: (1) rigorous risk analysis requires a precise specification of the probabilities associated with the various physical and social events and processes analyzed; (2) these probabilities are known only very approximately and their "exactness" implied by mathematical formulae describing them is illusory; (3) uncertainty about these probabilities is irreducible and cannot be removed by increasing the mathematical rigor of their description and estimation from empirical data; (4) under the present rapidly changing conditions, a credible estimation of these probabilities is becoming ever more elusive; (5) as economist F. Knight pointed out earlier this century, there is more to risk than mere probability – and the often bizarre external circumstances that mold the real-life uncertainties and play havoc with "rational probabilities" have not become any more rational, controllable, and predictable.

The result is a paradox, which is as dangerous as it is ironic: As the theory becomes more refined and rigorous, its assumptions, which it so painstakingly manipulates, are becoming less and less credible. What originated as an ambition to bring more objectivity, wisdom, and realism into the treatment of real-life uncertainties has developed into an ostentatious cleverness of games played with probabilistic chimeras conjured up "for practical purposes" and, lately, with "scenarios" often disguising a mix of ignorance and banalities as deeply scientifically reasoned "possibilities."

3.2 FOR PRACTICAL PURPOSES . . .

It is axiomatic that theoretical problems, however difficult, are more conducive to theoretical analysis than practical ones. A theoretical problem is entirely in the hands of the analyst. He formulates it, specifies the axioms and assumptions that will circumscribe his approach, and defines the rules of the game. With a practical problem things are more difficult since the analyst has little control over it. The problem is posed by the outside world – by nature, by society. Its solution has to abide by the rules of this outer world – by the laws under which the natural and social processes operate and which the analyst cannot arbitrarily create and impose but, at best, only discover – a fact that is often conveniently overlooked or ignored.

This discovery is the task of the empirical sciences. Cumulatively, these discoveries create the knowledge which must be respected and its total sets the limits to the theory of any real-life, or practical, problem. Here we reach the crossroads at which the road of speculation splits off from that of science. Here a prominent sign should be erected displaying the following maxim of Confucius, a maxim that should be the basis of the code of ethics of all science:

> If you know something, hold that you know it;
> if you do not, admit the fact – this is knowledge.

As is true of other maxims, this too is only an ideal, a light to illuminate the way rather than a summit to be reached, but it is a useful guide to honest scientific practice and a safeguard against its abuse.

Perhaps nowhere is such a safeguard more necessary than in the theoretical analyses of uncertainties inherent in real-life situations and of the risks (to human life, property, environment, security, etc.) that our responses to these situations entail. Because here we are attempting to tackle a very ambitious if not impossible task: a quantitative assessment of our degree of ignorance. To know how much we do not know, to measure and weigh it, is this not in itself a self-delusion, an arrogant pretense, an oxymoron? Unless we know the totality of what is knowable, how can we subtract from it the knowledge already attained to obtain the amount of the ignorance still outstanding – which is necessary to specify the degree of the uncertainty and quantify the risks?

In the realm of water resources systems, this Herculean task is often trivialized by invoking the magic formula

> "For practical purposes . . ."

Considering the extreme care with which mathematicians tread before venturing into any rigorous analysis despite the fact that they usually claim no practical utility for their results beyond an "aesthetic value" (Hardy 1967), one would assume that if somebody invokes practical purposes for an analysis where human life, economic welfare, and environmental quality may depend on its results, then he would be doubly careful. One would expect to hear something like "While the theory of risk analysis is most interesting, it is of little value for practical purposes since its basic building blocks cannot be reliably quantified and often not even defined . . . ," or "For any practical purposes one would have to make so many unverifiable simplifying assumptions that no rigorous analysis is warranted and decisions involving real-life risk situations must be guided by other considerations," or the like. Not so, at least not yet, in risk and uncertainty analysis of water resources systems. Here, the magic formula "for practical purposes . . ." is invoked to justify, in the same breath, two glaring contradictions.

On the one hand, it serves as a license for the most cavalier treatment of the fundamental issues, in particular for replacing the real unknown probabilities by some arbitrarily postulated fictitious ones, thus sweeping under the carpet the very uncertainties that are supposed to be analyzed.

On the other hand, it is regarded as a license for indulging in the most thorough and refined mathematical massaging of these fictitious probabilistic chimeras, using the "most rigorous," "most efficient," "unbiased," "maximum likelihood," etc., procedures for "extracting the most information" from them. One may naively ask: Information? About what? Perhaps about the imagination of the analyst? Of course not! About exactly those things that had been swept under the carpet, we are led to believe.

Clever! Unbearably clever! Magic? No, bluffing.

Surprisingly, transparent as it is (one does not have to be a Confucius to see through it), the bluff if seldom called. Perhaps there are no children around to say "The king is naked!" – or perhaps they took their Ph.D.s in risk analysis before they learned to speak.

It is true and usually inevitable that simplifying assumptions must be made in attempting to solve a complex practical problem. But by making them one must follow through and respect the consequences they imply. First and foremost, they do not reduce the inherent uncertainty of a problem, just the opposite: they increase it by introducing new unknowns into it. Therefore they dictate the use of a coarser yardstick, not a finer one. And, believe it or not, this is rather well known outside academic circles! For example, when simplifying the shape of a log to a cylinder for the practical purpose of estimating its volume, the logger would measure its diameter and length once, maybe twice, and take the averages. It would be quite difficult to persuade him that he should make twenty measurements with a micrometer, estimate the two parameters by maximum entropy, and calculate the volume to seven decimal places to "extract the most information" from the data.

To be fair, bluffing by hiding a lack of substance behind an overpolished form and inflated appearances is nothing new, only the use of "theory" for this purpose is a relatively new twist in the old game – Don Quixote made do with an old helmet which,

> when he had cleaned and repaired it as well as he could, he perceived there was a material piece wanting; for instead of a complete helmet, there was only a single head-piece: however, his industry supplied that defect; for, with some pasteboard, he made a kind of half-beaver, or vizor, which being fitted to the head-piece, made it look like an entire helmet. Then, to know whether it was cutlass-proof, he drew his sword, and tried its edge upon the pasteboard vizor; but, with the first stroke, he unluckily undid in a moment what he had been a whole week a-doing. He did not like it being broke with so much ease, and therefore to secure it from the like accident, he made it anew, and fenced it with thin plates of iron, which he fixed in the inside of it so artificially, that at last he had reason to be satisfied with the solidity of the work; and so, without any experiment, he resolved it should pass to all intents and purposes for a full and sufficient helmet. (Cervantes 1605)

If the performance of this helmet can be any guide, the current risk analysis models, fenced so artificially with all the rigor, are also destined to make history.

3.3 CLIMATE-CHANGE-IMPACT SCENARIOS: FROM BLUFFING TO METABLUFFING

Scenarios are a very serious business. They attempt to describe, in quantitative terms, variants of future events that can reasonably be expected based on present knowledge and understanding. This is an ambitious task, more exposed to uncertainties and therefore much more difficult than is reconstruction of plausible histories and causes of present events – the main objective of traditional science. Moreover, since the purpose of scenarios is to provide guidance for action (as opposed to that of science which is merely accumulation of knowledge), the extreme difficulty is here combined with extreme social responsibility.

This has been well understood in action-oriented disciplines like engineering or military operations where decisions (design and operation parameters, contingency plans, etc.) based on an unrealistic scenario may lead to embarrassments and disasters well within the lifetime of its originators who may then face unpleasant consequences including loss of jobs, fines, prison terms, even death sentences. Understandably, construction of scenarios has traditionally been approached with utmost care and entrusted only to teams of the most experienced and knowledgeable professionals.

However, the status and credibility of scenarios have been considerably eroded during recent years and, in my opinion, much of the blame must go to their widespread abuse and trivialization in climate-change-impact modeling. Here this was brought about by the collusion of several factors, namely (1) the long time-horizons involved (typically over fifty years) which practically absolve the originators of any personal risks associated with eventual falsification of their creations, thus diluting their sense of responsibility and encouraging a cavalier approach; (2) willingness, whether motivated by sincere good will or by calculated self-interest, to yield to political or management pressures for providing answers to problems far beyond the current state of knowledge (as discussed by Rogers [1983] and Klemeš [1991]); and (3) the ease with which scenarios involving "impacts" of any arbitrarily imposed "climate changes" can now be produced even by amateurs and dilettantes whose grasp of problems does not extend beyond an ability to insert "DO-loops" into the various models to which they may have gained access.

This has led to metabluffing where, in contrast to ordinary bluffing described above, it is not just the various questionable approximations of real (historic) events that are meticulously polished and presented as rigorous science, but concoctions produced by arbitrary and often physically incongruent changes of model parameters, process realizations that may be unrealizeable under the known physical laws.

On the other hand, even responsible and professional attempts to construct scenarios may be counterproductive if the scientific basis is inadequate vis-a-vis the complexity of the problem at hand. Such efforts may result in nothing more than stating the obvious in a spuriously convoluted manner indistinguishable from deliberate bluffing.

The latter point can be illustrated by the following example: It has often been pointed out (e.g., Klemeš 1982, 1991, 1992; Rogers 1991; Kennedy 1991) that not much more can be said about the hydrological effects of a possible climate change beyond the fact that it introduces another source of uncertainty into water management. Specifically, I summarized the achievements of a decade of "climate-change-impact" modeling in these words: "Basically, the only credible information obtained from the complex hydrological modelling exercises related to climate variability and change is that, if the climate becomes drier, there will be less water available (and opposite for a wetter climate). It appears to me that not much science is needed to ascertain as much" (Klemeš 1990). Five years and untold numbers of scientific scenarios later, the Intergovernmental Panel on Climate Change (1995) tells us that it took an "enormous intellectual and physical effort . . . of scientists and other experts world-wide" to arrive at conclusions like these: "More intense rainfall would tend to increase runoff and the risk of flooding . . . A warmer climate could decrease the proportion of precipitation falling as snow . . ."

O, sancta simplicitas!

3.4 IN PRAISE OF THEORY AND ROBUST RESULTS

When, thirty years ago, I was translating into Czech the famous book on stochastic storage theory written by the late Professor P. A. P. Moran (1959), I was struck by his repeated caveats regarding its practical applicability. No later than in the brief Preface, one keeps coming across statements like "It being difficult to obtain explicit solutions for the finite dam we attempt to simplify the problem. . . . The assumption of an input which is a type III additive process is certainly far removed from anything plausible in practice . . . This simple assumption would not often be true in practice . . . It is unlikely that in practice the assumptions underlying the model are sufficiently well verified . . . ," etc., expressing only a modest hope "that the theory given in this book may be helpful, if only as background knowledge, to those engaged in practical problems. . . ."

Professor Moran's modesty notwithstanding, it is precisely this background knowledge that can be a very valuable asset for the practitioner. The value of good theory is in facilitating a broad and deep general understanding, not in supplying a detailed recipe for solving each and every specific problem. Good theory helps the practitioner to develop a feel for the problem and provides a robust framework for guiding his judgment, for sorting the wheat from the chaff.

Moran's book itself offers several examples. For instance, one of the theoretical results led Moran to a conclusion that the distribution of reservoir storage "does not depend on the exact form of the [whole] distribution [of inflow] . . . ," so that ". . . the probability behaviour of the dam is chiefly dependent on the shape and location of the main part of the distribution of [input] and not on its tail" (p. 43).

Another example relates to the estimation of the probabilities of extreme floods – a fundamental problem in water resources risk analysis. On this, the background knowledge provided by the theory is summarized by Moran in these words: ". . . there is no theoretical reason why [the commonly fitted simple distributions] should fit observed series . . . It is important to realize that in this procedure there are two types of error, one arising from the fact that we do not know the true analytical form of f(x), and the other from the fact that the estimation of the parameters is based on a finite (and usually small) number of observations. *These errors are essential and cannot be avoided.* . . . the form of the tail of f(x) . . . is usually outside the range of the observations and can only be guessed at. *No amount of mathematical prestidigitation can remove this uncertainty*" (pp. 94–95; emphasis added).

Such conclusions, based on a purely theoretical analysis, are of a first-rate practical import: They give the lie to all the bluffing about the need for the mathematical prestidigitation in fitting distribution models to streamflow data in practical risk analyses of reservoir performance: When the risk of a reservoir running dry is of concern, there is no point in it since many different theoretical distributions will fit nicely the body of the empirical flow distribution as long as they preserve the mean and the variance of the data. When, on the other hand, the risk of flooding is of concern, it is irrelevant since the distributions being polished are only guesses anyway.

Over the past forty years or so, theoretical research has produced a number of similar robust results applicable in the risk analysis of water resources systems. One of the latest examples I am familiar with is due to Professor E. H. Lloyd, well known to hydrologists for his fundamental contributions to the stochastic reservoir theory. Recognizing the fact that we usually do not know the distribution type F of hydrological extremes, Lloyd (1996) has derived a simple formula for the confidence interval of the exceedance probability P_1 of the largest, that is, rank-one (annual) observation from an n-year record on the assumption that F is unknown. Retaining only the assumption of independence (which usually is not contradicted by the data), he established the lower and upper limits of a (central) confidence interval of P_1 as $1 - (1 - \theta)^{1/n}$ and $1 - \theta^{1/n}$, respectively (the width of the interval being $1 - 2\theta$). Thus, for example, for $n = 50$, the 95% confidence interval of P_1 (conventionally estimated by the expected value $1/(n + 1)$ of its distribution, i.e., $P_1 = 1/51 = 0.0196$ in this case) goes from 0.0005 to 0.071. In other words, we can specify the "true" return period of our "observed 50-year event" no better than by saying that there is a 95 percent probability that it will be somewhere between about fourteen and two thousand years. Anything beyond this kind of specification is speculation, notwithstanding any mathematical legerdemain by which it could have been obtained.

Robust theoretical results like these have little chance of being cited (not to say recommended!) by those who have built their careers on rigorous risk analysis, and for good reasons: Give or take a dozen of such results and the risk analysis experts could pack up and go home. Indeed, robust theoretical results pose a great risk to the burgeoning business of academic risk analysis. Therefore, let us end this section with the secret version of the academic risk analyst's *Pater Noster:*

Our Father,
Give us today our daily job security,
Lead us not into temptation to face reality,
And deliver us from robust results,
Amen.

3.5 A REALITY CHECK

The most formidable force which operates in the domain of water-related risk and which crushes under its feet the pretentious bluffing and the sound theory alike is Naked Reality. Moser and Stakhiv (1987) call it more deferentially "complex social problems" but this gallant label subtracts nothing from its raw and brutal nature and destructiveness.

Let it be made clear at the outset that, for theoretical purposes, the following observations about this force and its effects are of little significance. They cover just one forty-year record of a nonhomogeneous and highly nonstationary process with many missing data. But they may have some value for purely practical purposes.

Shortly after World War II, I came across a book describing what was called the Seven Wonders of the Modern World. Among them were the Taj Mahal, the Empire State Building, the Panama Canal, but the one that impressed me most (it still does) was the Hoover Dam.

Ten years later I was working in the Dams Division of the Water Resources Development Centre in Brno on a preliminary project of my first dam. At its ten-or-so metres of height, it was not quite a new wonder of the world but, then, it was not built either. I did not understand why since the cost-benefit analysis was very favorable, the water needs that it was supposed to serve seemed to me genuine and undisputable, and the risks that it mitigated against seemed clear and beyond reasonable doubt, so to speak. However, from one day to the next and by one stroke of a pen somewhere in the highest strata of the bureaucracy, all the risks and benefits were apparently conjured away and the work on the dam was stopped: There simply was a "change in the priorities" and I was assigned to work on another dam.

By that time I had read what I still consider to be one of the best books ever written on the design and operation of storage reservoirs, the *Water Resources Computations* by Kritskiy and Menkel (1952), where the concept of risk resulting from hydrological uncertainty was first explicitly related to the economics of the project. They suggested that it would be possible to arrive at optimal reliability levels for reservoir operation by relating the losses due to water shortages to the costs of their prevention, and pointed out that efforts in this direction were in their infancy. Here was an opportunity to advance the scientific level of risk analysis and I was eager to take it!

Since the hydrologic uncertainties seemed not much of a problem (hydrology textbooks of the day often presented, for example, the Pearson III distribution as something of a physical law governing the annual flows), all that seemed to be needed was an accurate determination of water-shortage-related losses in different economic sectors. I set out to do this for the region which our office was serving and thought it would be a rather straightforward exercise: Our office had the data on water requirements of individual users (they were the basis for our proposals for new dams) as well as the historic flow records, so I knew exactly where and when water shortages occurred. With these data in my briefcase, I went to steel mills, coal mines, chemical factories, fossil-fuel power plants, etc., trying to identify the losses that these shortages must have caused.

After several months of such detective work I came out empty-handed. The losses either did not occur (the official water requirements were inflated "to be on the safe side"); or were claimed to be "within normal fluctuations of output" caused by other factors like power failures, absenteeism, equipment failure, etc.; or were masked by ad hoc arrangements like emergency recirculation of cooling water (causing a slight but acceptable overheating of equipment), shutting down a part of the plant and doing the "scheduled maintenance" in it, sending employees on their annual leave ("they had to take it one time or other"); or were covered up by ingenious accounting procedures in order not to jeopardize bonus payments for a "100% fulfilment of the production plan"; or the losses actually reported were readily denied and explained away when I asked whether the plant would be willing to share the cost of their mitigation by a new dam (dams were normally financed from the state budget); or I was denied information under various excuses when some shady practices had been involved or some adverse effects of my survey were suspected.

My report gave a considerable headache to my superiors exactly because it shed light on some real uncertainties and, if followed up, could undermine the very way our office was doing its business. Even more important, it could bring into the open the many questionable practices inherent in "our advanced socialist system" which it was safer to keep behind closed doors. To mitigate against these real risks, my report was stamped SECRET and filed away – for all practical purposes it simply ceased to exist, in particular for me since I did not have clearance for working with classified documents. Only later did I squeeze a paragraph on my findings into my "Candidate of Science" dissertation (Klemeš 1963), pointing out the "often very narrow and wrong views of economic efficiency."

And so I learned my first practical lesson about the real risks and uncertainties in water resources management.

Let the reader not be misled! The above story is not just an irrelevant fossil from a bizarre and now vanished world. Similar situations keep arising, in different disguises, all around us, hidden away and not existing "for practical purposes." To give just one more example: Floodplain maps indicating the extent of twenty-year and one-hundred-year inundations were prepared in New South Wales between 1977 and 1984 to identify hydrological risks posed to development. However, as a result, landowners in those areas found it increasingly difficult to secure development loans and, by 1984, their pressures became so strong that the State Premier in his pre-election speech indicated that no maps existed and any that did had been withdrawn. And so they had. They are not available from the State and officially no longer exist (the university has a near complete set). And the loans are available again and, at another university just a few miles away, state-sponsored (I presume) research on Bayesian, robust, most efficient, etc., methods for flood probability estimation is making further breakthroughs.

Going back from floods to droughts again, one of my next dams was to augment low flows for a variety of water needs, each supported by claims of drastic economical consequences if the requirements were not met. The dam, to be constructed in a narrow gorge opening up into a wide valley, had an outstanding "economic effectiveness" based on the benefit-to-cost ratio but it also had the most devastating intangible effects, including a profound disruption of the whole social structure of the area due to the severe reduction of its agricultural base, relocation of several hundreds of inhabitants from five villages, and the crippling of the hundreds of years old local cross-valley road network, the flooding of unique aragonite caves, and destruction of a highly regarded thermal spa.

Although everybody knew the dam should not be built at that site, there was no way to stop it since economic effectiveness was the only criterion the communist regime officially recognized, a dogma. Challenging it by citing intangibles would be labeled "reactionary idealism" and treated as "sabotage of socialism" – and I had already learned my lesson in practical risk analysis. My feeble protest that there would be a risk of a catastrophe with enormous consequences if the water in the dam were ever contaminated with phenols in case of some failure in a huge tar-processing plant which was to be built immediately upstream of the reservoir was dismissed: "There is no such risk, comrade – the plant will be built according to Soviet specifications and there will be no failures, period."

Bizarre as it may sound, I killed the dam by a "conspiracy" improvised on the spur of the moment. When once, as often before, a Health Ministry official, an old medical doctor, came to my office to cry on my shoulder over the fate of the spa, imploring me "to do something" to save it, I snapped back in frustration: "If you certify it is worth 300 million I will save it" – for this was roughly the figure needed to make the dam's benefit-cost ratio unattractive. When he protested the impossibility of putting a value on human life, I told him the whole affair was just a game of numbers needed to mitigate against an irresponsible action of the government and offered him advice on how he could help me play it. I said, "Look, the economy will lose so and so many tons of coal or steel if a sick miner or steel worker will be convalescing several weeks longer because of the nonavailability of the spa. A ton of coal or steel costs so and so much money, there are so and so many patients treated in the spa and the lifespan of the dam is so and so many years; so you multiply this number with that and juggle the numbers of weeks and tons and interest rates, etc., until you come up with a present value of 300 million." He was impressed. Within a month I had a letter signed by the deputy minister of health giving the loss to the socialist economy due to the loss of the spa at (if I remember correctly) 356 million crowns.

The true risk of this analysis was that the doctor and I could have spent the rest of our lives in the Gulag if our "conspiracy" came to light. But the dam has not been built yet. The tar-treatment plant has.

Another socially complex situation arose in the late 1950s when a respectable explanation had to be found for the growing food shortages caused by the forced collectivization of agriculture. It was found in "shortage of water" and a massive program of irrigation was declared to be the savior. I was assigned to prepare an investment proposal for one of the large reservoirs for the irrigation water supply. Located in a flat alluvial plane in my native south Moravia, it was an awkward project which would be justified only in the most dire circumstances. But there was no way out of it because it had been identified "elsewhere" (meaning the highest circles of the communist party) as the key project of the program; and questioning the program would of course be a "counter-revolutionary act of sabotage."

Using the reservoir as a proxy vehicle for political dissent, biologists and ecologists mounted a fierce attack on it, predicting all kinds of risks, disasters, and catastrophes it would precipitate. Ironically, at the same time they were extolling ecological virtues of a similar and only slightly smaller system of reservoirs built in the fifteenth century just 10 kilometers away, requesting that these reservoirs be declared a protected nature reserve.

While I privately sympathized with the environmentalists' efforts to kill the dam for which I saw no need, my strategy was to do it by embellishing its design by all conceivable remedial environmental measures, outdoing even the environmentalists' own proposals (to their great frustration since this was weakening their arguments against the dam) in the hope of driving the cost of the project so high that it would kill its economic effectiveness. However, neither my nor their strategy worked and the dam was eventually built. So far, none of the foretold ecological disasters has materialized but one which nobody had predicted has – and it had a very positive effect on the mitigation of one risk which was none of the dam's business: It forced the implementation of an extensive wastewater treatment program upstream since the polluted waters from the whole basin, when dumped into the reservoir, made it a cesspool bathing the whole countryside in the stench of thousands of fish rotting in it under the hot south Moravian sun and attracting unwanted attention of local as well as foreign tourists and the media.

A much more spectacular, and by now internationally famous, example of a situation where a water resources system has become a proxy target of political dissent disguised as environmental movement is the Gabčíkovo-Nagymaros project on the Hungarian – Slovak section of the Danube River. It helped bring about the collapse of the communist regime in Hungary, strained the Hungarian–Slovak relations in the post-communist era, and the merits of the disputes about its fate and its water-related risks are to be decided not by the academic risk analysts or by the practicing water resource engineers or ecologists but by lawyers at the International Court of Justice in the Hague (Molnár 1996).

My last example is drawn from a risk analysis I attempted under the benign conditions of Canadian democracy over twenty years ago, shortly after joining the Inland Waters Directorate of the federal government. It involved a challenge of the native population against the James Bay hydroelectric development in Quebec, citing environmental and other risks and pointing out the hydrological uncertainties due to the sparse data on which the claims of the power output, flood protection, etc., were based. I set to work and was soon satisfied that the claimants had a strong case against Quebec Hydro. However, I also soon encountered puzzling obstacles and a lack of cooperation of the officials of the federal Department of Indian and Northern Affairs who had originally approached me with the request. One day I was quietly told that my "digging" into the matter was no longer desirable. The water management and environmental uncertainties, together with the risks of disruption of the native lifestyle, etc., had apparently all but vanished from one day to the next, following a generous cash-and-land settlement offered to the natives by the Quebec government and announced shortly after: The 16,000 affected native inhabitants were awarded $225 million (in 1975 $) plus 14,000 km^2 of land plus exclusive fishing, trapping, and harvesting rights over another 150,000 km^2 (a territory the size of half of Italy) plus an income security program plus some other government-guaranteed benefits (Smith 1995). This time my report was not stamped SECRET and filed away – I was discouraged from writing one.

The litany could go on and on, for instance, contrasting the scientific battles fought in the United States over the best flood frequency distribution (US Water Resources Council 1977) with the real uncertainties affecting the floods and with the forces shaping the American flood control policies (Kazmann 1995), but let me stop.

The moral of these anecdotes is this: The greatest uncertainties and risks in water resources systems arise from the fact that water resources are often used as proxies over which political battles, motivated by greed and lust for power, are waged. And the greatest risk to an honest analyst of these risks is that he may easily be crushed between the millstones grinding one against the other in these battles.

3.6 CONCLUSIONS, OR A TALE ABOUT UNKUNKS, KUNKS, AND SKUNKS

"Unkunk" was a label that the U.S. Air Force was using for unpredictable problems, or unknown unknowns (Linstone 1977). Accordingly, "kunk" can be defined as a known unknown; and, as is well known, skunk is a known which stinks.

Only kunks justify analyses by rigorous mathematical methods. These unknowns are precisely specified by known probability distributions, sampling rules, operating rules, etc. Their analysis requires the most efficient methods since their every point and state contains a piece of pure information about the behavior of the system. Their analyses and theories often provide illuminating insights and robust results which might be helpful in practical applications – especially if we lived in a more honest and reasonable world.

Real-life uncertainties and risks have the nature of unkunks, with the salutary exception of the kunks one faces in lotteries and other games of chance. Ironically, short of deliberate fraud, gambling is the most honest risk game in town. Compared with the academic risk games involving water resources systems, gambling is as innocent and straightforward as a charity bake sale in your local church. Indeed, in the casinos of Las Vegas or Monte Carlo, it would be a criminal offense if unkunks were presented to clients as kunks for the practical purposes of betting on them. Not so in academe – but this changes nothing in the fact that unkunks represented as kunks become skunks.

For practical purposes, the conclusions can thus be formulated as the following recommendations:

Kunks should be treated with rigor.
Unkunks should be treated with care.
Skunks should be avoided.

Because, as historian Daniel Boorstin once noted (Koch 1996), "The greatest obstacle to discovery is not ignorance, it is the illusion of knowledge."

REFERENCES

Bernstein, P. L. (1996) *Against the Gods*. John Wiley, London.

Cervantes Saavedra, M. de (1605) *Don Quixote*. Madrid (English translation: Wordsworth Classics, Ware, Hertfordshire, 1993).

Fiering, M. B. (1976) Reservoir planning and operation. In *Stochastic Approaches to Water Resources*, Vol. 2, Chap. 17, pp. 1–21. Edited and published by H. W. Shen, Fort Collins, Colorado.

Hardy, G. H. (1967) A *Mathematician's Apology*. Cambridge University Press, London.

Intergovernmental Panel on Climate Change (1995) *Climate Change 1995: Impacts, Adaptations, and Mitigation*. Summary for Policymakers. WMO and UNEP.

Kazmann, R. G. (1995) Flood control and the real world. *The Louisiana Civil Engineer* (May), pp. 4, 5, 17.

Kennedy, D. N. (1991) The political and institutional constraints of responding to climate change. In *Climate Change and Water Resources Management*, IWR Report 93-R17. Edited by T. M. Ballentine and E. Z. Stakhiv, pp. I-19–24. National Technical Information Services, Springfield, Virginia.

Klemeš, V. (1963) *A Water-Resources Assessment of Streamflow Changes Caused by a Storage Reservoir* (in Czech), CSc. Thesis. Slovak Technical University, Bratislava.

(1982) Effect of hypothetical climatic change on reliability of water supply from reservoirs. In V. Klemeš (1985), *Sensitivity of Water Resources Systems to Climate Variations*. WCP-98, Annex B, WMO, Geneva.

(1990) Sensitivity of water resource systems to climatic variability. In *Proceedings, Canadian Water Resources Association 43rd Annual Conference*, Penticton, B.C., pp. 233–42.

(1991) Design implications of climate change. In *Climate Change and Water Resources Management*, IWR Report 93-R17. Edited by T. M. Ballentine and E. Z. Stakhiv, pp. III-9–19, National Technical Information Services, Springfield, Virginia.

(1992) Implications of possible climate change on water management and development. *Water News* (April), S2–3, Canadian Water Resources Association.

Koch, C. (1996) Market-based management. *Imprimis* 25(8): 4–8, Hillsdale College, Hillsdale, Michigan.

Kritskiy, S. N. and Menkel, M. F. (1952) *Vodokhozyaystvennye Raschoty*. Gidrometeorologicheskoe Izdatelstvo, Leningrad.

Linstone, H. A. (1977) Questioning the methodology; Introduction. In *Futures Research*, Part I. Edited by H. A. Linstone and W. H. C. Simmonds, pp. 133–43. Addison-Wesley, Reading, Massachusetts.

Lloyd, E. H. (1996) A note on the exceedance probability of the sample maximum and the exceedance probabilities of other order statistics. Unpublished report.

Molnár, P. (1996) The Gabčíkovo-Nagymaros scheme on the Danube River – a review. In *Proceedings of the Sixteenth Annual American Geophysical Union HYDROLOGY DAYS*, edited by H. J. Morel-Seytoux, pp. 329–40. Hydrology Days Publications, Atherton, California.

Moran, P. A. P. (1959) *The Theory of Storage*. Methuen, London.

Moser, D. A. and Stakhiv, E. Z. (1987) Risk analysis considerations for dam safety. In *Engineering Reliability and Risk in Water Resources*, Proceedings of NATO ASI held at Tucson, Arizona, May 19–June 2, 1985, edited by L. Duckstein and E. J. Plate, pp. 175–98. M. Nijhoff Publishers, Dordrecht.

Rogers, P. P. (1983) Review of *Analyzing Natural Systems: Analysis for Regional Residuals – Environmental Quality Management*, edited by D. J. Basta and B. T. Bower. Resources for the Future, Washington, D.C., 1982. *Eos*, 64, p. 419.

(1991) What water managers and planners need to know about climate change and water resources management. In *Climate Change and Water Resources Management*, IWR Report 93-R17. Edited by T. M. Ballentine and E. Z. Stakhiv, pp. I-1–13. National Technical Information Services, Springfield, Virginia.

Smith, M. H. (1995) *Our Home or Native Land?* Crown Western, Victoria B.C.

U.S. Water Resources Council, Hydrology Committee (1977) *Guidelines for Determining Flood Frequency Information*. Bulletin 17A. Washington, D.C.

4 Aspects of uncertainty, reliability, and risk in flood forecasting systems incorporating weather radar

ROBERT J. MOORE*

ABSTRACT

Uncertainty in flood forecasts is dominated by errors in the measurements and forecasts of rainfall used as input. Error magnitudes are influenced by raingauge network density (possibly used in combination with weather radar), by rainfall intensity, and by the method of rainfall forecasting employed. An empirical approach to quantifying uncertainty associated with rainfall measurements and forecasts is taken, supported by data from two dense raingauge networks, together with weather radars, in southern Britain. The impact of uncertainty in rainfall on flood forecasts is examined through a comprehensive case study within the Thames basin in the vicinity of London. This study also allows the relative effect of model and catchment on flood forecast uncertainty to be better appreciated. Reliability of flood forecasts is considered in the context of the complexity of region-wide flood forecasting systems and the need to ensure that forecasts are made under all situations, including the possible loss of significant telemetry data. The River Flow Forecasting System's Information Control Algorithm is outlined as a solution to providing reliable forecasts, coping with both complexity and data loss. Risk is considered here in the context of when to issue a flood warning given an uncertain flood forecast. The use of both informal and more formal methods of ensemble forecasting is introduced as a means of quantifying the likelihood of flooding implied by a flood forecast.

4.1 INTRODUCTION

Flood forecasting systems function in real-time to transform telemetered field measurements (principally relating to river level and rainfall) and external forecasts (especially of weather) to forecasts of river level and flow, possibly along with settings associated with river control structures and reservoirs. Elements making up the system, including inputs and models, are associated with a variety of forms of uncertainty. The configuration of flood forecasting systems for region-wide areas is typically complex. Systems need to be reliable in the way they cope with complexity and possible data loss. Forecasts are ultimately used to decide on whether to issue a flood warning, which involves assessing the risk of being wrong given an uncertain forecast. Selected aspects of uncertainty, reliability, and risk associated with flood forecasing systems are considered in this chapter, with an emphasis on empirical results and operational systems.

4.2 UNCERTAINTY IN FLOOD FORECASTS

4.2.1 Introduction

Uncertainty in flood forecasts derives from a number of causes. These include the inadequacy of a model as a representation of a much more complex reality, the form of configuration of models making up a river network based forecasting system, errors in calibration of models using historical measurements, and uncertainty of the inputs to the forecasting system. This last cause is generally regarded as the most influential, particularly as they relate to measurements of rainfall as estimates of basin average rainfall and to forecasts of rainfall required for extended lead-time river flow and level forecasts using rainfall-runoff models. It is this source of forecast uncertainty that will be considered here with reference to both conventional raingauge networks and weather radar. An empirical approach to understanding rainfall input uncertainty, and its effect on flood forecasting uncertainty, is taken from the standpoint that errors are both highly heterogeneous (as rainfall error properties vary with intensity and weather dynamics), and nonlinear (for example in terms of the catchment reponse to rainfall).

4.2.2 Rainfall measurement uncertainty

The variability of rainfall in space and time makes it difficult to measure over basin-wide areas, particularly using point

* Institute of Hydrology (now Centre for Ecology and Hydrology), Wallingford, OX10 8BB, United Kingdom.

measurement techniques such as a tipping-bucket raingauge. Weather radar has the advantage over gauges in providing a spatially integrated measure, typically for 2 by 2 km grid square areas out to a range of 76 km. However, the sampling space actually relates to a volume of the atmosphere that increases in size and height above the ground with increasing range from the radar. Adjustments are thus required to infer ground-level rainfall for flood forecasting purposes, introducing further uncertainty. It is common today to combine physically-based radar adjustments with a final correction using a network of telemetering raingauges (Moore et al. 1994a), capitalizing on radar's ability to provide spatial measurements of the rainfall field and the raingauge's strength in providing more accurate point measurements of rainfall at the ground.

Effect of network density

It has been possible to quantify the uncertainty associated with rainfall estimates derived from raingauge networks, from weather radars, and from a combination of the two. Particularly important is how the uncertainty of rainfall estimates varies with the number of raingauges employed. An operational raingauge network, used for flood warning over London, provided the source of data underpinning the investigation. A set of forty-four gauges over a 60 by 60 km area provided concurrent data for thirteen storm events over the period June 1990 to April 1991. Data from the London weather radar at Chenies were also available for this period. Previous work (Moore et al. 1994a) had developed a methodology for combining the two data sources based on multiquadric surface fitting to calibration factor values, defined as a ratio of gauge to coincident radar estimates of rainfall. This work demonstrated, across twenty-three storm events, that an average accuracy improvement of 22 percent could be achieved, relative to an uncalibrated radar, when using up to thirty raingauges over this same area.

The approach used selected alternative "design networks," containing progressively fewer gauges, from the set of forty-four available and assessed their accuracy in turn. For a given network, the multiquadric surface calibration method was employed to estimate rainfall at raingauge locations not used in the calibration process and the error in estimation for these gauges pooled within and between events to form a performance criterion measure. Gauges were selected for deletion to create the network at the next density level based on first identifying the pair of gauges with the smallest inter-gauge distance and deleting the one that is nearest to any of the others. The result is a set of forty-three design networks containing from forty-four to two gauges and varying in terms of "mean distance between gauges" from 6.64 km to 78.6 km.

The procedure for forming the estimation errors at each fifteen-minute time-step, which are pooled to calculate a performance measure for each network design within and between events, is as follows. Starting with an n gauge network, one gauge is selected for omission and regarded as the true value at that point, R_g. The remaining $n - 1$ gauges are used to fit the multiquadric calibration factor surface, which is applied to the radar grid values to derive a recalibrated rainfall field and specifically an estimate of the rainfall, r, at the location of the raingauge selected for omission. The estimation error between R_g and r is calculated as the log error adjusted for quantization errors associated with the tipping-bucket raingauge; that is

$$e = \log\{1 + e'/(10 + r)\} \tag{4.1a}$$

$$\text{where } e' = \begin{cases} \varepsilon - \delta_g & \varepsilon > \delta_g, \\ 0 & |\varepsilon| \le \delta_g, \\ \varepsilon + \delta_g & < -\delta_g, \end{cases} \tag{4.1b}$$

$$\text{and} \quad \varepsilon = R_g - r. \tag{4.1c}$$

Here data units are mm h^{-1} and δ_g is the rainfall intensity corresponding to one tip of the raingauge in a fifteen-minute time-step. This procedure is repeated n times, for each gauge omission, and n errors obtained for the time-step. This is repeated for all time-steps in the event and for all events and the resulting n errors finally pooled in the root mean log error squared performance measure

$$rmse = (n^{-1}\Sigma e^2)^{\frac{1}{2}}. \tag{4.2}$$

If this same procedure was invoked for all design networks, then there would be progressively fewer errors making up the performance measure resulting in a lack of comparability in performance across networks. This problem was circumvented as follows. Using the network of $n - x$ gauges, where x gauges have been removed, then $n - x$ errors can be calculated as above and associated with a network of $n - x - 1$ gauges (the number of gauges used for calibration). Now use the $n - x - 1$ network to form another calibration factor surface and use this to estimate the rainfall at the x omitted gauges, giving a further x errors at each time-step. In this way there are a total of n errors calculated for a given time-step, irrespective of the number of gauges, x, omitted to form the design network. Hence, the pooled *rmse* will be made up of the same number of errors for each design network, thereby ensuring comparability.

Figure 4.1 shows the results obtained when the *rmse* criterion is pooled across the thirteen events and plotted against the number of gauges in the design network. A similar multiquadric surface fitting method has been used to infer the rainfall field from the raingauge network alone, this time fitting the surface directly to the point raingauge measurements of rainfall. The *rmse* resulting from this "raingauge-only" estimate as a func-

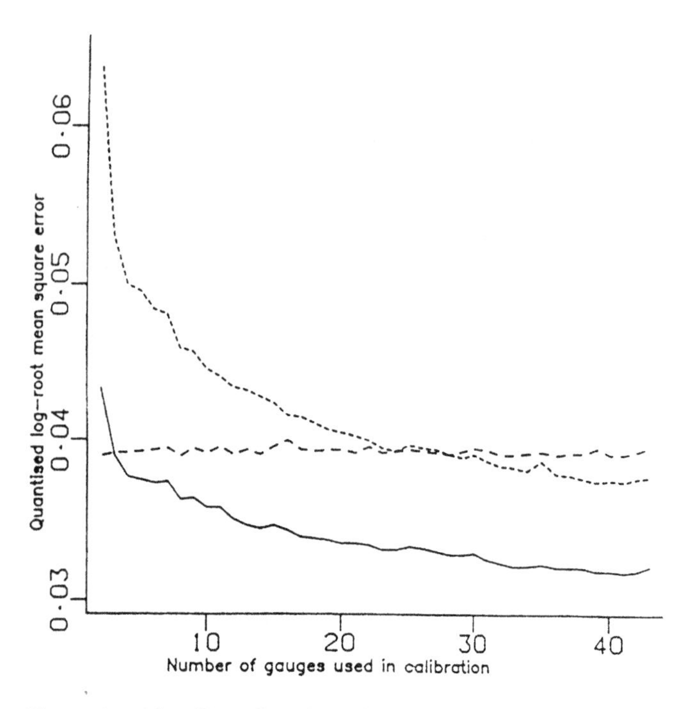

Figure 4.1. The effect of number of gauges on rainfall estimation accuracy. Results averaged over 13 events. Continuous line: calibrated radar; short dashes: raingauge-only; long dashes: uncalibrated radar.

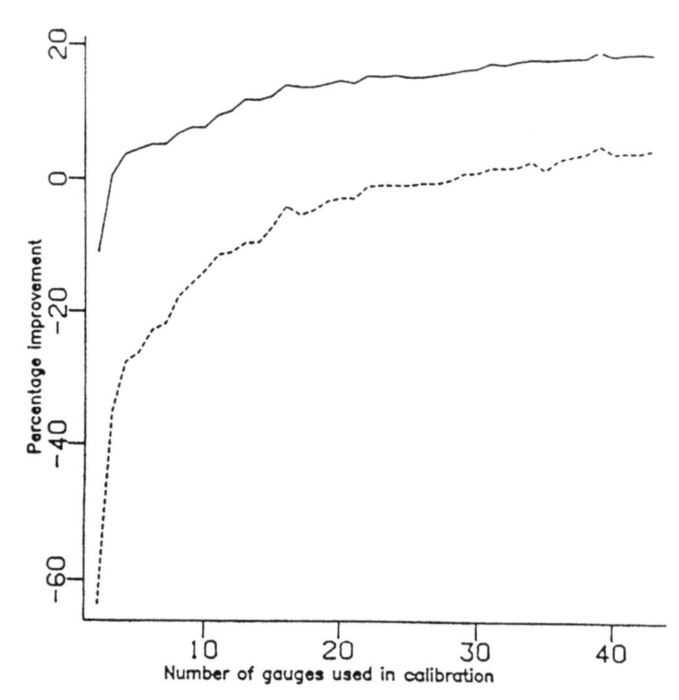

Figure 4.2. Percentage improvement in performance of raingauge-only (dashed line) and calibrated radar (continuous line) estimates of rainfall, relative to uncalibrated radar, as a function of number of raingauges in network. Results averaged over 13 events.

tion of number of gauges in each design network is shown on the same figure. Finally, the *rmse* calculated using the "radar-only" rainfall estimate, without using raingauges for calibration, is also displayed. Figure 4.2 shows the results expressed in terms of percentage improvement relative to uncalibrated radar data. This illustrates that as many as thirty gauges are required before the accuracy of the uncalibrated radar can be achieved, equivalent to a spacing of 9.2 km between raingauges. The accuracy of the calibrated radar is seen to continue to increase with increasing numbers of raingauges but at a progressively diminishing rate. This information might be used to support a choice of raingauge network to be used in combination with a weather radar.

A more detailed analysis on an event-by-event basis reveals that there is considerable variation about the average results arrived at by pooling across the thirteen events. As a result, any recommendation concerning an optimal number of calibrating raingauges must be highly circumspect. Transient factors influencing the optimal number include the effect of bright-band due to melting precipitation and the problem of beam-infill (where a higher radar beam elevation is used in place of the lowest when this is blocked by local obstructions) associated with shallow rainfall-forming clouds. In such conditions the value of radar is least and the complementary role of raingauges greatest.

Effect of rainfall intensity

The above analysis provided on-average estimates of rainfall uncertainty without regard for any variation with rainfall intensity. A special raingauge network designed for research purposes – as part of the NERC Special Topic HYREX (HYdrological Radar EXperiment) – has provided the opportunity to explore such variability in detail. The HYREX research facility is focused on a special network of forty-nine recording raingauges over the 135 km^2 Brue catchment, gauged at Lovington in Somerset, south-west England. The catchment is scanned continuously by two network C-band radars at Cobbacombe Cross and at Wardon Hill. The basic design comprises gauges in each 2 km^2 to provide broad catchment coverage, two SW–NE lines of four squares each containing two gauges for orientation studies (orthogonal to the topography and aligned with the prevailing storm direction), and two dense networks with eight gauges per square in areas of low and high relief to support gauge spacing studies. An arrangement for the eight-gauge-within-a-square networks was chosen to provide the "best" estimate of the mean rainfall over the square. This resulted in a diamond-within-a-square configuration, with sides .778 and 1.38 km respectively, within the 2 km^2.

Empirical measures of accuracy have been used to explore the dependence of rainfall estimation accuracy on rainfall magnitude. A simple approach has been pursued in which the average of the values from the eight gauges in a 2 km^2 is used as the "true pixel rainfall" for the square. Departures from this at each gauge are then used to compute the standard error associated with a single gauge estimate. Plotting the standard error against the true rainfall as individual points for each fifteen-minute wet period yields an empirical relationship, in the form of a scatter plot, between gauge accuracy and rainfall magnitude. This can also be done for radar data, using the same true rainfall, and the accuracy relationship compared. In practice the scatter plots are reduced to a smoothed "best line" along with 90 percent confidence bands indicating how well the best line has been estimated.

A sample of the results obtained are shown in Figure 4.3 for the Wardon Hill radar. The plots show the standard errors, as confidence envelopes, as they vary with rainfall intensity. The rainfall estimators being assessed in Figures 4.3(a) and 4.3(b) are the 2 and 5 km resolution radar data and a typical gauge, for fifteen-minute and hourly interval data respectively. The estimate of concern is the rainfall over the 2 km^2 containing the eight-gauge network in the area of low relief; the typical gauge represents a gauge in the same square.

Standard errors of catchment average estimates can be obtained in a similar way. A weighted average of all the gauge values in the Brue catchment provides a "true catchment rainfall" against which raingauge and radar estimates can be judged. The results, corresponding to those previously described for the pixel estimators, are shown in Figures 4.3(c) and 4.3(d).

Note that fifteen-minute totals from a single raingauge have errors of 25 percent at 4 mm in estimating the grid square rainfall, whereas radar has errors of circa 50 percent for a range of rainfall magnitudes. However, a typical single raingauge in the Brue will provide a worse estimate of the catchment average rainfall than that provided by radar. As the time-scale of interest coarsens to one hour, the performance of a single gauge and radar becomes comparable, at least on average.

4.2.3 Rainfall forecast uncertainty

The uncertainty associated with radar rainfall forecasts used in flood warning systems are arguably the greatest source of error by far, particularly for higher lead times from two to six hours ahead. Beyond six hours, Mesoscale Numerical Weather Prediction models can provide predictions which may be quantitatively useful for flood forecasting over larger scale areas (16 km grid lengths, or more) and for coarser time intervals of aggregation.

Moore et al. (1994b) present a local radar rainfall forecasting method that infers the storm speed and direction from two radar images fifteen minutes apart, which is then used to advect the current image forward. The forecast field is constructed from this advected image by smoothing it toward the field average value of the image, with more smoothing at increasing lead time. (Further details are given later in the chapter.) An assessment has been made of this local forecasting method against the UK Frontiers national radar rainfall forecasting system using data from fourteen storm events within the 76 km radius field of the London Weather Radar. Frontiers employs a grid of velocity vectors derived from the steering winds in a Mesoscale Meteorological Model in forming advection forecasts and uses man/machine interface techniques to suppress radar anomalies (Brown et al. 1994).

The results indicate superiority of the local forecast method at all lead times when judged on the main criterion for assessment, the log root mean square error of forecasts, using both radar and raingauge data as "truth" [Figure 4.4(a)]. However, the dependence of the local forecast method on data from a single radar means that the field it is able to forecast diminishes with increasing lead time, most markedly for higher storm speeds. On the evidence of the fourteen storms investigated, this is not a serious problem for forecast lead times up to two hours. Forecast performance has also been judged using three categorical "skill" criteria. The False Alarm Rate, the Critical Success Index, and the Probability of Detection all indicate a cross-over in superiority in favor of the Frontiers forecasts at lead times beyond $1^1/_2$ to 3 hours [Figures 4.4(b), (c), and (d)]. These results seem to support the original conjecture that the development of a local forecasting method, providing high resolution forecasts for 2 km grid squares, updated at fifteen-minute intervals, for lead times up to two hours, would provide a valuable complement to the Frontiers product, which has national coverage on a 5 km grid for lead times up to six hours, updated every thirty minutes. The local rainfall forecasting system has been in operational use by the UK Environment Agency (formerly the National Rivers Authority) to support flood warning over London and the Thames basin since November 12, 1991. The local calibration and forecasting procedures, together with anomaly correction algorithms, have most recently been integrated with a PC animated weather radar display system and given the acronym HYRAD (Moore et al. 1994c).

4.2.4 Flood forecast uncertainty

The main interest here is the effect of input uncertainty – especially that associated with rainfall measurements and forecasts

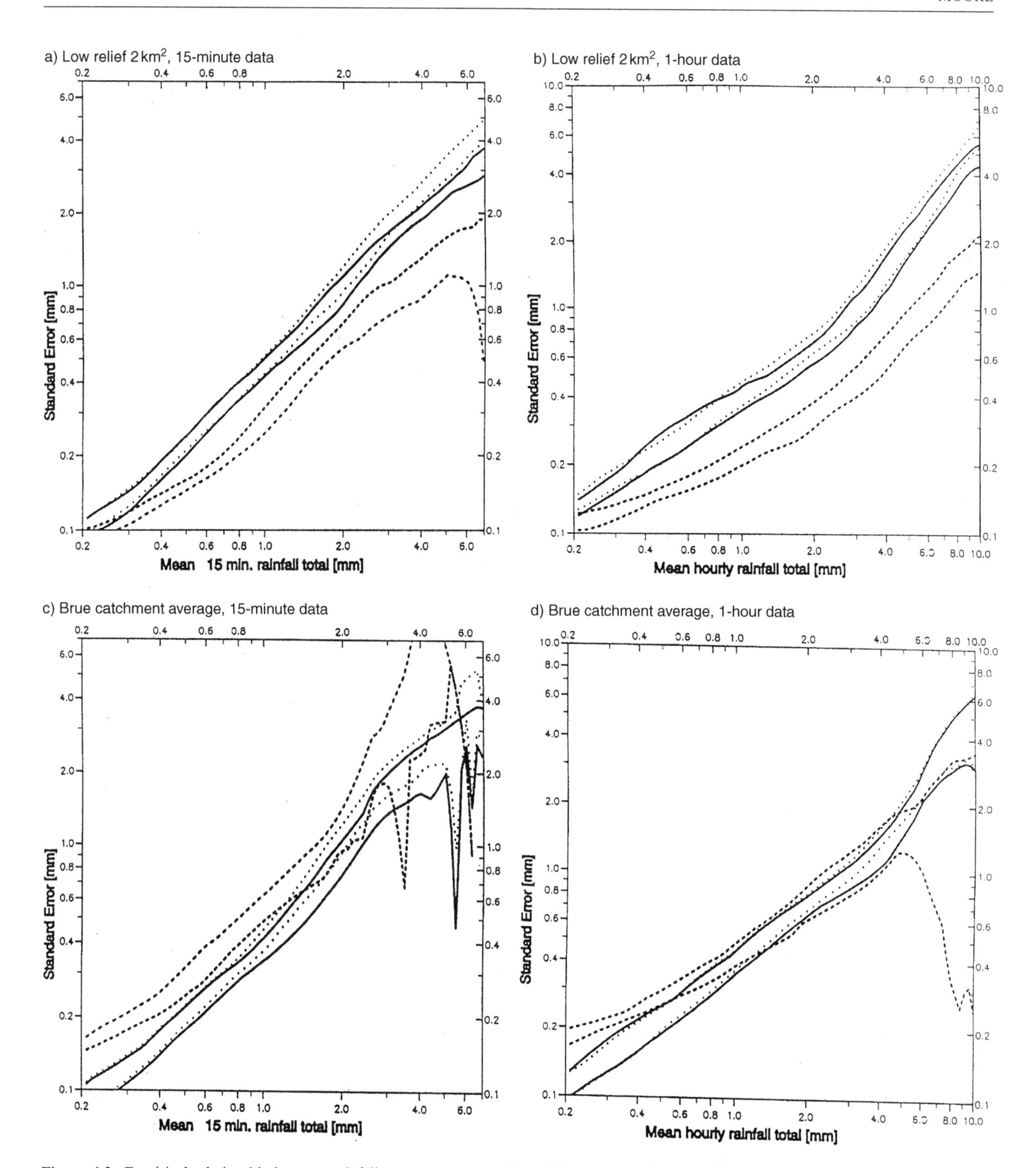

Figure 4.3. Empirical relationship between rainfall sensor accuracy and rainfall magnitude. 90% confidence envelopes are shown for the following estimators: single gauge (continuous line); 2 km radar (dashed line); 5 km radar (grey dotted line).
(a) Low relief 2 km square, 15-minute data. (b) Low relief 2 km square, 1-hour data. (c) Brue catchment average, 15-minute data. (d) Brue catchment average, 1-hour data.

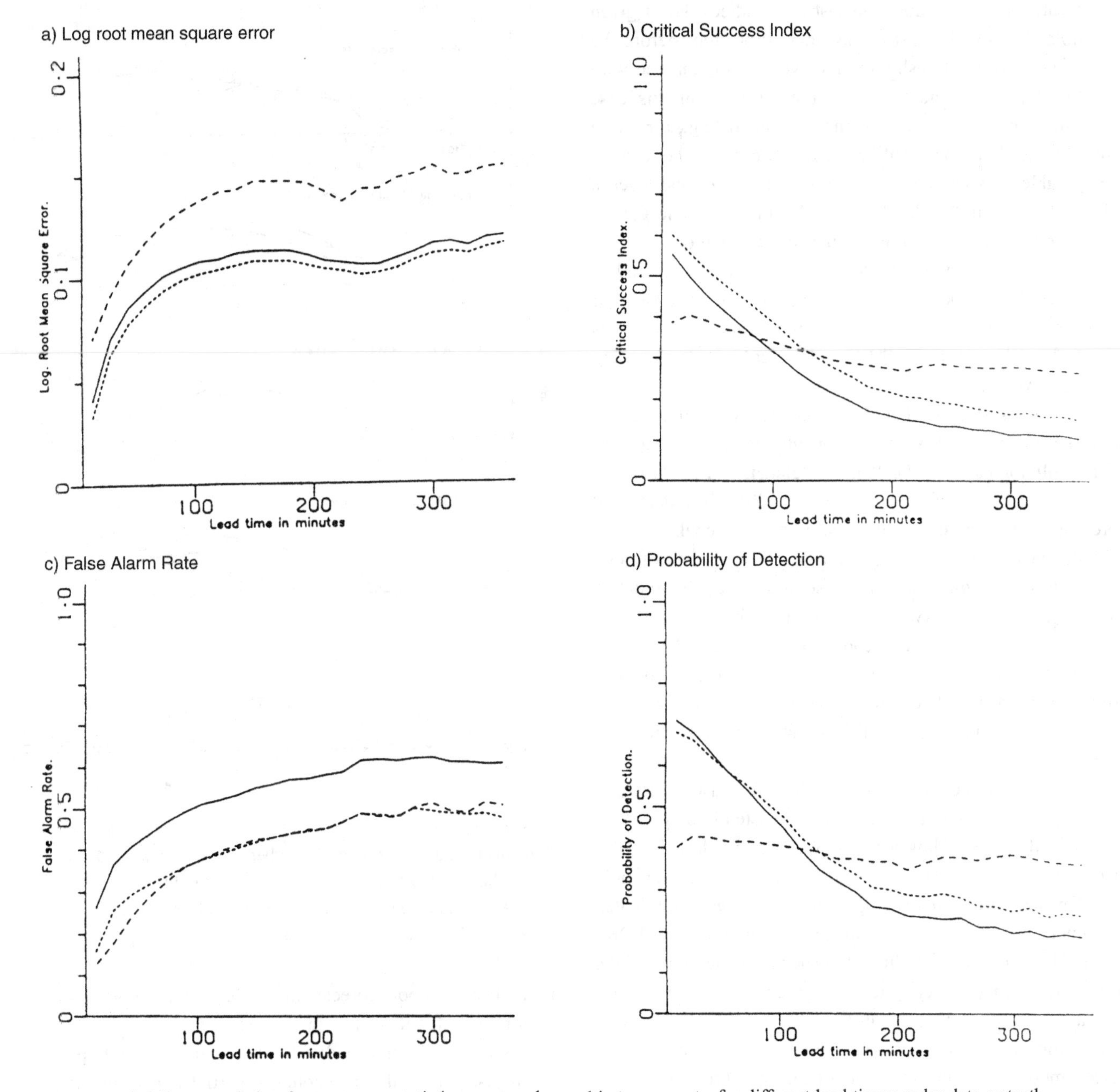

Figure 4.4. Rainfall accumulation forecast error statistics, averaged over 14 storm events, for different lead times: radar data as truth (continuous line: 2 km local forecast; short dashes: 5 km local forecast; long dashes: Frontiers forecast).
(a) Log root mean square error. (b) Critical Success Index. (c) False Alarm Rate. (d) Probability of Detection.

– on flood forecast performance. However, the study to be reported also provides the opportunity to judge the relative effect of choice of model on forecast uncertainty and its variation with the catchment of interest.

A comparison has been made of flood forecasts obtained from three lumped conceptual rainfall-runoff models using four types of rainfall forecast: perfect (observed), conditional (simple Markov state-transition), local radar and Frontiers. The models, of differing complexity, have been calibrated for nine catchments of various sizes and types in the Thames basin. They have been used to forecast a separate set of flood events over a range of lead times, while maintaining a water balance between events. Observed rainfall and updating based on observed flows have been used up to the forecast origin, with forecast rainfall thereafter, and various measures of the flow forecast performance then calculated.

The performance measures used in the assessment included the R^2 statistic, taken as a median estimate across events and a set of categorical threshold forecast skill indices. For a given threshold flow the forecast originating τ time steps before the first crossing of the threshold in the event was compared with this threshold over the period of the forecast (in this case six hours) and the type of contingency noted (e.g., threshold exceeded in both observations and forecasts). The contingency table was incremented across all events and over a wide selection of threshold values. Finally, the table entries were used to calculate three skill scores: the probability of detection, the correct alarm rate, and the critical success index. These categorical skill indices were chosen because of their relevance to flood warning procedures in which exceedences of set alarm levels are used to invoke a color-coded warning of flood severity.

The results suggest that the R^2 performance of forecasts up to six hours ahead is not significantly affected by the type of rainfall forecast for the larger catchments (140–740 km^2). For the smaller catchments (down to 29 km^2) radar-based forecasts appear to be superior to simple conditional forecasts, with the local forecast system performing best. The use of radar rainfall forecasts consistently increases the probability of detection of an exceedence (warning), with recalibrated local radar forecasts performing best in conjunction with the PDM rainfall-runoff model (Moore 1993). A set of forecasts produced from the PDM model for the Silk Stream at Colindeep Lane using the five types of rainfall forecast as input is shown in Figure 4.5.

The effect of choice of model on flow forecast uncertainty was not clear cut, with no one model consistently outperforming the other two. However, the simplest model, the IEM (Isolated Event Model), was worst overall and was especially poor for the more complex responding large, chalk-dominated catchment. The model of intermediate complexity, the PDM, provided sufficient flexibility of response while proving the most robust in producing less extreme poor performances. Updating was important for all models. The range of forecast uncertainty was greater across catchments of different type than across models for the same catchment. Peak errors are more affected by the rainfall estimate used than the type of model. Further details are available in Moore et al. (1993) and Austin and Moore (1996).

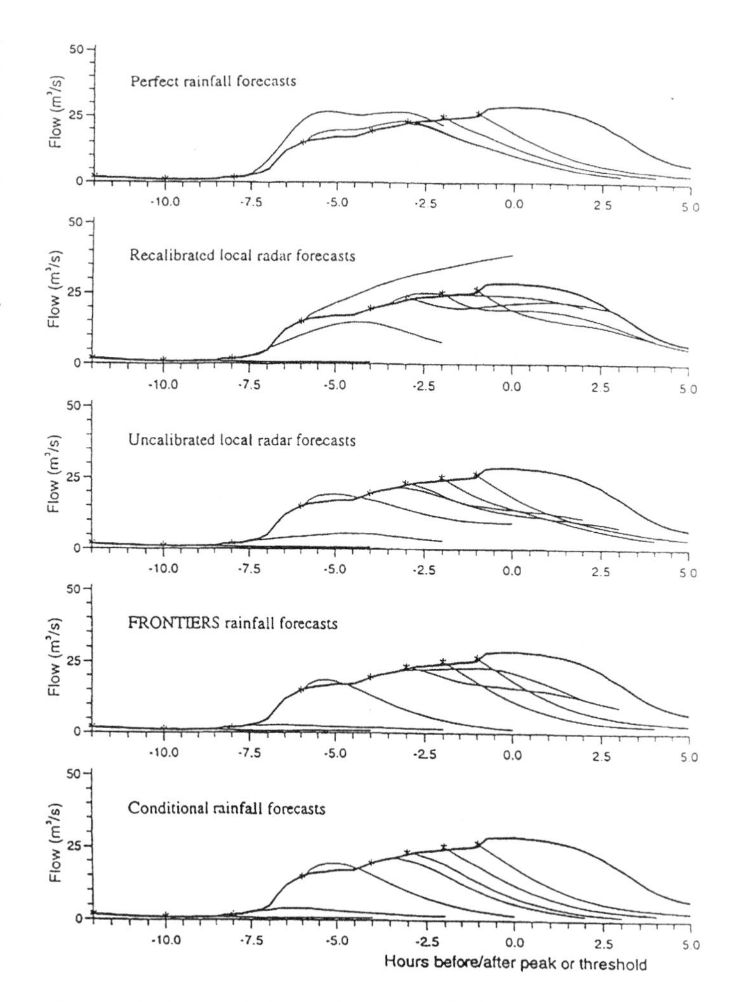

Figure 4.5. Forecast hydrographs for the Silk Stream at Colindeep Lane made at different time origins before the peak at 5:00 A.M., September 23, 1992. Flows computed using the PDM model using HYRAD recalibrated data and forecast rainfall from the 5 sources indicated ("local" are forecasts from HYRAD).

4.3 RELIABILITY AND SYSTEM COMPLEXITY

4.3.1 Introduction

Reliability will be understood here to have its literal connotation of that which may be relied upon, and not its technical association with probability. We are thus concerned with the reliability of flood forecasting systems with regard to the requirement for them not to fail. They may prove inaccurate through loss of telemetry data but they must not fail to produce a forecast, especially on account of partially missing data.

Flood forecasting systems are often required to be configured to meet a highly complex set of requirements, involving forecasts of river levels and flows across networks of river basins and the automated control of reservoirs, river gates, and tidal barrages. Modeling of such complex systems requires a well thought out software control architecture if the reliability of forecast construction is to be achieved. The River Flow Forecasting System (RFFS) hydrological kernel, developed by the Institute of Hydrology, employs the Information Control Algorithm or ICA to achieve this reliability of forecast construction (Moore and Jones 1991; Moore 1993, 1994; Moore et al. 1994c).

4.3.2 The RFFS Information Control Algorithm

The ICA tackles a possibly complex overall forecasting problem through division into a number of simpler subproblems, or "Model Components." The results of one or more model components are fed as input series into a subsequent Model Component. These input series will in practice be made up of observed values in the past, model infilled values in the past where observations are missing, and model forecast values in the future. The infilled or forecast values will usually be constructed using preceding Model Components. At the extremities of the information network, beyond the river network, are special Model Components which can provide missing values for their input series through the use of backup profiles. Rainfall series completion is a typical example. Thus, a particular forecasting problem may be viewed as a node and link network in which each node is associated with a Forecast Requirement (e.g., to forecast flow at site x) and each link to a Model Component which contains a set of Model Algorithms used to construct the forecasts.

A particular configuration of forecast points within a river system is described within the ICA by a set of description files. These files take two main forms:

i. a Model Component file that defines the form of model structure, through the specification of Model Algorithms to be employed, and the data inputs to be used to make forecasts for a particular location or set of locations; and
ii. a Forecast Requirement file that defines for each forecast point the Model Component to be used to construct the forecast for that point, the type of forecast (e.g., river level, flow, snowmelt) and the connectivity with other model components.

A Model Component is typically made up of a number of Model Algorithms, for example for snowmelt modeling, rainfall-runoff modeling, and real-time updating. Model Algorithms comprise forecasting procedures which can be used to create forecasts for several sites in the region, a particular procedure typically only differing from site to site through the model parameters used. The Model Algorithms to be used are defined within the Model Component file description, together with the parameter values appropriate to the site(s) for which the component is to provide a forecast. Figure 4.6 illustrates a typical Model Component and its associated Model Algorithms, and Figure 4.7 illustrates the connectivity between model components. This connectivity allows the ICA to represent river systems with complex dendritic structures including bifurcations. The specifications for a Model Component can be replaced with those for a modified version without disturbing the specification of the rest of the Model Component network and, essentially, without having even to consider more than the local part of the rest of the network. This provides for flexibility and ease of modification of an existing configuration.

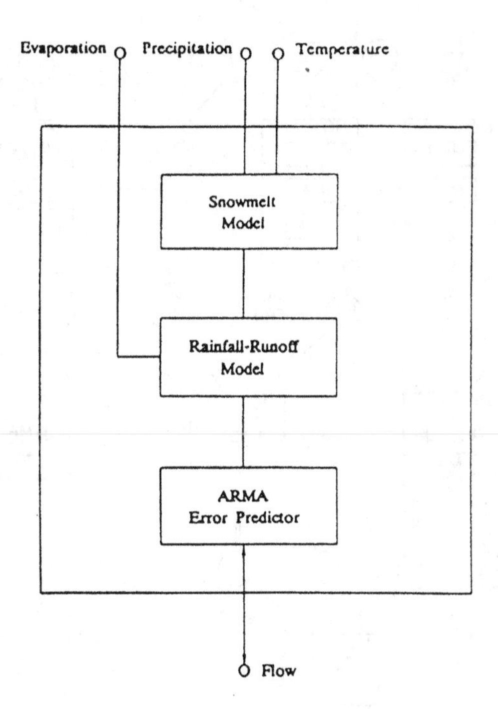

Figure 4.6. A Model Component and its associated Model Algorithms.

On completion of a forecast run, the "states" of the models required to initialize a subsequent run are stored. The states typically will be the water contents of conceptual stores within snowmelt and rainfall-runoff models or the river levels and flows of channel flow routing models. A subsequent run at a later forecast time origin will start forecasting forward from the time of storing the states from a previous run. The set of model states for an algorithm, for a particular time point, contains a sufficient summary of all data prior to that time point to allow the algorithm to continue executing as if it had been run over a warm-up period up to that time point.

Operationally in non-flood conditions the system might be run automatically once a day, say at 7 A.M. following routine data gathering by the Telemetry System. This means that the model states are available to provide good initial conditions from which to run the model for a flood event occurring later the same day, thus avoiding the need for a long "warm-up" period for model initialization. During flood events the system might be run frequently under the control of the operator. For small, fast responding systems a fully automated mode of operation might be preferred.

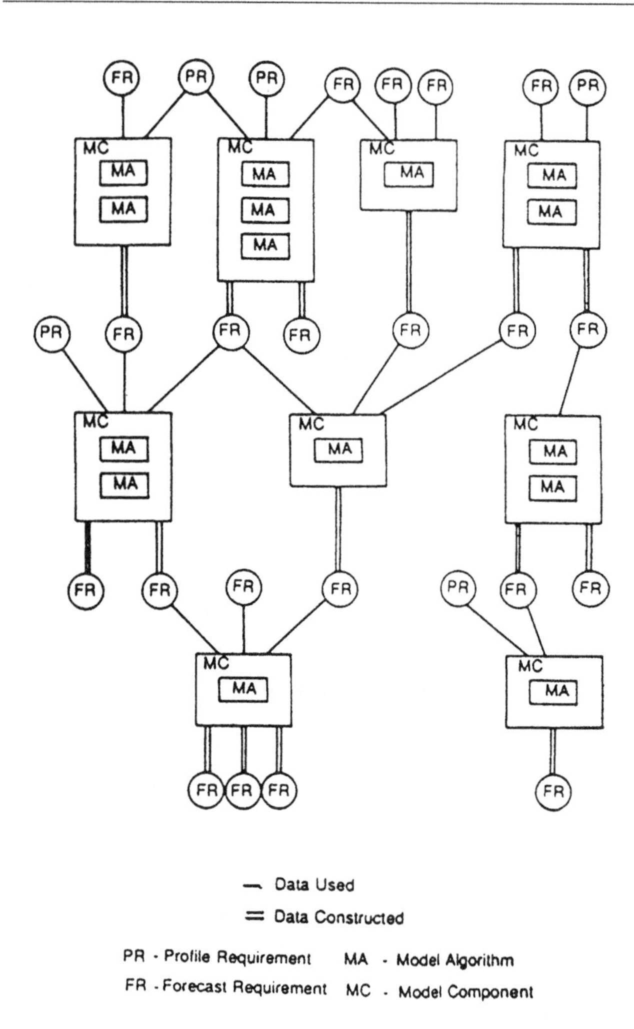

Figure 4.7. Connectivity between Model Components.

4.3.3 Data loss, system resilience, and profile data

The ICA has been designed so that the generation of flow forecasts can be resilient to data loss. This is accomplished for a point "internal" to the network by ensuring that the Model Component which constructs forecasts for the point will also infill missing values in the past data. For "external" points, typically rainfall and other forms of climate data, Model Algorithms are used to merge data time series from a variety of sources. In the event that no data are available, provision is made to supply a backup profile. A hierarchy of priority of data source can be imposed in the case of data being available for a given time from more than one source. For example, in the case of rainfall the priority for a given catchment rainfall might be radar data, raingauge data from *n* raingauges, and then any combination of less than *n* (allowing for raingauge system malfunction), and a backup rainfall profile. For future times when no observation data are available the priority might be a Local Radar Rainfall Forecast from HYRAD, a FRONTIERS forecast and a synoptic forecast (provided automatically to the modeling computer by a computer/telex facility from a Weather Center) and finally a backup rainfall profile. The rainfall profiles can be selected to be seasonally dependent and categorized into light, moderate, and heavy with the option of invoking a selection at run time. They may also be subdivided into different synoptic regions over a region.

Another important use for profiles is to provide boundary conditions for a tidal hydraulic model. Astronomical tidal predictions would be made available in profile form and augmented with tide residual forecasts supplied by a Storm Tide Warning Service. Other uses for profiles are for potential evaporation and temperature to support rainfall-runoff and snowmelt modeling.

4.4 RISK AND ENSEMBLE FORECASTING

The assessment of flood risk in real-time can be assessed with reference to the "best estimate" forecast alone. If the forecast is much greater than the flooding threshold, and provided the input/modeling system has been well configured, then it might be reasonable to categorize the risk as high. However, in more marginal cases, particularly where the cost of false-warning or failure-to-warn may be high, then a probabilistic risk assessment is ideally required. Discussion of uncertainty in flood warning systems has already implied that quantification of risk in probabilistic terms will not be easy, due to the inherent non-stationarity and nonlinearity present. It can be made easy by making inappropriate assumptions concerning linearity and stationarity and this is often done in practice, using standard methods drawn from the theory of time series analysis. More informal methods of risk assessment have attractions. We have already highlighted the source of greatest uncertainty as often being associated with the rainfall input to flood forecasting systems. Thus an informal ensemble of rainfall inputs, derived for example from raingauges, from weather radar, from a typical rainfall and from a no-rainfall and extreme-rainfall, might be used to obtain an ensemble of forecasts whose envelope is likely to provide an impression of the flood risk posed. Figure 4.8 shows a very simple ensemble forecast produced by a raingauge measurement and a "typical rainfall," in this case not too different from no-rain; the observed hydrograph is also shown, although it would not be available for future times in real time.

The Institute of Hydrology's River Flow Forecasting System already incorporates informal ensembles through providing a logical "What if?" switch used to "switch out" radar estimates of catchment rainfall in favor of raingauge estimates, when the

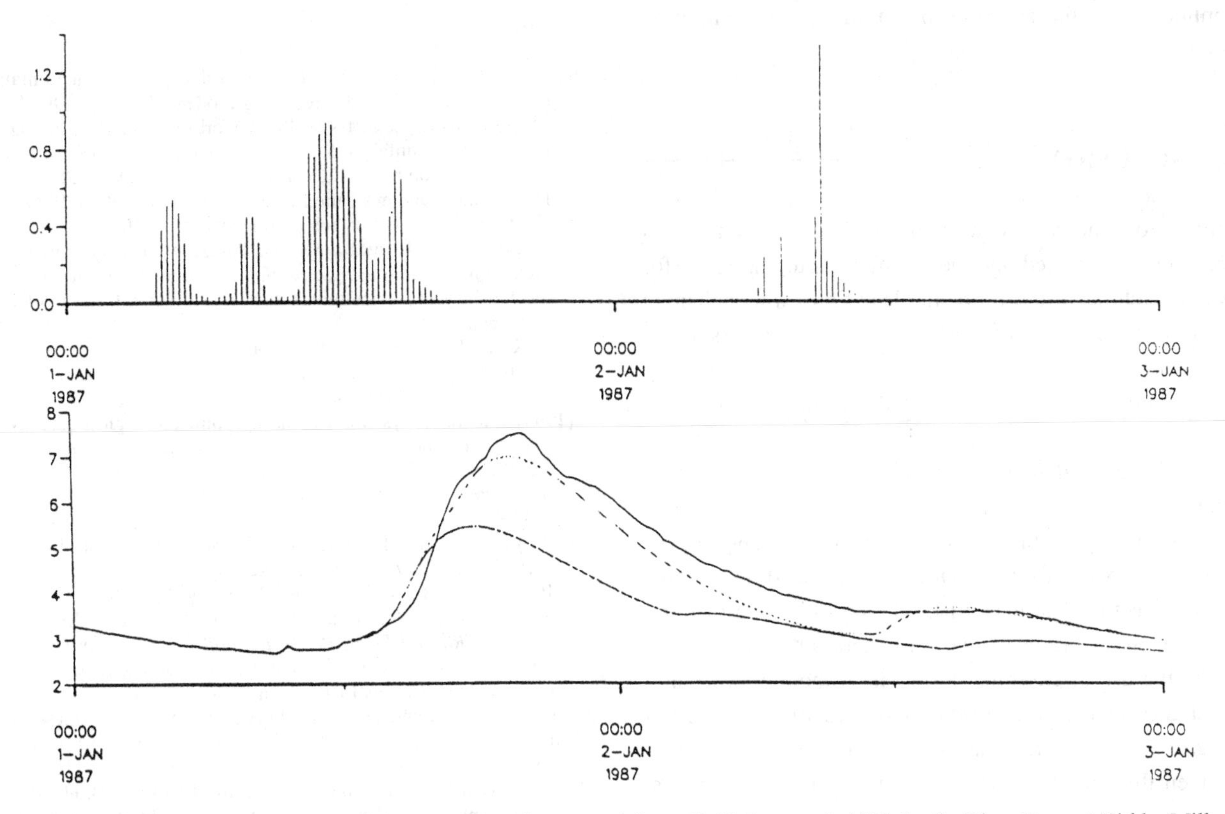

Figure 4.8. Operational flow forecast made using the RFFS; forecast origin at 12:00 January 1, 1987 for the River Dove at Kirkby Mills. Bold line: observed; upper dashed line: forecast assuming perfect foreknowledge of rainfall; lower line: forecast using backup rainfall beyond forecast origin.

latter are known to be unreliable, or when the sensitivity of flood forecasts to the alternative rainfall inputs is of interest. In addition, provision is made for multiplying rainfall forecasts by a given factor. Both of these can be of use. A more conventional ensemble forecast might derive from the use of ensemble rainfall forecasts as inputs to a catchment model. This might take the form of a dynamic rainfall model with radar data assimilation, whose initial conditions are stochastically perturbed (Moore 1995).

A simple approach would be to take the advection forecasts from HYRAD, IH's local radar rainfall forecasting scheme, and to introduce variability into the quantities that are used in forecast construction. Specifically, forecasts are constructed for radar grid square (i, j) as a weighted average of the field average value, $\bar{R}$ and the advection forecast for this square, $\hat{R}_{i,j}$, such that

$$\tilde{R}_{ij} = \begin{cases} \bar{R} + a\left(\hat{R}_{ij} - \bar{R}\right) & \hat{R}_{ij} > 0 \\ 0 & \hat{R}_{ij} = 0 \end{cases} \tag{4.3}$$

where the weight $a = f^{\tau}$ is a shrinkage factor, reducing with increasing lead time τ, with f a constant value. Thus, the forecast is based on an advection forecast which is progressively shrunk toward the field average value with increasing lead time. The advection velocity used in calculating $\hat{R}_{ij}$ is inferred from a form of correlation analysis between the current radar image and that fifteen minutes ago. Equation 4.3 provides smoothing, which is appropriate to the requirement for a good central forecast. In the case of an ensemble forecast some additional random variation is required, rather than smoothing, and this could be provided by randomly varying the velocity vector and/or by adding random noise. The correlation response function, obtained for a grid of velocity vectors at the velocity inference step, could be used in the generation of sets of velocity vectors and corresponding $\hat{R}_{i,j}$ values. The set of $\hat{R}_{i,j}$ values, possibly with the addition of some random noise, could be used to generate ensembles of rainfall forecasts from which flood forecast ensembles could be derived. Weighting of ensembles to derive approximate confidence bands on forecasts provides further possibilities and opportunities for research.

Recent research has considered a more complex method of deriving uncertainty estimates as part of a Bayesian scheme for model updating and forecasting (Jones 1996). However, the assumptions made at present lead to these estimates being overconfident, and the approach has a significant computing overhead.

4.5 CONCLUSION

The nature of uncertainty in flood forecasting systems is complex and dominated by the rainfall input, at least for headwater catchments and for higher lead times where channel flow routing models are concerned. Empirical studies using radar/raingauge networks have been used here to quantify this uncertainty and its effect on flood forecasts for different models and catchments. Reliability of flood forecasting systems is paramount if loss of life and/or property is to be minimized. The River Flow Forecasting System's Information Control Algorithm has been presented as reliably coping with the complexity of region-wide flood forecasting systems and the prospect of data loss. Risk relating to issuing of flood warnings given uncertain forecasts is important and might be helped by the use of ensemble forecasts. Although informally constructed ensembles are already proving useful, more formal methods giving confidence bands are seen as a research challenge given the complex role of uncertainty in flood forecast construction.

Acknowledgments

The strategic support of the MAFF Flood Protection Commission in this work is acknowledged along with the Commission for the European Communities under the EPOCH project "Weather radar and storm and flood hazard" and the Environment Programme project "Storms, floods and radar hydrology." Support by the Environment Agency (formerly the National Rivers Authority) and by NERC under the HYREX Special Topic is also acknowledged.

REFERENCES

Austin, R. M. and Moore, R. J. (1996) Evaluation of radar rainfall forecasts in real-time flood forecasting models. *Quaderni Di Idronomia Montana* 16 (Special Issue, Proc. Workshop on "Integrating Radar Estimates of Rainfall in Real-Time Flood Forecasting"), 19–28.

Brown, R., Newcomb, P. D., Cheung-Lee, J., and Ryall, G. (1994) Development and evaluation of the forecast step of the FRONTIERS short-term precipitation forecasting system. *J. Hydrol.* 158: 79–105.

Jones, D. A. (1996) *An evaluation of a Bayesian scheme for model updating and forecasting using a rainfall-runoff model.* Contract Report to the Ministry of Agriculture, Fisheries and Food, Institute of Hydrology, 36pp.

Moore, R. J. (1993) Real-time flood forecasting systems: perspectives and prospects. *British-Hungarian Workshop on Flood Defence*, September 6–10, VITUKI, Budapest, 121–66.

(1994) Integrated systems for the hydrometeorological forecasting of floods. In *Proceedings of the International Scientific Conference EUROPROTECH, Part I: Science and Technology for the Reduction of Natural Risks*, edited by G. Verri, May 6–8, CSIM, Udine, Italy, 121–37.

(1995) Rainfall and flow forecasting using weather radar. In *Hydrological Uses of Weather Radar*, edited by K. Tilford, BHS Occasional Paper No. 5, British Hydrological Society, 12: 166–86.

Moore, R. J., Austin, R. M., and Carrington, D. S. (1993) *Evaluation of FRONTIERS and Local Radar Rainfall Forecasts for Use in Flood Forecasting Models.* R&D Note 225, Research Contractor: Institute of Hydrology, National Rivers Authority, 156pp.

Moore, R. J. and Jones, D. A. (1991) A river flow forecasting system for region-wide application. *MAFF Conference of River and Coastal Engineers*, Loughborough University, England, July 8–10, 12pp.

Moore, R. J., May, B. C., Jones, D. A., and Black, K. B. (1994a) Local calibration of weather radar over London. In *Advances in Radar Hydrology*, edited by M. E. Almeida-Teixeira, R. Fantechi, R. Moore, and V. M. Silva, Proc. Int. Workshop, Lisbon, Portugal, November 11–13, Report EUR 14334 EN, European Commission, 186–95.

Moore, R. J., Hotchkiss, D. S., Jones, D. A., and Black, K. B. (1994b) Local rainfall forecasting using weather radar: The London Case Study. In *Advances in Radar Hydrology*, edited by M. E. Almeida-Teixeira, R. Fantechi, R. Moore, and V. M. Silva, Proc. Int. Workshop, Lisbon, Portugal, November 11–13, Report EUR 14334 EN, European Commission, 235–41.

Moore, R. J., Jones, D. A., Black, K. B., Austin, R. M., Carrington, D. S., Tinnion, M., and Akhondi, A. (1994c) RFFS and HYRAD: Integrated systems for rainfall and river flow forecasting in real-time and their application in Yorkshire. In *Analytical techniques for the development and operations planning of water resource and supply systems*, BHS National Meeting, University of Newcastle, November 16, BHS Occasional Paper No. 4, British Hydrological Society, 12pp.

5 Probabilistic hydrometeorological forecasting

ROMAN KRZYSZTOFOWICZ*

ABSTRACT
The U.S. National Weather Service has supported the development of an integrated probabilistic hydrometeorological forecasting system. The system produces probabilistic quantitative precipitation forecasts that are used to produce probabilistic river stage forecasts; these in turn are input to optimal decision procedures for issuing flood warnings, operating waterways and barges, or controlling storage reservoirs. The system is designed based on Bayesian principles of probabilistic forecasting and rational decision making. This chapter outlines the system concept.

5.1 INTRODUCTION

5.1.1 Systems approach to hydrometeorological forecasting

That forecasts should be stated in probabilistic rather than categorical terms has been argued from operational (Cooke 1906) and decision-theoretic (Murphy 1991) perspectives for almost a century. Yet most operational systems produce deterministic forecasts and most research in physical and statistical sciences has been devoted to finding the "best" estimates rather than probability distributions of predictands. Undoubtedly, the leap from a deterministic frame of thought to one that not only admits our limited knowledge and information, but also quantifies uncertainty about future states of the environment, requires a vast and coordinated effort at two levels: engineering – to design probabilistic forecasting systems, and organizational – to alter the institutional mindset and *modus operandi.*

The U.S. National Weather Service (NWS) has embarked on making such a quantum change (Zevin 1994; Krzysztofowicz 1998). The goal is to increase the value of service to users by developing and implementing an integrated probabilistic hydrometeorological forecasting system. The system is built of three components, as depicted in Figure 5.1:

- Probabilistic Quantitative Precipitation Forecasting (PQPF) system for a Weather Forecast Office (WFO);
- Probabilistic River Stage Forecasting (PRSF) system for a River Forecast Center (RFC); and
- Flood Warning Decision (FWD) system for a WFO.

The overall objective of the system is threefold: (i) to integrate processing of information, beginning with the preparation of meteorological forecasts of precipitation amounts at the WFO, through the production of hydrological forecasts of river stages at the RFC, to decision making concerning the issuance of flood watches and warnings by the WFO; (ii) to quantify uncertainty in forecasts of precipitation amounts and river stages; and ultimately (iii) to provide additional information to users of river forecasts and flood warnings.

5.1.2 Potential benefits of probabilistic forecasts

A time series of precipitation is the major input to hydrologic models that produce forecasts of river stages. It is generally agreed that by extending the time series of precipitation already observed with a forecasted time series, river stage forecasts can be improved in two ways: (i) the lead time of forecasts can be increased and (ii) the bias toward underestimation of river stages during rises (the phenomenon known as stair-stepping) can be reduced. But there is a challenge to realizing these improvements: precipitation amount is one of the most difficult meteorologic predictands and the major source of uncertainty in river forecasting.

In contrast to a deterministic QPF that gives a single estimate of the precipitation amount, a PQPF allows the forecaster to convey the *degree of uncertainty* about the amount. A quantification of this uncertainty has three potential benefits (Krzysztofowicz et al. 1993): it aids the forecaster in making unbiased judgments, it provides input necessary to produce PRSFs, and it provides additional information to forecast users. To wit, the PRSF allows users to consider a trade-off between the forecast uncertainty and the lead time. In effect, risks can be explicitly accounted for in decision making, and this addi-

* Department of Systems Engineering, University of Virginia, Charlottesville, VA, USA.

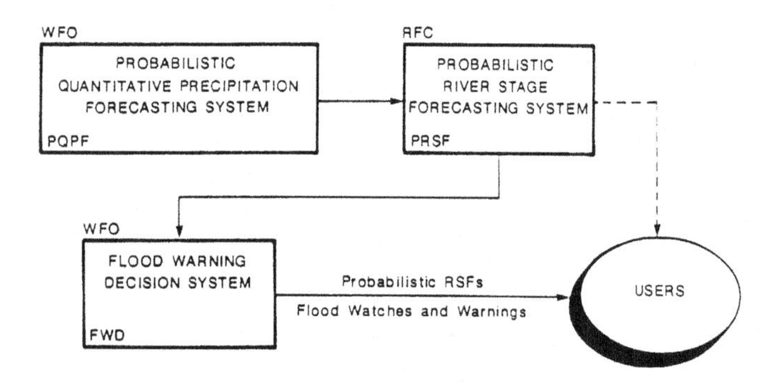

Figure 5.1. An integrated probabilistic hydrometeorological forecasting system for a Weather Forecast Office (WFO) and a River Forecast Center (RFC).

tional information is expected to increase economic benefits of forecasts. The premise of this assertion derives from Bayesian decision theory (Krzysztofowicz 1983; Alexandridis and Krzysztofowicz 1985): when used optimally, probabilistic forecasts are always at least as valuable (economically) as deterministic forecasts. Moreover, the larger the prior uncertainty about the predictand, the greater the potential economic benefits from probabilistic forecasts, relative to deterministic forecasts. These benefits may be especially significant to emergency management agencies, storage reservoir managers, waterways and barge operators during high rises and flood events. Likewise, a PRSF will allow the NWS to cast flood watches and warnings in a probabilistic format, which is more informative and effective than the categorical format in conveying predictions of rare and severe weather events (Murphy 1991).

5.2 PROBABILISTIC FORECASTS

5.2.1 Probabilistic quantitative precipitation forecast

Let W denote the *basin average precipitation amount* accumulated during a fixed period, long enough to cover an entire weather event or a significant portion thereof. Let W_i denote the basin average precipitation amount accumulated during the ith subperiod, $i \in \{1, \ldots, n\}$, where n is the number of subperiods such that $W_1 + \ldots + W_n = W$. For each i, define a variate $\Theta_i = W_i/W$, representing a *fraction* of the total amount accumulated during subperiod i, so that $0 \leq \Theta_i \leq 1$ and $\Theta_1 + \ldots + \Theta_n = 1$. Fractions $(\Theta_1, \ldots, \Theta_n)$ are assumed to be stochastically independent of the total amount W, conditional on timing of precipitation within the diurnal cycle. This property, termed the *conditional disaggregative invariance*, was tested and confirmed for river basins in Pennsylvania via statistical analyses of 24-hour precipitation amounts disaggregated into four 6-hourly amounts (Krzysztofowicz and Pomroy 1997).

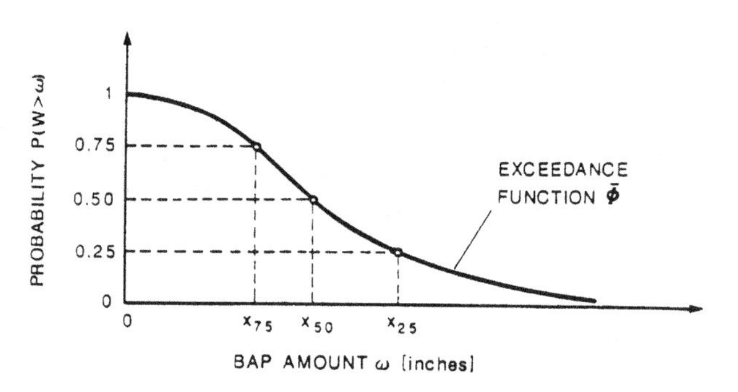

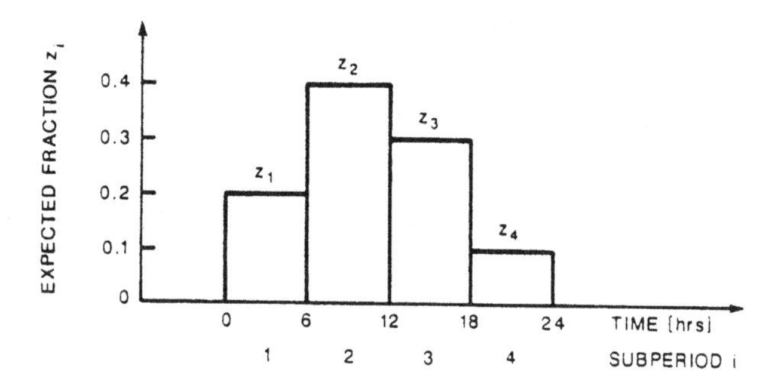

Figure 5.2. Example of a probabilistic quantitative precipitation forecast (PQPF). (a) Exceedance function of the 24-hour basin average precipitation amount. (b) Expected temporal disaggregation of any 24-hour amount into four 6-hourly amounts.

The forecast consists of two parts, as illustrated in Figure 5.2: the first part is a probabilistic forecast of W; the second part is a deterministic forecast of $(\Theta_1, \ldots, \Theta_n)$. This dichotomy and the particular format of the forecast, which consists of $(3 + n)$ estimates, is a result of a compromise between theoretical considerations (whose objective is to maximize the information contents of the forecast) and operational considerations (whose objective is to minimize the complexity of judgmental tasks required of a forecaster and the overall effort involved in forecast preparation).

The first part of the forecast specifies three *exceedance fractiles* (x_{75}, x_{50}, x_{25}) of the total basin average precipitation amount W. With p denoting a probability number, $0 < p < 1$, the $100p$% exceedance fractile of W is an estimate x_{100p} such that the exceedance probability is $P(W > x_{100p}) = p$. The three exceedance fractiles are used to estimate a parametric *exceedance function* $\bar{\Phi}$ such that

$$\bar{\Phi}(\omega) = P(W > \omega).$$

The second part of the forecast specifies n *expected fractions* $(z_1, \ldots, z_n)$, which are estimates of $(\Theta_1, \ldots, \Theta_n)$. Specifically, $z_i = E(\Theta_i)$, where E denotes the expectation, $0 \leq z_i \leq 1$ for every subperiod $i \in \{1, \ldots, n\}$, and $z_1 + \ldots + z_n = 1$, which implies that $(n-1)$ estimates are sufficient. The estimates $(z_1, \ldots, z_n)$ define the *expected temporal disaggregation* of any total basin average precipitation amount W accumulated during a period, into amounts $(W_1, \ldots, W_n)$ accumulated during the n subperiods. For any hypothesized total amount, say $W = \omega$, one can obtain the conditional expectation of W_i as

$$E(W_i \mid W = \omega) = z_i \omega, \quad i = 1, \ldots, n.$$

A complete characterization of uncertainty about the temporal disaggregation is obtained by using the forecast $(z_1, \ldots, z_n)$ to update a prior (climatic) joint distribution of $(\Theta_1, \ldots, \Theta_n)$ into a posterior distribution from which a conditional joint distribution of $(W_1, \ldots, W_n | W = \omega)$ may be derived.

In summary, an operational PQPF for a river basin specifies $3 + n$ estimates

$$(x_{75}, x_{50}, x_{25}; z_1, \ldots, z_n),$$

where the first $3 + (n - 1)$ estimates are sufficient because of the constraint $z_1 + \ldots + z_n = 1$. In the operational testing of the forecasting system conducted by the WFO Pittsburgh since August 1990, W is a 24-hour basin average precipitation amount, which is disaggregated into four 6-hourly amounts. Forecasts are issued twice a day for 24-hour periods beginning at 0000 and 1200 UTC (Universal Time Coordinate).

5.2.2 Probabilistic river stage forecast

Let t denote the time measured on a continuous scale from the beginning of the period for which a PQPF is prepared. Let $H(t)$ denote the *river stage* in a gauging station at time t. The predictand for a gauging station is a sequence of river stages $\{H(t): t = t_1, t_2, t_3, \ldots, t_r\}$. Epochs $t_j, j = 1, 2, 3, \ldots, r$, coincide with epochs for which a hydrologic model calculates the discrete hydrograph; for instance, $t = 6, 12, 18, \ldots, T$ hours.

The PRSF for a gauging station consists of a *sequence of exceedance functions*

$$\{\overline{\Psi}(h \mid t): t = t_1, t_2, t_3, \ldots, t_r\},$$

such that

$$\overline{\Psi}(h \mid t) = P(H(t) > h),$$

which specifies the probability of the river stage $H(t)$ exceeding level h, for any h between the gauge datum and infinity. Figure 5.3 shows an example.

Informative PRSFs may be produced for lead times up to T hours, where $T \leq$ (projection time of the PQPF) + (response time of the river basin). Beyond this horizon, the forecast exceedance functions $\overline{\Psi}(h|t)$, if they are well calibrated, should tend toward a climatic exceedance function of the river stage.

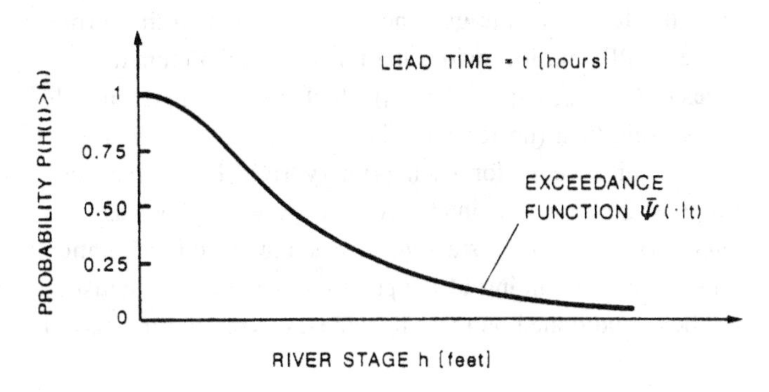

Figure 5.3. Example of a probabilistic river stage forecast (PRSF) for a gauging station.

5.3 PRECIPITATION FORECASTING SYSTEM

The PQPF system is designed for a WFO and consists of three subsystems: forecasting methodology, local climatic guidance, and forecast verification.

5.3.1 Forecasting methodology

The purpose of the forecasting methodology is to aid a field forecaster in coupling principles of weather forecasting with principles of probabilistic reasoning, and to automate all algorithmic processing tasks. Its implementation takes the form of a human-computer system. Its main components are judgmental tools and estimation procedures (Krzysztofowicz et al. 1993).

The purpose of the *judgmental tools* is to guide the forecaster's reasoning. With training and experience, these judgmental tools can be internalized by the forecaster to become part of his cognitive skills. There are two such tools: (i) an information processing scheme, which the forecaster follows to transform information from multiple sources (numerical model outputs, centrally produced guidance, local climatic guidance) into judgmental estimates; and (ii) a protocol for assessing exceedance fractiles, which the forecaster follows to verify probabilistic properties of judgmental estimates.

The purpose of the *estimation procedures* is to elicit judgmental estimates from the forecaster and then to transform this input into the final product, ready for transmission to the RFC. The estimation procedures are interactive and mostly graphical.

5.3.2 Local climatic guidance

The purpose of the local climatic guidance is threefold: it allows the forecaster to become familiar with climatic statistics of the predictand; it reinforces the frame for his judgments, since the format of the guidance is identical with the format of the PQPF; and it provides the forecaster with climatic estimates of (x_{75}, x_{50}, x_{25}; z_1, z_2, z_3, z_4) which play an important role in assessing final (posterior) estimates.

The guidance is for each primary river basin and each month of the year. It is interactive and allows the forecaster to enter various hypotheses about the basin average precipitation amount, and the timing of the precipitation; climatic statistics are then conditioned on these hypotheses (Krzysztofowicz and Sigrest 1997).

5.3.3 Forecast verification

A PQPF specified in terms of seven estimates (x_{75}, x_{50}, x_{25}; z_1, z_2, z_3, z_4) is verified against the actual 24-hour basin average precipitation amount and its temporal disaggregation, which are calculated by the RFC based on raingauge observations. The forecast is verified with respect to four attributes: (i) *calibration* of the exceedance fractiles (x_{75}, x_{50}, x_{25}), (ii) *informativeness* of the exceedance fractiles, (iii) *bias* of the expected fractions (z_1, z_2, z_3, z_4), and (iv) *informativeness* of the expected fractions (Krzysztofowicz and Drake 1992; Krzysztofowicz 1992, 1996; Krzysztofowicz and Sigrest 1999).

A comparison of the latest forecasts with actual amounts is available at 1200 UTC each day. Verification statistics are produced and reviewed with the forecasters at the end of each month. The group receives verifications for the latest month, three months, and twelve months. In addition, each forecaster receives, for his eyes only, verifications of forecasts he prepared during the latest three months and twelve months.

5.4 STAGE FORECASTING SYSTEM

The PRSF system is designed for a RFC and consists of four subsystems: precipitation forecast processor, river forecasting methodology, river climatic guidance, and river forecast verification.

5.4.1 Precipitation forecast processor

The RFC receives PQPFs from all WFOs located within its service area. Processing these forecasts involves tasks such as: (i) quality control, (ii) reconciliation of any significant discontinuities between PQPFs for adjacent river basins lying within service areas of different WFOs, (iii) composition and review of a PQPF for the entire service area of the RFC, and (iv) computation of time series of basin average precipitation amounts that will be input into a hydrologic model.

5.4.2 River forecasting methodology

The methodology takes the form of a hydrologic-statistical forecasting system. Its purpose is to transform a PQPF for a river basin into a PRSF for a station at the basin outlet. The system embodies three premises. First, it may be implemented with any deterministic hydrologic model. Second, it decomposes the total forecast uncertainty into *hydrologic uncertainty* and *precipitation uncertainty*, which are independently estimated and then integrated. Third, it is based on principles of Bayesian inference (Krzysztofowicz 1985, 1999).

The hydrologic uncertainty refers to the posterior uncertainty about the actual river stage, conditional on the hydrograph output by a hydrologic model when the future precipitation input is known. This uncertainty is characterized in terms of a family of conditional probability density functions of the form $\eta(h|s, t)$, where h is a magnitude of the unknown actual river stage $H(t)$ at time t, and s is an estimate of this stage produced by the hydrologic model under the hypothesis that the precipitation input is known perfectly.

The precipitation uncertainty refers to the uncertainty about the river stage estimate output from the hydrologic model when the future precipitation input is a random variable whose distribution is specified by a PQPF. This uncertainty is characterized in terms of a density $\pi(s|t)$, where s is a magnitude of the river stage $S(t)$ at time t that is output from the hydrologic model.

Of ultimate interest is the predictive (Bayesian) density of the actual river stage $H(t)$ at time t. This density is specified by the total probability law:

$$\psi(h \mid t) = \int_{-\infty}^{\infty} \eta(h \mid s,t)\pi(s \mid t)ds.$$

The exceedance function $\overline{\Psi}$ of the river stage $H(t)$ at time t is then readily obtained as:

$$\overline{\Psi}(h \mid t) = P(H(t) > h) = \int_{h}^{\infty} \psi(u \mid t)du.$$

The processors that output densities η and π are *Bayesian post-processors*, estimated partly off-line via a simulation experiment and partly on-line via a contingency analysis conducted on a set of river stage time series produced by the hydrologic model from the PQPF.

5.4.3 River climatic guidance

The purpose of the river climatic guidance is to provide prior statistics and distributions of river stage time series. This information is needed by the Bayesian uncertainty processors in the forecasting subsystem. It is also needed by the verification subsystem. The guidance is for each river station and each month of the year.

5.4.4 River forecast verification

A PRSF, specified in terms of a sequence of exceedance functions for epochs $t = 6, 12, 18, 24, \ldots, T$ hours, is verified against a time series of actual river stages. For every epoch t, the exceedance function is verified with respect to two attributes: *calibration* and *informativeness*. Each verification statistic is next analyzed as a function of the forecast lead time t.

5.5 FLOOD WARNING DECISION SYSTEM

The FWD system is designed for a WFO. Its purpose is to provide optimal rules for deciding flood watches and warnings based on PRSFs. The system is structured according to a decision-theoretic methodology (Krzysztofowicz 1993; Kelly and Krzysztofowicz 1994).

The methodology conforms to the *Bayesian postulates of rationality*: It seeks a rule that minimizes the expected disutility of outcomes, given a PRSF. Two elements of the forecast enter the decision rule: the forecast lead time t and the probability of the river stage $H(t)$ exceeding a threshold stage h_0 (for example, the flood stage) at time t. This probability is specified by the exceedance function as $P(H(t) > h_0) = \bar{\Psi}(h_0|t)$. The general structure of an *optimal decision rule* is as follows:

If $P(H(t) > h_0) > p^*$ and $t < t^*$, then issue a flood alarm.

The alarm may be either watch or warning. The *probability threshold* p^* and the *lead time threshold* t^* have two sets of values, one for the watch and another for the warning. These values are determined via the decision model, and they depend upon the disutilities of outcomes. Together, p^* and t^* characterize a trade-off between the probability of flooding and the lead time of a flood alarm. This trade-off is optimal from the viewpoint of users, or decision makers, whose disutilities have been represented in the model.

In summary, PRSFs and the decision-theoretic methodology will make it possible to replace the predominantly arbitrary hydrologic criteria for issuing flood watches and warnings with rational decision criteria, defined and estimated from the viewpoint of users.

5.6 CLOSURE

A conceptual design has been presented of an integrated probabilistic hydrometeorological forecasting system. The system is built of three main components: PQPF system, PRSF system, and FWD system. Research, development, and testing of system elements have been progressing since 1990. A demonstration of a prototype of the total system is planned next. The premise of this effort is that the integration of meteorologic forecasting, hydrologic forecasting, and decision making into a total system, and according to Bayesian principles of rationality, will harness recent advances and ultimately increase societal benefits of hydrometeorologic science and services (Krzysztofowicz 1995).

Acknowledgments

This chapter was written in 1996 while the author was on assignment with the U.S. National Weather Service, Eastern Region, under an Intergovernmental Personnel Act agreement. References were updated in 1999. Research leading to this chapter was also supported in part from a subaward under a cooperative agreement between the National Oceanic and Atmospheric Administration (NOAA) and the Cooperative Program for Operational Meteorology, Education and Training (COMET) of the University Corporation for Atmospheric Research (UCAR).

REFERENCES

Alexandridis, M. G. and Krzysztofowicz, R. (1985) Decision models for categorical and probabilistic weather forecasts. *Applied Mathematics and Computation* 17: 241–66.
Cooke, W. E. (1906) Forecasts and verifications in Western Australia. *Monthly Weather Review*, 23–24, January.
Kelly, K. S. and Krzysztofowicz, R. (1994) Probability distributions for flood warning systems. *Water Resources Research* 30(4): 1145–52.
Krzysztofowicz, R. (1983) Why should a forecaster and a decision maker use Bayes theorem. *Water Resources Research* 19(2): 327–36.
(1985) Bayesian models of forecasted time series. *Water Resources Bulletin* 21(5): 805–14.
(1992) Bayesian correlation score: a utilitarian measure of forecast skill. *Monthly Weather Review* 120: 208–19.
(1993) A theory of flood warning systems. *Water Resources Research* 29(12): 3981–94.
(1995) Recent advances associated with flood forecast and warning systems. *Reviews of Geophysics, supplement*, 1139–47, July.
(1996) Sufficiency, informativeness, and value of forecasts. *Proceedings*, Workshop on the Evaluation of Space Weather Forecasts, Space Environment Center, Boulder, Colorado, June, 103–12.
(1998) Probabilistic hydrometeorological forecasts: Toward a new era in operational forecasting. *Bulletin of the American Meteorological Society* 79(2): 243–51.
(1999) Bayesian theory of probabilistic forecasting via deterministic hydrologic model. *Water Resources Research* 35(9): 2739–50.

Krzysztofowicz, R. and Drake, T. R. (1992) Probabilistic quantitative precipitation forecasts for river forecasting. *Preprints*, Symposium on Weather Forecasting, 72nd Annual Meeting of the American Meteorological Society, Atlanta, Georgia, 66–71, January.

Krzysztofowicz, R., Drzal, W. J., Drake, T. R., Weyman, J. C., and Giordano, L. A. (1993) Probabilistic quantitative precipitation forecasts for river basins. *Weather and Forecasting* 8(4): 424–39.

Krzysztofowicz, R. and Pomroy, T. A. (1997) Disaggregative invariance of daily precipitation. *Journal of Applied Meteorology* 36(6): 721–34.

Krzysztofowicz, R. and Sigrest, A. A. (1997) Local climatic guidance for probabilistic quantitative precipitation forecasting. *Monthly Weather Review* 125(3): 305–16.

(1999) Calibration of probabilistic quantitative precipitation forecasts. *Weather and Forecasting* 14(3): 427–42.

Murphy, A. H. (1991) Probabilities, odds, and forecasts of rare events. *Weather and Forecasting* 6(2): 302–307.

Zevin, S. F. (1994) Steps toward an integrated approach to hydrometeorological forecasting services. *Bulletin of the American Meteorological Society* 75(7): 1267–76.

6 Flood risk management: Risk cartography for objective negotiations

O. GILARD*

ABSTRACT
Flood damage in France and Europe in recent years has shown that there is still a long way to cope with this problem. It seems that the conceptualization of the risk by dividing it between a socioeconomical dimension (vulnerability) and a hydrological-hydraulic dimension (hazard) is a promising means of investigation. Moreover, recent hydrological synthetic models, called flow-duration-frequency models, allow one to propose quantification of these two parameters of risk, that is, vulnerability and hazard. Estimating its spatial characteristics is very useful in the process of objective negotiation where land use managers take into account flood risk and socially acceptable risk. Representative maps, such as those proposed by the "inondabilité" model, can be forwarded to decision makers in order to help them use hydrological and hydraulic results in a more efficient way. These new concepts and methods should improve risk mitigation and lead to better acceptability of the risk level in the potentially flooded area.

6.1 INTRODUCTION

Extreme floods have been particularly numerous in France in the recent past. They caused severe economic and human damage. Among disastrous recent floods were those of Vaison la Romaine (1992), Corsica (1993) and Camargue area (1994), north and west of France (1994 and 1995), Var (1994), and so on. These events showed that flood risk management, and especially land use management in flood plains, is not sufficient to cope with the problem.

A risk policy should be based on three different aspects, as shown in Figure 6.1: these are prevention in the phase of land use management, flood forecasting and crisis management, and individual risk culture to improve citizen reactions to flood risk (Gilard and Givone 1993). Even if much work has been devoted to flood forecasting, it seems that more efforts should be made in the area of prevention in order to avoid human risk. Recent progress in hydrology allows one to propose a cartographic method based on an objective quantification of flood risk. It is useful to organize a real negotiation around flood risk.

Using all the diversity of land use, such methods should lead to more balanced solutions using both structural and nonstructural works, adapted to a basin scale and more respectful of the environment.

6.2 DEFINITIONS AND CONCEPTS

6.2.1 First step of a risk model

When speaking of the risk situation, some complex concepts appear. In fact, one intuitively makes a distinction between a campsite localized near a river with slow flow and another campsite near a torrent: the former situation is more acceptable than the latter. One also makes the same intuitive distinction between development of a house and of an industrial area near a river.

Analyzing this intuitive risk comprehension, one assumes that the risk can be analyzed by breaking it up into two independent components: one based on the socioeconomic perception and the other depending on the hydrologic and hydraulic knowledge of the hydrological regime (Figure 6.2).

The first factor, which will be called vulnerability, represents the sensitivity of the particular land use to the flood phenomenon. Consequently, it depends only on the land use type and on the social perception of risk. It can be different from one area to another, even for the same type of land use, and can also vary in time. One can assume for instance that the same campsite has the same vulnerability independent of where it is located, in a flood plain or on top of a hill. The difference in the risk level is due to the hazard of occurrence of flood, which is obviously different in the two situations.

The second factor is called hazard and depends only on the flow regime of the river, independent of the land use of the flood plain. That is to say, the same flow will flood the same area with the same physical parameters such as maximum water depth, duration of inundation, etc., independent of the land use.

* Division Hydrologie – Hydraulique, Cemagref, Lyon, France.

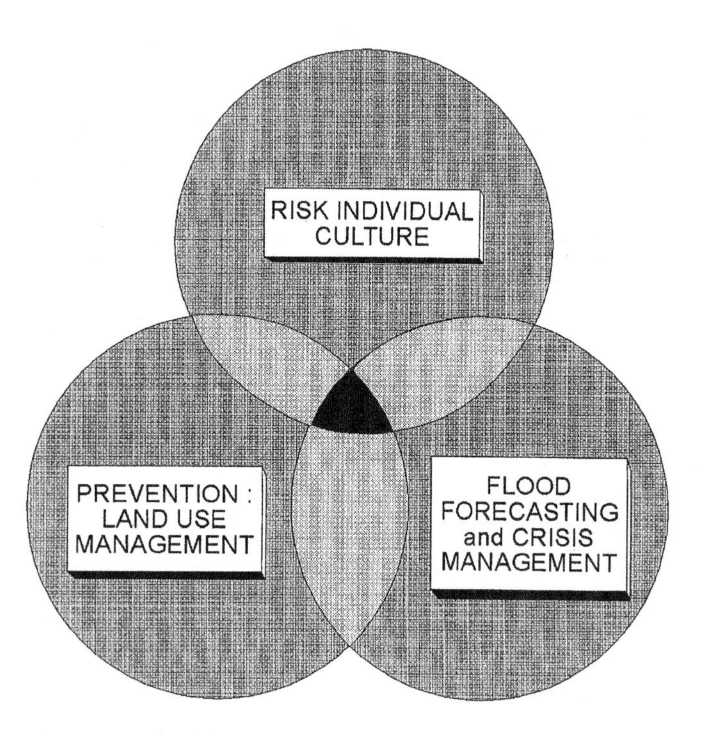

Figure 6.1. The three components of a risk management policy.

The flood risk level is a combination of these two basic factors: hazard and vulnerability. This decomposition of risk into two components is a first simplification or conceptualization of a complex reality, which should allow one to achieve a better comprehension of the problem.

6.2.2 New rules for river management

Until now, most risk studies have been limited to a hydraulic analysis of the river in order to design structural flood protection works. Moreover, they are limited to estimating design flow and do not take into account the possibility of an out-bank flow. Investigations are carried out for a defined reach of the river where the problem is most accurate even if it is not the best scale for flood management. Such an approach leads to accelerating the water in its movement downstream by structural means such as river embankments or by channelization, as the only way to mitigate the risk. Using the concepts defined previously, that means that one assumes an *implicit* vulnerability level and studies mostly the hazard factor. Because of the generalization of such works, their efficiency decreases because of cut-off of natural polders. The problems are conveyed downstream. At the same time, such structures induce a modification of land use, and of course of vulnerability, regardless of the real hazard level because people think they are perfectly protected. So, in case of extreme floods (for example, when the dikes are overflowed) damages could be very significant.

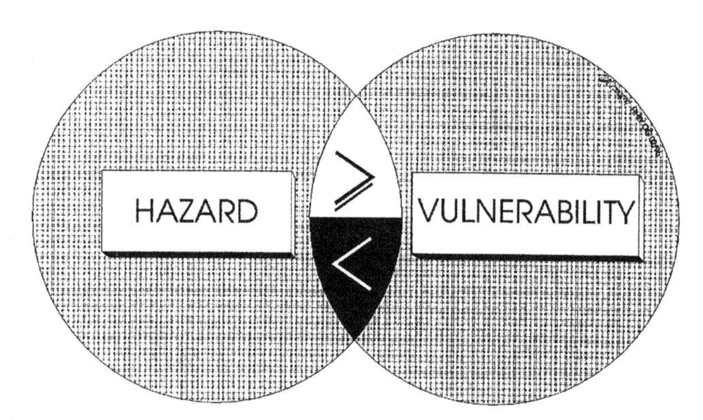

Figure 6.2. Risk conceptualization.

Today, according to recent French law (water law of 1992, environment law of 1995), one has to cope with the problem of flood management at the catchment scale, taking into account the whole river length and environmental objectives at the same time. At least, increasing needs in water supply and water quality lead to a broader perspective in terms of water resources. According to these new objectives, it seems that the generalization of the dynamic slowing-down concept could be a pertinent response to the problem. It requests enhancing water storage during flood periods wherever possible, favoring natural means, and fitting hydraulic constraints (frequency, duration, depth, and velocity) to a more realistic need of protection linked to each land use type. Consequently, each type of land use will be protected differently. A similar idea was proposed for a new floodplain management of the Mississippi basin in 1993 after the last big floods (Rasmussen 1994). Work on this concept is ongoing and is aimed at application on the catchment scale.

In order to do so, one needs an objective and quantitative estimation of both vulnerability and hazard. If the hazard is more or less imposed by natural conditions (hydrology and hydraulics), a real negotiation should be launched with inhabitants (or their representatives) of flood plains to define the adapted vulnerability (need of protection or acceptable maximum risk). The "inondabilité" method has been developed to provide an understandable reference based on hydrological, hydraulic, and socioeconomic knowledge in order to facilitate the negotiation process.

6.3 "INONDABILITÉ" METHOD

This method is based on an effective quantification of the two components of the risk – vulnerability versus hazard – as put in evidence by the previous concept. In order to compare objectively two largely different factors, it is necessary to quantify each one with the same units. The synthetic hydrological models called flow-duration-frequency (QdF) allow such transformation.

Their theoretical basis is the assumption that in each point of the area of concern, a local depth-duration-frequency relationship is available or, likewise, a flow-duration-frequency relationship. It has been demonstrated that most French and European river regimes can be described by three regional models with two local parameters, that is, the instantaneous maximum flow of a ten-year return period (QIXA10), and the characteristic duration for the catchment (D). Using a regional model with two local parameters, it is quite easy to produce a graph such as the one in Figure 6.3 determining maximum value of flow for any duration between 1 second and thirty days and for any return period between 0.5 year and 1,000 years. Further information on these models is given in Galea and Prudhomme (1993).

These QdF models compare favorably to the traditional "reference flood event" used in works design because of the following properties:

- synthetic description of the whole flow regime of the river,
- ability to transfer hydrological parameters all along the river and including ungauged sites,
- coherence in terms of peak flows as well as volumes,
- possibility of straightforward construction of synthetic hydrographs (called mono-frequency synthetic hydrographs),
- continuity in the scale of duration as well as frequency.

Theoretical and practical studies on numerous catchments put in evidence these properties at different scales (from less than 1 to more than $2{,}000\,km^2$).

As shown in a general flow chart of the method (Figure 6.4), these hydrological models integrate both hydraulic results (hazard modeling) and socioeconomic analysis (vulnerability modeling) to produce synthetic maps visualizing the risk of a flood plain of concern (Gilard 1995).

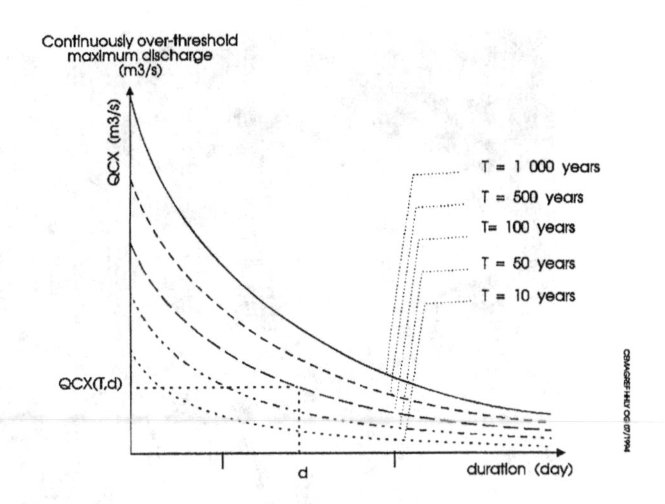

Figure 6.3. Example of local QdF abacus.

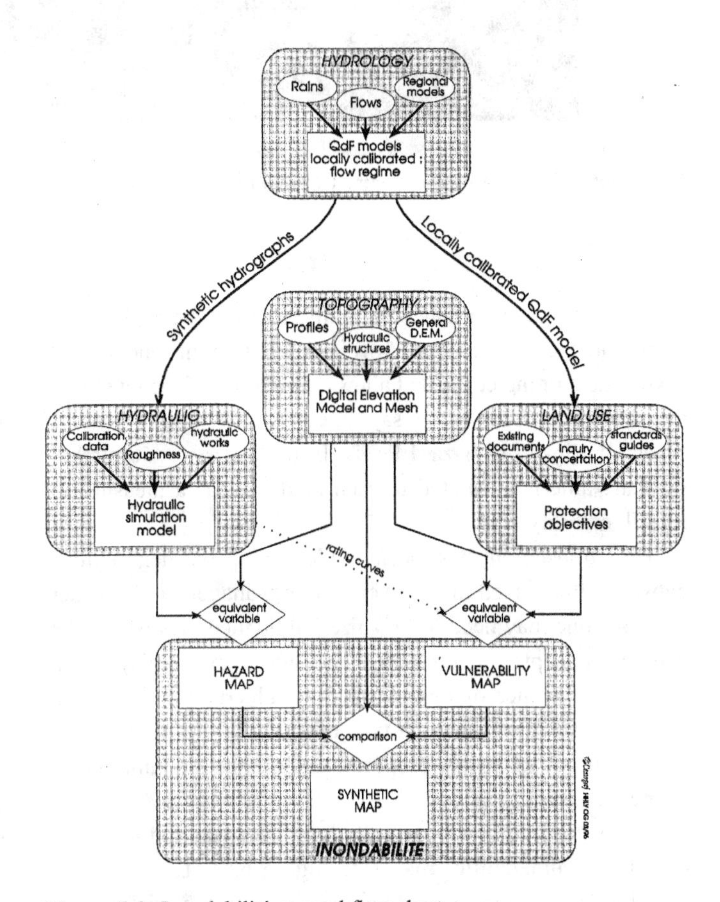

Figure 6.4. Inondabilité general flow chart.

6.3.1 Vulnerability mapping

The vulnerability maps are in fact land use maps in the whole flood plain of the river, which is supposed to be the maximal extension of an extreme flood (return period of 1,000 years and even more, if necessary). Each land use type will be affected with a "need of protection" expressed in terms of hydrological variables such as the frequency of the tolerable flood, its duration, its depth, and eventually its velocity.

Each plot of land is characterized by a couple (duration, frequency) or a triplet (depth, duration, frequency) or eventually a quadruplet (velocity, depth, duration, frequency). Locally calibrated QdF curves make it possible to transform these in an equivalent measure, expressed in equivalent return period, which means the return period for the flood equivalent to the protection objective. Figure 6.5 shows the practical way for this conversion. To take into account the depth, which

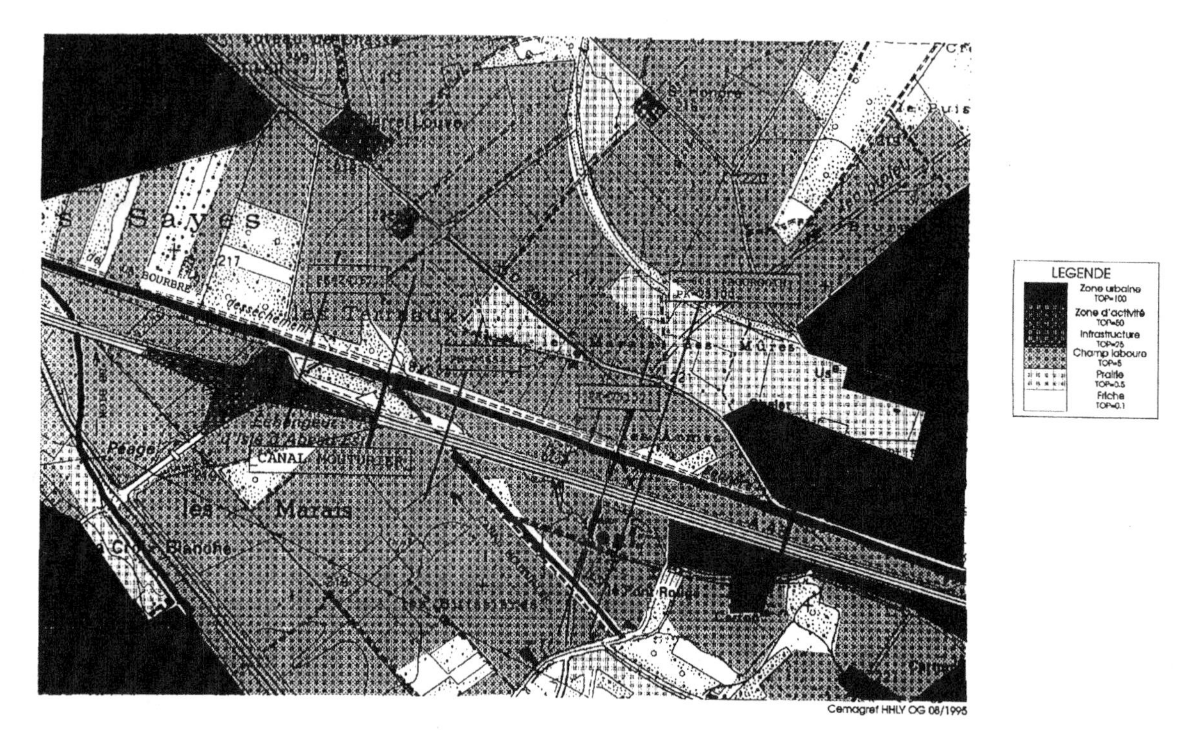

Figure 6.5. Vulnerability map (extract from a real example).

could be very important, especially in rural areas because some damages occur only above a defined depth, one should have a local rating curve that allows translation of the level into the discharge. That is to say, one needs hydraulic results, especially linked to hazard levels. In the end, a single mathematical value is attributed to each land plot, as a measure of its vulnerability.

This translation has been made based on negotiations between all involved in the river system management. In order to do so, one may need to organize public meetings where the general concepts are presented and where every player can express his or her opinion about the vulnerability level that should be attributed to each type of land use. To improve the process, usual values are proposed and used as the initial point of the discussion.

This value can be mapped all along the river, and it can be called the vulnerability map (see Figure 6.5). To evaluate a "need of protection" (see Figure 6.6) from the triplet (frequency, duration, depth), to a land use plot, a general survey in the basin is needed. It is useful to conduct meetings with representatives of the inhabitants to negotiate the final result as an iterative process between the theoretical model and the various wishes of the population. This process leads to a consistent result when all the representatives are aware of all the problems laid on the table.

6.3.2 Hazard mapping

The hazard level of a land plot is assumed to be linked to the return period of the first flooding flow. Similarly, assuming the flood amplitude constant, one can have a continuum of possibilities ranging from frequent exceedances of short duration to rare exceedances of long duration. So the return period of the first flooding flow summarizes the whole hydrological and hydraulic regime of the river, taking into account all the usual flood parameters such as frequency, duration, depth and, no doubt, velocity (through the QdF model assumption of cause).

This hazard level usually can be calculated with a mathematical Saint-Venant model solving hydrodynamic equations in transient mode with "mono-frequency synthetic hydrographs" (built on QdF curves; see Galea and Prudhomme 1994) as upstream conditions in order to be coherent with the QdF model of the river and with the way to quantify vulnerability. Such a model takes into account storage processes and is useful to test the consistency of hydraulic works.

In effect, maps of the flooded areas are available for many return periods ranging from one year to one thousand years (and even more if necessary). That is to say, one can determine the return period of the first flooding flow for each land plot of the concerned area (Figure 6.7) and this return period is taken as

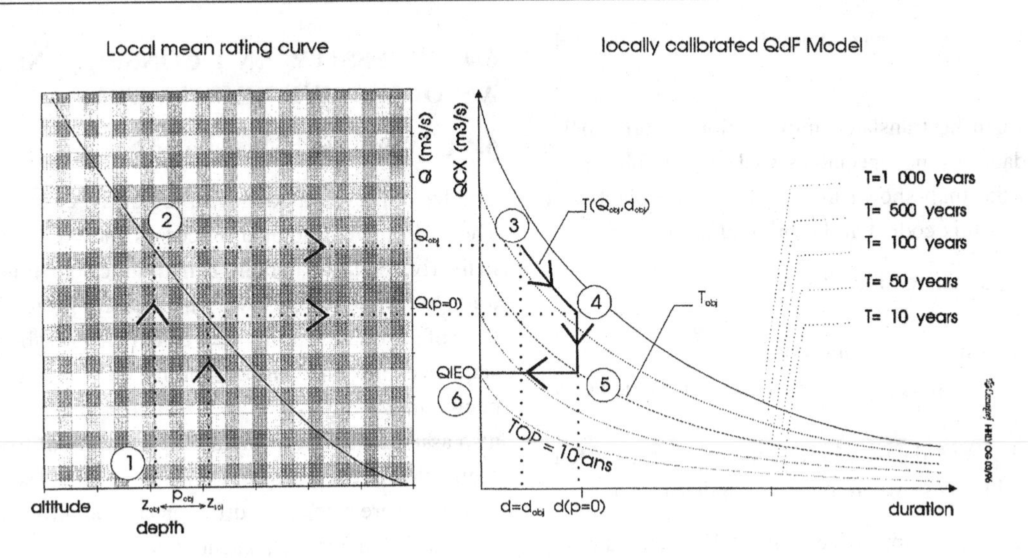

Figure 6.6. Need of protection equivalent variable calculation.

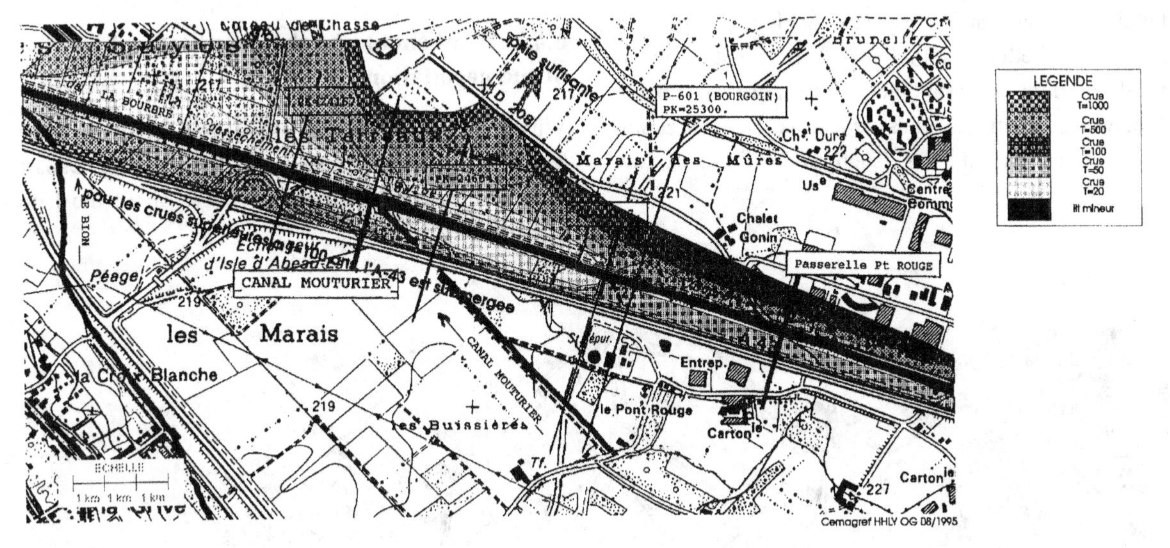

Figure 6.7. Hazard map (extract from a real example).

the hazard measure of this plot, expressed in years and thus comparable to the vulnerability assessment.

6.3.3 Synthetic "inondabilité" maps

With these two maps of vulnerability and hazard, comparison is available as both of them use the same unit and lead to the following possibilities:

- First, the hazard level, expressed as a return period, is undefined. It is higher than the maximum simulated return period, that is, the plot is outside of the maximum simulated flood; its risk level is undefined too and, as its objective of protection is finite, this plot can be considered to be overprotected;
- Second, the hazard level is higher than the vulnerability level; that means that the occurrence of a flooding flow is smaller than tolerated; the plot is overprotected with a security margin of the difference between the hazard level and the vulnerability level;
- Third, the hazard level is smaller than the vulnerability. That means that this is a situation of risk where local need of protection is not satisfied. An estimation of the underprotection or risk magnitude is given by the

difference between the hazard level and the vulnerability level.

These results can be translated into a color code to make them understandable by nonspecialists such as a public decision maker, and the map shown in Figure 6.8 is a black and white adaptation of this code. On the original color map:

- Vulnerability is mapped in yellow.
- Hazard is blue.
- Risk is mapped in a three-color code:
  - yellow for the first case where plots are not flooded even in the worst event;
  - green in the second case where plots are overprotected;
  - red in the third case where plots are underprotected.

Such synthetic maps summarize the whole knowledge of a risky situation, including socioeconomy and hydrology. In the same way, one can test each structural work scheme in order to verify that the problem is not being conveyed downstream and also test the impact of land use development. This can be translated into maps and compared. Such a comparison should lead to a better comprehension of what kind of nonstructural solutions are available and to more sustainable solutions between structural ones (such as channelization or dikes) and nonstructural ones (such as land use of low vulnerability).

6.4 NECESSITY AND CONSEQUENCES OF AN OBJECTIVE NEGOTIATION

6.4.1 What is to be negotiated?

The hazard level, as defined previously, cannot be negotiated. The hydrometeorological processes and the hydraulic transfers in the riverbed lead to definite hydraulic constraints. On the contrary, the vulnerability level (the need for protection) could be a term of negotiation as it seems evident to everybody that the consequences are different depending on whether a phenomenon occurs in a rural area or an urban area. In a rural area, one can also assume that the vulnerability is different from a forest to a crop field, from a grass land to an orchard, and so on. In the same way, there are also some differences in an urban area between a stadium and a campus, a school, a hospital, a housing development, or an individual house. So, the first step of the negotiation could be the elaboration of vulnerability level in accordance with each land use type when building vulnerability maps.

6.4.2 What should be the quantitative base of the negotiation?

With the "inondabilité" method, one is able to quantify both hazard and vulnerability in the same units. Usually, people try to convert information they have about the land use into

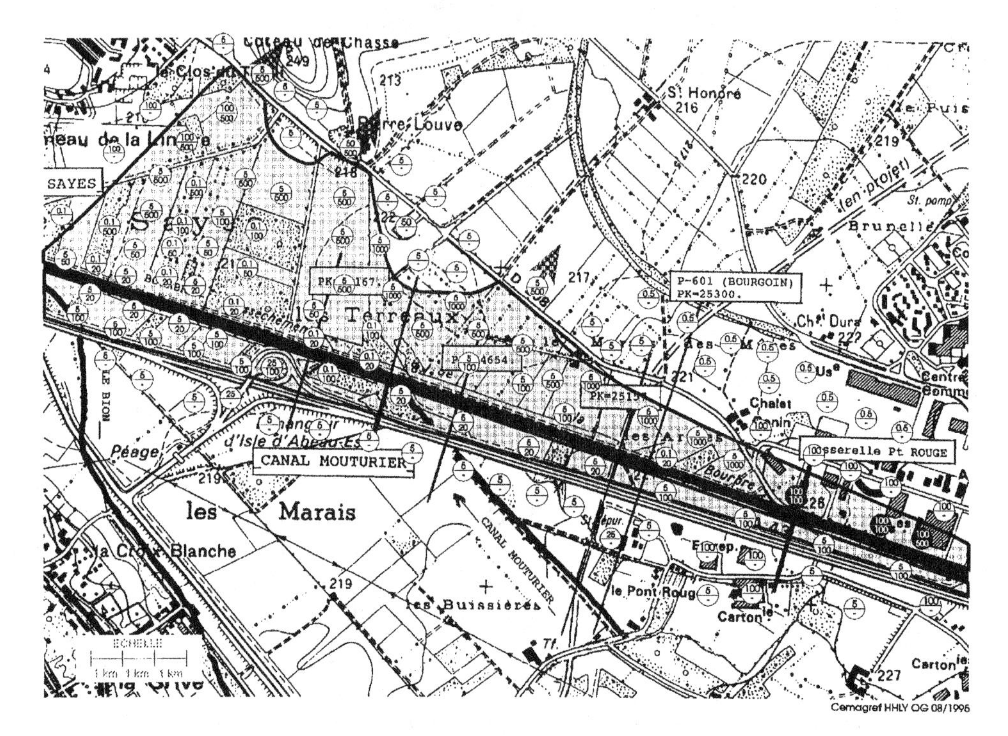

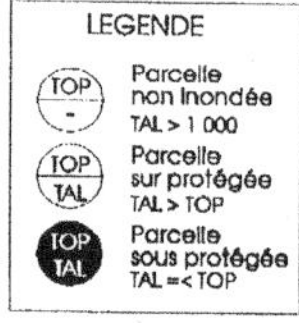

Figure 6.8. Synthetic risk map (extract from a real example).

economic value. This leads to converting flood knowledge into damage evaluation with a significant problem of accounting the probability of each event. In fact, when planning land use development or works on a river, one should bear in mind that the area is not exposed to a single well-known flood event but rather to the whole hydrologic regime, with many various manifestations such as big and small floods as well as severe droughts. Moreover, flood damage includes not only direct material damages; there are also many indirect damages which cannot be easily translated into monetary units (Green et al. 1995).

To reverse the usual order of things, with "inondabilité" method one chooses to translate economic concepts such as flood damages and economic stakes linked to land uses all along a river into hydrologic parameters. An objective way was devised to quantify both vulnerability and hazard in a coherent unit on each land plot. So it becomes possible to base negotiation on a quantitative estimation of risk. In fact, due to the continuity equations, every hydraulic work results in transfer of a water volume in time or in space. This transfer can be seen as an exchange of "risk value" and it can be efficiently negotiated between riverside residents (individually or collectively).

6.5 CONCLUSIONS

Extreme floods of the 1990s demonstrate the limitations of a traditional approach to flood protection, that is, the one based on hydraulic structural works. Today it seems possible to develop a real negotiation process, based on a quantitative description of risk all along a river. The latter will provide an efficient base of land use planning for the decision makers and representatives of inhabitants in charge of the development of their basin and management of the land use of the river's flood plain. The negotiation process will be effective once results of quantitative risk assessment are available and exchange between under- and overprotected zones is possible, such as via exchanges of vulnerability levels leading to future land uses. This way to design and to use a "risk market" at a basin scale proved to be an efficient way to better mobilize all the diversity and the potential of the landscape (natural polders, forests), to promote a better individual "cultural risk" and finally, to avoid damage in implementing a better land use management policy. The "inondabilité" method has been used in several basins in France, of areas from 20 up to 1,000 km^2 (Oberlin et al. 1989–1995). Results prove that this quantitative method helps decision makers to design better policies of flood management.

Acknowledgments

This chapter reports the initial results produced within the European Research Project on "FLOODAWARE" (Environment and Climate, DGXII, 4° PCRD). Part of this work was also supported by the French Ministry of Environment with the research program on natural hazards.

REFERENCES

De Laney, T. A. (1995) Benefits to downstream flood attenuation and water quality as a result of constructed wetlands in agricultural landscapes. *J. Soil and Water Conservation*, 620–26.

Galea, G. and Prudhomme, C. (1993) *Characterisation of ungauged basins floods behaviour by upstreaming QdF models*. Second international conference on FRIEND "Flow regimes from international experimental and network data," October 11–15, Technische Universität Braunschweig, RFA, IAHS Publication 221: 229–40.

(1994) The mono frequency synthetic hydrographs concept: definition, interest and construction. International conference, Alpine Regional Hydrology – A1: Regional synthesis of hydrological regimes (AMHY III), Stara Lesna, Slovakia, September 12–16.

Gilard, O. (1995) The "INONDABILITÉ" method. *Sistema Terra*, 4(3): 61–64.

Gilard, O. and Givone, P. (1993) Flood events of September 1992 South of France: reflections about flood management. Conference of coastal and river engineering, Loughborough University.

Green, C. et al. (1995) Vulnerability refined: analysing full flood impacts. In *Floods across Europe (flood hazard assessment, modelling and management)* edited by E. Penning-Rowsell and M. Fordham, European research program EUROFLOOD.

Oberlin, G., Gilard, O. et al. (1989–1995) Différentes études d'inondabilité (Bon-Nant, Bourbre, Rival, Egrenne . . .). Rapports d'études, Cemagref HHLY.

Rasmussen, J. L. (1994) Floodplain management into the 21st century: a blueprint for change – sharing the challenge. *Water International* 19: 166–76.

7 Responses to the variability and increasing uncertainty of climate in Australia

JONATHAN F. THOMAS* AND BRYSON C. BATES**

ABSTRACT

This chapter describes recent developments in the methods used to cope with uncertainty in water systems in Australia. The very high historical variability of Australian rainfall and runoff means that climate change simply may amplify a preexisting problem. Variability of flows to urban, irrigation, and environmental uses is considered, as well as issues of infrastructure robustness in the face of an increasing probability of large rainfall events. In general, the thrust has been to develop practices for flow allocation and demand management that are based on relative water availability, stochastically interpreted, rather than on absolute quantities. Operating rules are being adopted that allow decentralized decision making to take place within the constraints imposed by variability. This allows individual agents to express their risk preferences, and where possible to exercise their own decisions about risk and reliability. There is a growing need for expressions of social risk preference to be built explicitly into the trade-offs that are implicit in system design and operation.

7.1 VARIABILITY IN AUSTRALIA'S CLIMATE AND HYDROLOGY

The hydrologic environments of Australia and of Southern Africa are fundamentally different from those of the Northern Hemisphere. Even within the same climatic zones, annual flow variability of these Southern Hemisphere continents is two to four times that of northwest Europe and North America (see Table 7.1). For Australia, part of the difference is that the continent lies at the western end of the El Niño/Southern Oscillation (ENSO) system (Ropelewski and Halpert 1987) and is affected by rain depressions resulting from tropical cyclones. There is growing evidence that ENSO effects extend to the west coast of Australia, influencing Indian Ocean temperatures in the higher latitudes via oceanographic and atmospheric dynamics exerted through the Indonesian archipelago (Meyers 1996; Meyers et al. 1995). Interannual fluctuations are modulated by decadal-scale and longer-term variations. Climate variability is magnified in runoff variability, and this is in part due to soil properties and the transpiration strategies of the native vegetation (Findlayson and McMahon 1988; Fleming 1995).

Groundwater levels in many Australian aquifers also show remarkable variations through time. See, for example, Jolly and Chin (1991) on aquifer responses in the dry tropics of northern Australia, and Macumber (1991) on aquifer response in the temperate zone in the state of Victoria.

In contrast to the natural variability of rainfall and runoff, a prime objective of Australian water supply systems is to provide reliable yields. It has been calculated, using the Gould-Gamma equation, that the average storage size required to provide equivalent yields is approximately six times greater in Australia than it is in Europe (yield being expressed as a given percentage of the mean streamflow with a given design probability of failure). This is before taking account of higher evaporation rates in Australia (Findlayson and McMahon 1988). For example, in 1982 Sydney had a water storage capacity of 932 m^3 per head of population served, as compared with 250 m^3 for New York, 73 m^3 for Glasgow, and 18 m^3 for London. The average volume of irrigation storage capacity in Australia is 15,200 m^3 per hectare of irrigation, compared to 7,600 m^3 in the United States; 3,800 m^3 for Egypt; and 1,500 m^3 in India (Burton 1982).

The problem of determining required storage capacity is exacerbated by the fact that mean annual flows and their standard deviations vary dramatically with the climatic period selected. The differences are enormous for the more variable streams (Fleming 1995). Developing reliable storage-yield relationships is extremely difficult, and the high variability of runoff also places serious limitations on the management of storages. Consequently, as well as providing large storage capacities, water suppliers have adopted methods for managing water demand and delivery systems during periods of drought.

* Resource Economics Unit, Perth, Australia.
** CSIRO Land and Water, Wembley, Australia.

Table 7.1. *Comparison of annual flow variability in catchments of 1,000 km² to 10,000 km², as between Australia/Southern Africa (ASAF) and the rest of the world (ROW): after McMahon and Finlayson (1991). C_v = coefficient of variation*

Climatic zone (Koppen System)	ASAF number of streams	ASAF Median C_v	ROW number of streams	ROW median C_v	Ratio of C_v
Aw	1	0.76	7	0.26	2.9
Bsk	9	0.93	9	0.36	2.6
Cfa	26	0.92	40	0.34	2.7
Cfb	18	0.87	32	0.25	3.5
Csa	2	1.33	6	0.45	3.0
Csb	2	0.61	1	0.50	1.2
Cwa	9	0.96	5	0.23	4.2
Cwb	12	0.64	7	0.41	1.6

7.2 CLIMATE CHANGE

In southern and eastern Australia, which have highly variable interannual rainfall, climate change projections suggest that there may be an increase in the frequency of large events and of droughts. However, the extreme variability of rainfall sequences makes it difficult to confirm nonstationarity. Water supplies have been severely reduced in much of Queensland and New South Wales in recent years by an extended drought. Meanwhile, the perceived higher probability of extreme rainfall events has also drawn attention to the structural integrity of dams and spillways.

The southwest of the continent has also experienced reductions in rainfall and streamflows since 1975. A reduction of the order of 40 percent in streamflow in many Darling Range rivers has occurred, comparing records from 1911 to 1974 with those from 1975 to 1994. Figure 7.1 presents data for one representative catchment. Here there is evidence of nonstationarity in the streamflow series. Groundwaters that have not been subject to pumping or land use change exhibit declining water tables over the same period, though factors other than rainfall may be involved. It is anticipated that there may be a long-term increase in the (currently low) level of summer rainfall and a reduction in winter rainfall. This would reduce system yields markedly, as much of the additional summer rainfall would be lost by evapotranspiration, while surface water storages only begin to accumulate after a wetting-up of the soil profile in early winter.

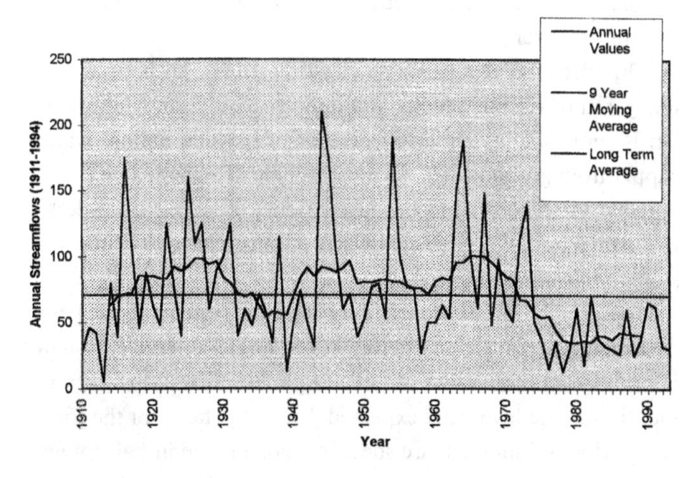

Figure 7.1. Serpentine annual streamflows (1911–94).

7.3 URBAN SYSTEMS

The reliability of Australian urban supply systems varies, depending on local conditions and past design choices. Perth has elected to adopt an "unreliable" system, with a need for water use restrictions on average once in every ten years. Sydney has a system that is designed to give at least 90 percent of average yield in 99 percent of years. In recent times it has been necessary to impose restrictions more frequently. Perth has had restrictions on water use in five of the last seven years, and Sydney has had to impose water use restrictions for the last two years, after a thirty-year period of no restrictions.

Ruprecht, Bates, and Stokes (1996) report that water utilities in Western Australia have responded to the increasing uncertainty about system yields by (i) seeking to improve the accuracy of medium-term meteorological forecasts, thus allowing improvement in operating decisions; (ii) changing the length of climatic time series used for system operation and augmentation calculations, so as to represent the weather pattern "currently" being experienced (where "currently" represents a judgment that will be reconsidered on a decadal time scale); (iii) advancing in time some or all of the planned source developments needed to meet anticipated demands; (iv) diversifying water sources (including groundwater development, water re-use and cloud seeding) so as to make the overall system more resilient to rainfall variability; and (v) devising more effective and efficient systems of demand management during periods of shortage.

7.3.1 Meteorological forecasting

In an attempt to improve decision making about operating rules and water use restrictions, attention is being focused on the prospects for improving medium-term rainfall forecasting; that is, at the time scale of one to two years. For example, in the west of the continent research funds are being sought to increase the availability of data on surface and depth temperatures of the Indian Ocean, to allow linkages to be made to regional climate models. Reducing the errors in dam inflow and groundwater recharge forecasts for the Perth system would mean that a decision *not* to impose water use restrictions at the beginning of the summer period could be made with greater confidence about the probability of adequate storage recovery during the following winter (see comments on restrictions policies below). At present the Bureau of Meteorology and other weather consultants provide advice to the WA Water Corporation each spring.

7.3.2 Selection of climatic time series for planning

Planning and management of the Perth supply system has been made more complicated over the last ten years by what appears to be nonstationary variability in the streamflow record. The data for one representative catchment in the southwest of Western Australia, shown in Figure 7.1, demonstrate a markedly different pattern of annual streamflows since 1975. In particular, the high flows that occurred at five- to ten-year intervals in the longer term data series have been absent since the mid-1970s. This has been associated with a marked tendency for rainfall to decline late in the winter period in most years since 1975. The utility's response has been the pragmatic and reasonable one of basing system planning and operation on the streamflow history of the last twenty years rather than on the longer term record.

7.3.3 Capital works rescheduling

This option, which has been described in Boland, Carver, and Flynn (1980), identifies the incremental cost of increasing supply capacity so that restrictions on use are no longer necessary. This policy is increasingly being adopted in Australian cities, but it can be rather costly.

7.3.4 Water use restrictions

Types of restriction

Although it is economically rational in times of shortage to tolerate wide fluctuations in the price of a commodity, this has been rejected by Australian state governments who control the urban water utilities. Instead, they have chosen an approach based on price stability plus a mixture of voluntary and mandatory restrictions on water use.

In urban areas, water use restrictions introduced at any given point in time are based on an ascending scale of severity, depending on the calculated probability of supply failure. Garden watering is the principal target. The form of restrictions varies, but the most common is to ban garden sprinkling between 9:00 A.M. and 6:00 P.M. each day. This allows householders the freedom to supply all the water their garden needs, while saving considerably on evaporative losses. Other forms of restrictions include alternate-day watering for odd- and even-numbered houses, and in severe droughts a total ban on sprinklers.

Stochastic analysis

Probabilities of the need for restrictions are explicitly considered in system design and in the choice of operating rules. For example, the metropolitan water supply system for Perth (population 1.1 million) is designed to require water use restrictions once in every ten years. This part of the continent experiences a Mediterranean climate, in which rainfall is highly concentrated in a four-month winter period. In springtime, when the winter's gain in storage is known, a decision must be made as to whether water use restrictions will be required in the forthcoming summer.

This decision requires a stochastic estimate to be made of likely future storage credits and deficits. The Perth Metropolitan Region obtains about 60 percent of its total annual water supply from groundwater. As this resource is managed on a sustainable yield basis, and because of environmental considerations, high pumping of groundwater during years of low surface storage has been resisted.

The simulation approach described by Nathan (1996) for developing a restrictions policy for a single reservoir system is broadly representative of general practice. First, the current month, storage level, and expected demand pattern for the forecast period (<9 months) are specified. For a given initial storage level and antecedent conditions the expected storage level in

n months' time is determined from a mass balance based on expected demands and losses, and the volume of inflows corresponding to a range of nonexceedance probabilities, using a monthly time-step. Two control variables are used: (i) trigger level for imposition of restrictions, and (ii) the severity of restrictions expressed as a percentage of normal demand. The consequences of different scenarios, in terms of exceedance probabilities and alternative levels of the trigger signal and severity of restrictions, can thus be evaluated, making extensive use of graphical output.

It has been our experience that models which use linear programming to determine operating rules for multireservoir systems often incorporate very crude representations of conjunctive-use systems involving groundwater as well as surface water resources and are not well suited to deriving short-term operating rules, for example, rules for real-time operation. The latter limitation is of great practical importance as water managers deal with uncertainty in real time, adjusting reservoir releases and groundwater extraction rates in response to observed conditions and climate forecasts.

Economics and social equity of restrictions policies

The importance of water use restrictions in Australian urban water systems has also focused attention on the economics of system designs which imply more or less restrictions (i.e., smaller or larger storage capacities). The implementation of restrictions will normally result in economic costs to both consumers and the water utility. Under this policy, water utility revenues fall dramatically during periods of restricted supply, while the consumers suffer from reduced water availability.

Dandy (1992) estimated the potential economic costs of water restrictions in Adelaide, South Australia, assumed to be equal to the loss of consumer surplus plus the loss of producer surplus. The "external" costs of restrictions in the form of the "browning" of the city during the drought period were not considered. A general expression for the economic costs of time restrictions on outdoor water use was presented:

$$L = pQ_0\{e/(1+e)\}\{1-(1-r)^{(1+e)/e}\} \qquad (7.1)$$

where

L = economic loss ($/unit time)
p = water price ($/m^3)
Q_0 = unrestricted outdoor water demand at price p (m^3/unit time)
r = mean proportional reduction in water use due to restriction policy
e = price elasticity of demand for outdoor water use

Using a simulation model of the southern Adelaide water supply headworks system, Crawley and Dandy (1996) concluded that the imposition of restrictions on water use is costly for both consumers and producers, and that operating rules should be such as to minimize the use of these policies where possible.

It is by no means certain, however, that consumers will be willing to pay slightly higher average long-term water prices in exchange for a lower probability that water use restrictions will be imposed. Questionnaire surveys of consumers in Perth, Sydney, and Canberra have suggested a general willingness to tolerate a higher frequency of restrictions rather than face long-term average price increases. Nancarrow and Syme (1989) asked water consumers the question: "In order to avoid water restrictions in times of drought or water shortage, would you be prepared to pay any of the following increases to your water rates or excess water charges (i.e., the extra money could be used to build more dams or consider desalination plants, etc.)?" Table 7.2 summarizes the results.

Such data must always be interpreted carefully. To the economist, the fact that approximately a third of consumers have a positive willingness to pay for increased system reliability suggests the existence of a potential Pareto improvement through higher charges. However, the public good aspect of water supply systems often means that in order for this minority of consumers to obtain the more reliable service that they are willing to pay for, higher charges must be imposed on the majority who would rather tolerate more frequent restrictions. For a politician who may be held responsible for the decision and for the affected community, there is clearly a problem of equity. An intermediate solution might be found, however, through a system of water delivery and metering that allows individual consumers to choose to pay a higher water price for increased service reliability. Alternatively, under a two-part tariff structure, the water utility might be able to devise a system of fixed charges that compensate low-income house-

Table 7.2. *Urban water consumers' willingness to pay to avoid water use restrictions*

Response	Perth % (n = 352)	Sydney % (n = 344)	Canberra % (n = 349)
Zero	70	66	63
up to $10	10	7	8
up to $30	5	11	10
up to $50	9	10	9
up to $100	6	5	7
>$100	1	1	2

holds for the higher variable charges needed to finance higher reliability.

Risk trade-offs

It seems that, despite our improving ability to simulate the behavior of water supply systems that are subject to stochastic loadings, subjective judgment is always involved in the final policy choice. Some researchers have responded to this by proposing methods of multi-objective and multi-attribute utility analysis that would make the decision-making procedure and the trade-offs that it implies more transparent.

One example is given by Thomas and Macpherson (1991), who developed an object-oriented system for evaluating alternative water resource development scenarios that were subject to varying levels of risk and reliability. For each possible new source, the system assumed expected (deterministic) flows based on a stochastic analysis of reservoir and groundwater yields. The source development problem was then defined as the selection of a subset of possible projects, including dams, wellfields, and trunk distribution pipes to serve a number of defined regional markets. The objective function was to maximize the net present value of producer and consumer surplus, or as a variant of this, to maximize the financial surplus of the water utility, through the optimal scheduling of development. A version of the Metropolis Algorithm, or "method of simulated annealing" (Press et al. 1986), was used to minimize the risk that a particular schedule of source developments could be a local rather than a global optimum.

The method was applied in a study of source development in the rapidly urbanizing southeast of Botswana, undertaken by an Australian firm of consulting engineers. The perceived risks surrounding particular source developments were numerous and extended well beyond issues of stochastic hydrology. For example, some source development options involved dam construction in another country, and hence international agreements on cost sharing and water allocation. The capacity of the entire schedule of developments to be financially viable had to be measured against the risks associated with any delay in augmenting storage. One of the trade-offs involved in this choice was between having small recycling projects and water use restrictions in the early years of the planning period, as against making an earlier start to major construction, indeed in advance of the capacity of the water market to pay for a big scheme. A multi-objective framework in which various scenarios were generated and evaluated seemed appropriate for the circumstances; however, it was found that there was reluctance among the stakeholders to lay bare the trade-offs that were inherent in the final choice. There is generally great pressure on advocates of big projects to present them as being risk-free, or at least risk-manageable. Similarly, the impression must be given that all the trade-offs have been equitably resolved. A frequent result is that all the back-room work on the true risks and trade-offs must be kept in the back room if the program is to proceed. This should not deter those who are responsible for project design from considering risks and trade-offs explicitly and within a standard framework.

7.4 IRRIGATION SYSTEMS

7.4.1 Overview

There are over two million hectares of irrigated land in Australia, including pastures and, increasingly, cropping and horticultural land. Most of Australia's agricultural irrigation from surface storages occurs within the Murray-Darling Basin. While using only 1 percent of total agricultural land, irrigation accounts for about 25 percent of national agricultural production and generates around $6 billion in gross national product, $4 billion in exports, and most of Australia's production of fruit, vegetables, dairy products, wine, cotton, and rice.

The Murray-Darling catchment, an area of 1.06 million km^2, has high rainfall in its headwaters in the Great Dividing Range, and low rainfall in the lower reaches in western Queensland, western New South Wales and South Australia. The river system is fully regulated, with large dams in the headwaters and some hundreds of weirs and other regulating structures throughout the basin. Approximately 7,650 Gl/year are supplied for irrigation use and a further 327 Gl/year for domestic, industrial, and commercial use. The basin experiences highly variable rainfall and runoff, as is apparent from the streamflow data shown in Figure 7.2.

Water allocation systems have been devised to cope with uncertainty. Generally, the system of water allocation to farmers comprises a fixed and a variable allocation, often referred to as a "surplus flow" allocation. The latter depends on available storage at the beginning of the summer irrigation season.

Three major policy developments within the last decade have been (i) the division of irrigation water supply systems into two components, namely a wholesale and a retail component; (ii) the privatization/corporatization of local irrigation delivery systems; and (iii) the introduction of transferable water entitlements, including inter-catchment transfers. These initiatives have mainly been a response to microeconomic reform, issues concerning the financing of infrastructure renewal, and the fact that in many regions water resources are now fully developed and committed. However, these new institutional arrangements increasingly define the context within which the management of system reliability, risk, and robustness has to proceed.

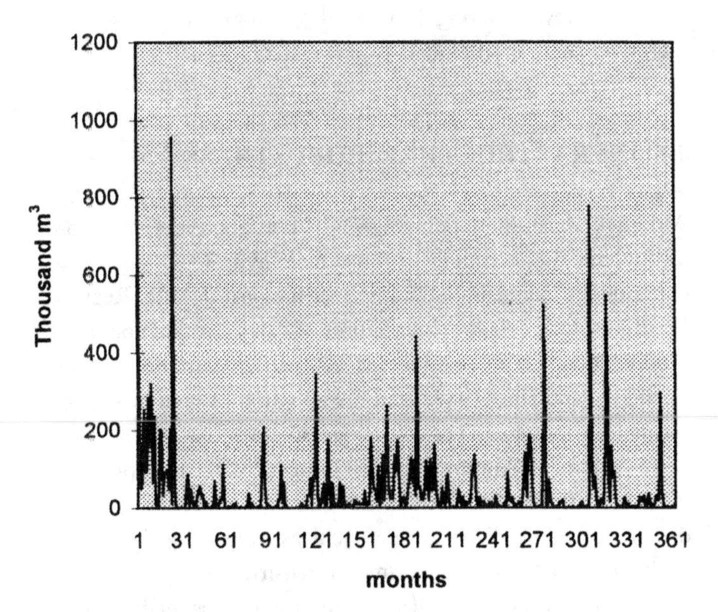

Figure 7.2. Namoi river flows at Keepit reservoir, New South Wales for 33 years (after Dudley 1991).

7.4.2 Stochastic analysis

Deterministic approaches based on observed climatic or hydrological sequences are generally appropriate for water systems with seasonal regulation and short planning horizon relative to the lengths of historical records. They are not appropriate in assessments of yield, reliability, or resilience when the lengths of reliable, concurrent climatic, and hydrological records are short relative to the planning horizon or to the lengths of long-term droughts. Stochastic approaches make uncertainty explicit and allow investigation of system response against the possible range of system input and output sequences. Computer simulation is the principal method used in estimating yields, storages, and the effects of different operating rules. Long sequences of data, usually up to 100 years of monthly rainfall, are required to obtain adequate samples of individual events and sequences. Synthetic streamflow data, and historical data that provide information about other variables such as cultivated area, or other climatic variables, are often used.

In large multiple-reservoir systems providing irrigation supplies, such as in the Murray-Darling system, there is a need to match short-run decisions (e.g., area of crop to plant) with medium-term operational decisions (e.g., reservoir contents to carry over to the next year) and with long-run decisions (e.g., total area planned to be irrigated, or the level of fixed water flow to be reserved for wetland inundation).

This has encouraged joint optimization approaches that address long-term and short-term decision variables simultaneously (Dudley 1990). Conducting a complete search by computer simulation for the optimal set of operational decisions, which can be thought of as an optimal path through a network, is a huge computational task, even for modern computers.

However, dynamic programming can be combined with simulation to obtain the "best" decisions to follow under uncertainty for every state that the system might be in at each decision point (Dudley 1991). Instead of beginning at the start of, say, a 100-year sequence with 1,200 decision intervals, Dudley's model takes just one decision interval (e.g., January) and simulates the effects of all decision alternatives over that interval, for all possible starting system states, using data for the 100 Januaries in the record. The effect is measured in end-of-period system states and economic returns. From these an expected value of returns is obtained and a frequency distribution of end-period system states, which is taken as the state variable transition probability distribution for that discrete decision and state. The process is then repeated for all other decision intervals, forming return matrices and state variable transition probability matrices for use by a dynamic programming model, which looks ahead many years when optimizing decisions. According to Dudley (1991) the state variables can be defined to reflect the degree of uncertainty faced by decision makers. This contrasts with many applications of stochastic simulation models which assume perfect foresight over the immediate decision interval.

7.4.3 Allocation under conditions of uncertainty

In relation to hydrological variability, there is a need to identify separately the ability to trade water in the short term as distinct from the trading of a perpetual right to receive water. The concept of "capacity sharing" (Dudley and Musgrave 1988) has been developed in order to decentralize decision making about releases from storage in regulated river systems. The capacity sharing model conceptually divides a single large reservoir into many compartments or subreservoirs, each controlled by a single decision maker (user). Each user is allocated a *percentage* share of reservoir capacity and a *percentage* share of net inflows to the reservoir (net inflow is inflow minus evaporation and seepage losses for the period). Thus, the emphasis is on coping with highly variable water availability rather than with guaranteeing a constant supply volume. All holders of water rights may trade.

The set of decision makers would include all the irrigation farmers and any other users of the resource such as towns and industries, but also agencies which perform a resource management function, for example a flood control agency or an environmental agency. The percentage shares of reservoir capacity do not need to be equal across users; nor is it necessary for the percentage share of net inflows allocated to an individual user to be the same as its share of total reservoir capacity. For example, if one of the reservoir users is a flood control agency, it might be best for it to own a share of reservoir capacity, but not any share of net inflow.

The management of the overall reservoir is reduced to little more than monitoring individual releases by the water-using decision makers, and calculating their available capacity and unused share of net inflow at any point in time.

The capacity sharing method has been extended to situations where there is a need to establish some priorities among the set of users, as to which is to have first call on available flow. For example, a town supply might take the first priority, even though its share of total flows is smaller than that of an agricultural irrigation area.

The concept of capacity sharing has been extended to include unregulated streamflows. For example, a user's total entitlement may be made up of a share of empty reservoir space, a share of reservoir inflows, and a share of downstream unregulated flows. The unregulated flow entitlement is used on an "as available" (no storage) basis (Dudley and Scott 1993).

There is a growing number of applications of the capacity sharing concept. Two large schemes, the Snowy Mountains Scheme and the Hume-Dartmouth Dam, are operated on a capacity sharing basis. The state of Victoria has adopted capacity sharing at the level of bulk allocations to water wholesalers. A capacity sharing scheme is being established for a dam on the Fish River in northern New South Wales. In this scheme reservoir capacity is shared between a thermal power generator, the Sydney Water Corporation, two local councils, and a fifth group representing miscellaneous users.

Interest in capacity sharing is greatest in areas of the greatest hydrological uncertainty, such as in northern New South Wales. Irrigation farmers are generally in favor of this type of scheme because they feel increasingly threatened by possible re-allocations of flow for environmental purposes. There is evidence to suggest that the greater the security of title held by resource users, the greater is the incentive for conservative use (Young 1992). A study currently under way at the Centre for Water Policy Research at the University of New England will use a streamflow model of the Namoi River Basin in order to show what level of reservoir capacity would have to be allocated to irrigators in order to guarantee them an equivalent sequence of flows as they have received under the current system.

The capacity sharing system also has potential ramifications for catchment land use management. For example, if proposed land use changes such as afforestation or farm dam construction would decrease expected flows, then the perpetrators of the land use change would need to buy flow rights from existing users. Conversely, the reservoir manager could offer to buy any increased flow resulting from land development and sell the increased share entitlements to downstream users.

7.5 INFRASTRUCTURE ROBUSTNESS

Although Australian dams have had a good safety record throughout the twentieth century, their hydrologic safety has come under increasing scrutiny over the past decade. Reasons for this include catastrophic failures of dams in other countries, modification of procedures for estimating extreme rainfalls (Kennedy and Hart 1984), and the adoption of nonlinear flood estimation models. These have led to revisions of estimates of probable maximum precipitation (PMP) and probable maximum flood (PMF). It is now accepted that the spillway designs of many dams are deficient as judged by the design probability of failure at the time of construction.

The high cost of upgrades has led to an interest in dam break analysis, hazard rating and mapping, and risk analysis procedures (Cantwell and Himsley 1987; Wellington 1987). Other initiatives include depth-velocity-loss studies, breaching characteristics of dam failures, hydraulic roughness of river valleys, valuation of human life, and society's perception of acceptable levels of risk and the ethics, accountability, and liability involved in decisions about dam safety (Bates et al. 1989).

The Australian National Committee on Large Dams (ANCOLD) has developed guidelines on design floods for dams (ANCOLD 1986) which define the PMF as a limiting flood that could "reasonably" be expected to occur. Conventional practice remains based on engineering design and cost-benefit analysis. The application of risk analysis is growing, but there has been little attention to the quantification and incorporation of social perceptions of risk in decision making. Bates et al. (1989) proposed a multi-attribute utility approach that would incorporate empirical expressions of stakeholder utilities into the decision-making framework. Stakeholders would include the community, professionals with expertise in water resources, government agencies, and politicians. The objectives were to enhance communication between interested groups, systematize evaluation procedures, develop socially legitimated standards of safety, and help resolve differences of views and interests between groups. There has, however, been relatively little progress in this direction.

Recent events may change this situation, as concern grows about a number of dam structures. It seems odd to call upon social psychologists or economists when the issue is the probability of failure of a large structure. But the most readily available responses to higher risk involve social, political, and economic decisions. For example, if structural stability of a dam is in question, then changes in operating rules, such as a higher level of releases and lower average storage, differentially affect various groups in society. Irrigators receive less water, on average, while recreational and environmental flows increase. Money spent on upgrading flood warning systems might be more cost effective than spillway upgrades, but is often not an attractive option for local politicians and members of the general public, who are likely to demand the "safest possible" structure.

7.6 CONCLUSIONS

The very high historical variability of Australian rainfall and runoff means that climate change simply may amplify a pre-existing problem.

Within Australia, attention is being focused on the development of rules and institutional structures which give explicit recognition to the economic, social, and environmental values of the people affected by decisions on risk, reliability, and robustness, and which link water entitlements to relative rather than absolute levels of resource availability. Operating rules are being adopted that allow decentralized decision making to take place within the constraints imposed by variability. This allows individual agents to express their risk preferences, and where possible to exercise their own decisions about risk and reliability.

REFERENCES

ANCOLD (1986) Guidelines on design floods for dams. *Australian National Committee on Large Dams*, Canberra.

Bates, B. C., Milech, D., Syme, G. J., and Fenton, M. (1989) Establishment of dam safety criteria and evaluation of dam safety: a multi-attribute utility approach. *Australian National Committee on Large Dams Bulletin No. 82*, 21–27.

Boland, J. J., Carver, P. H., and Flynn, C. R. (1980) How much water supply capacity is enough. *Journal of the American Water Works Association* 72(7): 367–74.

Burton, J. R. (1982) Environmental issues and water allocation. In *Irrigation water: policies for its allocation in Australia*, 77–94. Armidale: Australian Rural Adjustment Unit, University of New England.

Cantwell, B. L. and Himsley, N. J. (1987) Australian approaches to hazard rating. *Australian National Committee on Large Dams Bulletin No. 78*, 55–58.

Crawley, P. D. and Dandy, G. C. (1996) The impact of water restrictions costs on the selection of operating rules for water supply systems. In *Proc. 23rd Int. Hydrology and Water Resources Symp.* Canberra: Institution of Engineers Australia National Conference Publication 96/05, Vol. 1, 43–49.

Dandy, G. C. (1992) Assessing the economic cost of restrictions on outdoor water use. *Water Resources Research* 28(7): 1759–66.

Dudley, N. J. (1990) Alternative institutional arrangements for water supply probabilities and transfers. In *Transferability of Water Entitlements – an International Seminar and Workshop*, edited by J. J. Pigram and B. P. Hooper. Armidale, NSW: University of New England, Centre for Water Policy Research, 79–90.

(1991) Management models for integrating competing and conflicting demands for water. In *Seminar and Workshop Water Allocation for the Environment*, edited by J. J. Pigram and B. P. Hooper. Armidale, NSW: University of New England, Centre for Water Policy Research, 169–81.

Dudley, N. J. and Musgrave, W. F. (1988) Capacity sharing of water reservoirs. *Water Resources Research* 24(5): 649–58.

Dudley, N. J. and Scott, B. W. (1993) Integrating irrigation water demand, supply and delivery management in a stochastic environment. *Water Resources Research* 29(9): 3093–101.

Finlayson, B. L. and McMahon, T. A. (1988) Australia vs the world: a comparative analysis of streamflow characteristics. In *Fluvial Geomorphology of Australia*, edited by R. J. Warner. Sydney: Academic Press, 17–40.

Fleming, M. (1995) Australian water resources are different. *Australasian Science* 16/2, 8–10.

Jolly, P. B. and Chin, D. N. (1991) Long-term rainfall-recharge relationships within the Northern Territory, Australia: The foundations for sustainable development. In *Proc. Int. Hydrology and Water Resources Symp.* Canberra: Institution of Engineers Australia National Conference Publication 91/2, 824–29.

Kennedy, M. R. and Hart, T. L. (1984) The estimation of probable maximum precipitation in Australia. *Civ. Eng. Trans.* Vol. CE26, No. 1. Canberra: The Institution of Engineers Australia.

McMahon, T. A. and Finlayson, B. L. (1991) Australian surface and groundwater hydrology – regional characteristics and implications. In *Proc. Int. Sem. and Workshop Water Allocation for the Environment*, edited by J. J. Pigram and B. P. Hooper. Armidale, NSW: University of New England, Centre for Water Policy Research, 21–40.

Macumber, P. G. (1991) Interactions between groundwater and surface systems in Northern Victoria. Melbourne: Department of Conservation and Environment Victoria.

Meyers, G. (1996) Variation of Indonesian throughflow and the El Niño Southern Oscillation. *Journal of Geophysical Research* 101: 12555–263.

Meyers, G., Bailey, R. J., and Worby, A. P. (1995) Geostrophic transport of Indonesian throughflow. *Deep Sea Research*, Part I, 42: 1163–74.

Nancarrow, B. E. and Syme, G. J. (1989) Improving communication with the public on water industry policy issues. Research Report No. 6. Melbourne: Urban Water Research Association of Australia. 114 pp. plus appendices.

Nathan, R. J. (1996) Development of a drought forecasting procedure to aid water supply management. In *Proc. 23rd Hydrology and Water Resources Symposium, Hobart* Vol. 2, 615–21. Canberra, The Institution of Engineers Australia.

Press, W. H., Flannery, B. P., Teukolsky, S. A., and Vetterling, W. T. (1986) *Numerical Recipes: the Art of Scientific Computing.* Cambridge: Cambridge University Press.

Ropelewski, C. F. and Halpert, M. S. (1987) Global and regional scale precipitation patterns associated with the El Niño/Southern Oscillation. *Monthly Weather Review* 115: 1606–26.

Ruprecht, J. K., Bates, B. C., and Stokes, R. A. (1996) Climate variability and water resources workshop: a summary of outcomes. *Water Resources Technical Report Series* No. WRT5, Perth: Western Australia Water and Rivers Commission.

Thomas, J. F. and Macpherson, D. K. (1991) Economic maximisation procedure. In *Botswana National Water Master Plan Study* Vol. 12, Part III. Gaborone: Government of Botswana.

Wellington, N. B. (1987) A review of risk analysis procedures for dam safety assessment. *Australian National Committee on Large Dams Bulletin* No. 78, 45–54.

Young, M. D. (1992) Sustainable investment and resource use: equity, environmental integrity and economic efficiency. *Man and Biosphere Series* Vol. 9. Paris: United Nations Educational, Scientific and Cultural Organization.

8 Developing an indicator of a community's disaster risk awareness

NORIO OKADA*

ABSTRACT
The primary objective of this chapter is the development of an appropriate social indicator which represents society's robustness against disaster risk. With a focus on the risk of drought, a sociopsychological approach based on the concept of social representation is presented. As the working hypothesis, the society's perceived level of readiness against drought (SPRD) is defined in terms of the social message of relevant newspaper articles by its article area. Using actual drought cases for the cities of Takamatsu and Fukuoka, this working hypothesis has been examined and reexamined from two different analytical viewpoints – that is, a contextual analysis and an analysis of the water-saving phenomenon. It is shown that the results are basically positive in support of our working hypothesis.

8.1 INTRODUCTION

The Great Hanshin-Awaji Earthquake, which struck the heart of the Kobe-centered metropolis on January 17, 1995, has demonstrated the often forgotten fact that the citizens of a large city are bound to coexist with the risks of an urban disaster. Though not so catastrophic as this earthquake, several metropolitan regions in Japan experienced a drought of unprecedented scale during the preceding summer. Among them are the cities of Fukuoka (on Kyushu Island) and Takamatsu (on Shikoku Island), which underwent the most serious socioeconomic damage. This must have strengthened the awareness that in society, large cities are bound to coexist with the risks of urban disaster. In both cases, citizens seemed to have learned that the community's disaster risk awareness makes a difference when examining the criticality of the disaster damage.

The primary objective of this chapter is to provide a scientific vehicle to analyze this point. That is, an attempt will be made to discuss scientifically how to model the community's awareness against disaster risk. More specifically, the focus will be on the 1994–95 drought using the two comparative case study areas – the cities of Fukuoka and Takamatsu.

8.2 THE 1994–95 DROUGHT

In the summer of 1994, Japan was struck by a period of drought unprecedented in scale, having an occurrence probability of one in one hundred. Fukuoka City, the capital city of Fukuoka Prefecture and also the political and economic center of Kyushu Island, experienced this drought which resulted in restrictions of municipal water supply service from July until September 1994. A drought also struck Takamatsu City and was similar in terms of rainfall shortage, date of occurrence, and drought duration. There was a difference, however, in the way the two cities reacted. Specifically, the robustness of Fukuoka toward drought was clearly over that of Takamatsu. Interestingly, it appeared that the citizens and industries of Fukuoka were more "ready to accept" the inconveniences that accompanied the drought, such as reduced water supply service. If this can be accepted as true, then it becomes interesting to determine what the major "background factor" was which contributed to the "societal robustness" against the drought. What would be an appropriate "social indicator," if we intend to measure the level of this "societal robustness"? These are the very questions addressed in this chapter. Answering these questions should lead to reexamination of more effective and systematic countermeasures during drought. These countermeasures may combine to build up "society's readiness against drought" as a part of the invisible, soft infrastructure in conjunction with the development of "hardware" infrastructure.

8.3 MEASURING THE "INVISIBLES" OF SOCIETY

The point of departure will be interpreting the "societal robustness" against drought as "society's perceived level of readiness

* Disaster Prevention Research Institute, Kyoto University, Uji Kyoto, Japan.

against drought" (abbreviated as SPRD). The immediate difficulty that arises is measuring this entity, which is invisible by nature. In other words, SPRD is not a physical entity that is visible to the eye, nor directly measurable on a physical scale. To overcome this difficulty, a sociopsychological approach can be used, where any observable aspect of "society's readiness" is interpreted as the "social representation," which is accompanied by some measurable "social message." As the relevant social representative, newspaper articles in reference to a particular drought occurrence in that specific region in real time, or drought occurrences in the past which are quoted in association, is proposed. In this way, the amount of social message by the newspaper articles may be measured by the area of the related articles carried in the "representative newspaper" of a region.

This approach basically parallels the approach proposed and tested by Sugimori, Yamori, and Okada (1993) examining the society's perceived level of readiness against flood and combined mud slide disasters in the city of Nagasaki, where they frequently occur.

The approach is applied here to droughts, using Fukuoka and Takamatsu as the case study areas. The span of the time-series data is set so that it is identical to the period of the water restriction (termination of water supply and reduced faucet pressure). The time interval of concern is 1994 and a portion of 1995 when the drought struck, and the intent is to compare analytically development in two locations by using the data sets plotted over the duration of the drought. Further, another drought is analyzed, which struck Fukuoka. The period of this drought was from May 1, 1978, to May 31, 1979, which was unprecedented in scale at that time for Fukuoka, causing enormous economic and social damage to the city.

Figure 8.1 (Takamatsu, 1994), Figure 8.2 (Fukuoka, 1994–95), and Figure 8.3 (Fukuoka, 1978–79) are the results of the above study, illustrating the measured area of relevant newspaper articles which appeared in local representative newspapers. Table 8.1 itemizes the major events that took place during each drought. From Figure 8.1 we may make the following observations:

1. Assuming that the time-series data indirectly represent SPRD, the SPRD rises sharply immediately after a drought regulation is introduced by the city of Fukuoka.
2. The SPRD then reaches its peak and declines, then picks up to reach another lower peak and declines again, in a damped oscillating fashion.
3. Overall, it shows a decaying trend over time and we may interpret from this that the SPRD becomes activated to its fullest extent immediately after any significant sign of the occurrence of a drought. However, it hardly maintains its peaks, and tends to decrease with the passage of time.

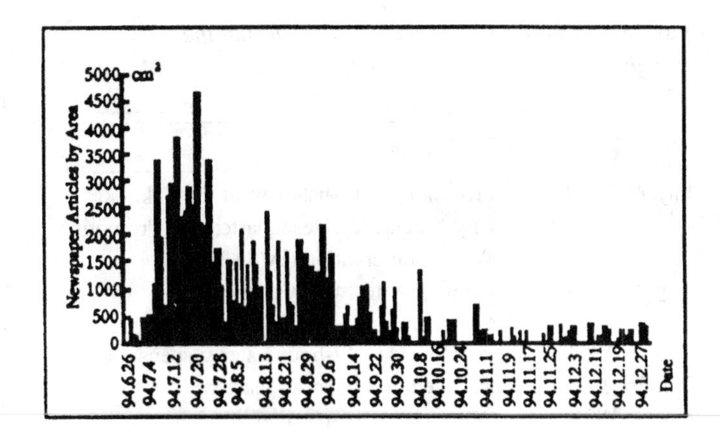

Figure 8.1. SPRD graph (Takamatsu, 1994).

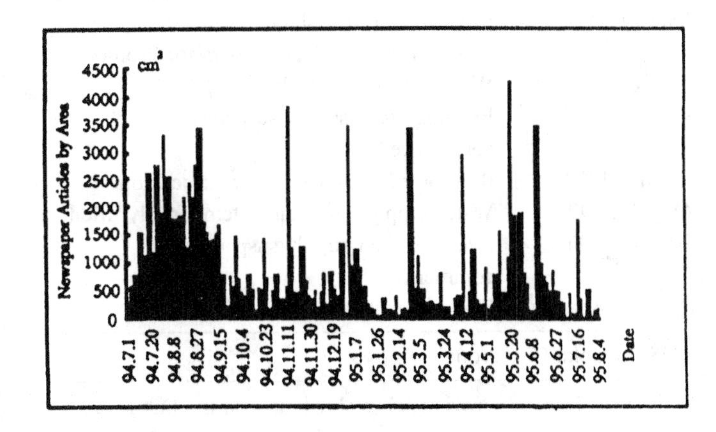

Figure 8.2. SPRD graph (Fukuoka, 1994–95).

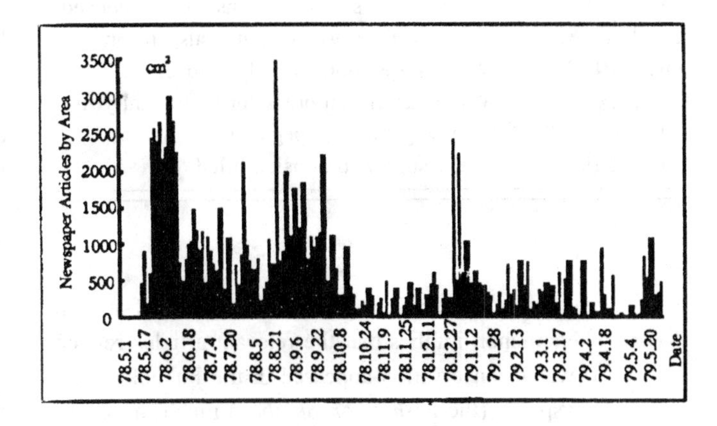

Figure 8.3. SPRD graph (Fukuoka, 1978–1979).

Comparison of Figure 8.1 with Figure 8.2 shows the following:

(a) The observations for Fukuoka are basically the same as that for Takamatsu.

(b) Takamatsu tends to show a much more rapid decay in SPRD than Fukuoka, which has a series of intermit-

Table 8.1. *Chronology of major events during the three droughts*

1994–95 Fukuoka City drought	
July 11, 1994	Provisional establishment of the Fukuoka City Emergency Headquarters for Promoting Water Conservation (by decree)
Aug. 8, 1994	Evening water supply suspension commenced
Sept. 1, 1994	Reinforced to 12-hour water suspension
Oct. 26, 1994	Reduced to 8-hour suspension
June 1, 1995	Water supply suspension lifted
1994 Takamatsu City drought	
June 27, 1994	Provisional establishment of the Takamatsu City Emergency Headquarters for Promoting Water Conservation (by decree)
July 11, 1994	Evening water supply suspension commenced
July 15, 1994	Reinforced to 5-hour water suspension
Aug. 19, 1994	Water supply suspensions temporarily lifted
Sept. 4, 1994	Evening water supply suspension commenced
Oct. 1, 1994	Water supply suspension lifted
1978 Fukuoka City drought	
May 15, 1978	Provisional establishment of the Fukuoka City Emergency Headquarters for Promoting Water Conservation (by decree)
May 15, 1978	9-hour water supply suspension commenced
June 1, 1978	Reinforced to 19-hour water suspension
June 11, 1978	Water suspension partially reduced
Dec. 20, 1978–Jan. 10, 1979	Water suspension break for traditional year-ending and year-beginning festivities
Mar. 25, 1979	Water supply suspension lifted

tent peaks and declines more slowly. This off-and-on phenomenon implies that the relevant articles related to water conservation appeared in the Fukuoka local newspaper (the *Nishinihon Shinbun*) much more frequently and periodically than the Takamatsu local newspaper (the *Shikoku Shinbun*). That is, Fukuoka tends to be more periodically activated against drought through the social message from its local newspaper than Takamatsu.

(c) This tendency of periodicity may be attributed to Fukuoka's past experiences that accumulated in society from the 1978–79 drought, which did not occur in Takamatsu.

(d) From (a)–(c), Fukuoka can be judged as "more robust" against drought than Takamatsu.

By comparing Figure 8.2 and Figure 8.3, we may well infer that Fukuoka has become more robust against drought when compared to the time before the 1978–79 drought.

8.4 A CONTEXTUAL ANALYSIS OF THE DROUGHT-RELATED SOCIAL MESSAGES OF NEWSPAPER ARTICLES

One analysis is based on the working hypothesis that the amount of social message of the newspaper articles represents the society's perceived level of readiness against drought (SPRD). Now, this working hypothesis is re-examined by carrying out a contextual analysis of the drought-related social messages of the newspaper articles. The relevant newspaper articles are classified according to the context in which the drought-related messages were sent out into society. That is, related articles are classified into the following two categories (see Suzuki, Okada, and Ikebuchi, 1997).

Category P (passive message): The readers receive these messages from a given newspaper in a passive manner. This category includes articles such as description of related facts (amount of rainfall, water level of a stream, commencement of water supply regulation, etc.), essays that intend to raise society's awareness against drought, and so forth.

Category A (active message): These articles are contributed, directly or indirectly, by the readers. "Opinions," "commentaries," and "letters from the readers" are typical examples that come under this category. Articles reporting people's active and visible attitudes or actions taken against the drought are also categorized as "A," since they motivate newspaper reporters to further inform the readers of related facts. In other words, this type of social message is interpreted as the active message which originates from society, is then filtered through the newspaper media, and is eventually conveyed back into society.

In the next approach, the amount of social message of the articles, either "A" or "P," is measured by newspaper article area. Figures 8.4, 8.5, and 8.6 show the results plotted over time for Takamatsu (1994), Fukuoka (1994–95), and Fukuoka (1978–79). Here the curves have been smoothed by the use of the moving average method so that an overall trend can be derived for each case. The time-span for this smoothing practice was set to ten days. The rationale for the ten-day time-span comes from preliminary studies which showed a more clear-cut pattern for the ten-day moving average over the seven-day and thirteen-day time-spans, which were also tested. From these results, we may make the following observations, which are common to all three cases:

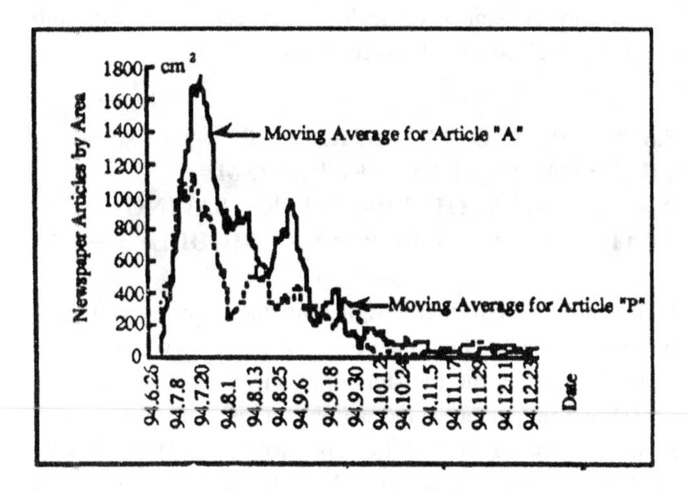

Figure 8.4. Comparison of curves for articles "A" vs. "P" (Takamatsu, 1994).

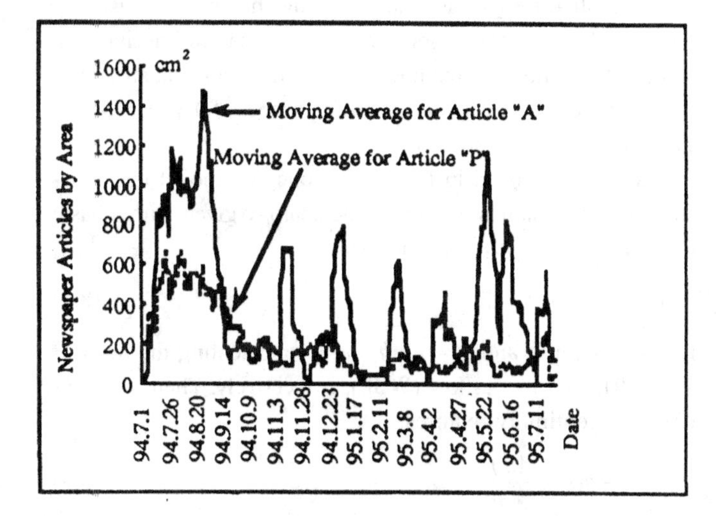

Figure 8.5. Comparison of curves for articles "A" vs. "P" (Fukuoka, 1994–95).

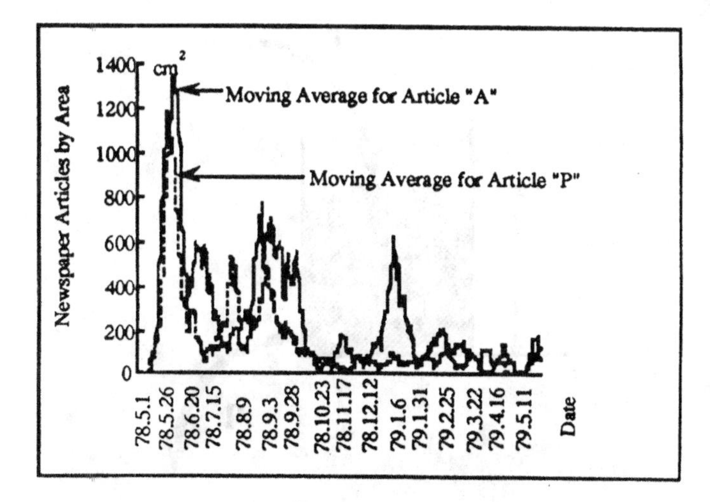

Figure 8.6. Comparison of curves for articles "A" vs. "P" (Fukuoka, 1978–79).

1. The curve for article "A" tends to be more dynamic (fluctuates) than the one for article "P."
2. More specifically, the former shows a sharp rise immediately after the commencement of water supply regulation, then reaches its peak, followed by a decline. Then it recovers again to a lower peak before declining again, and repeats this pattern until it gradually decays to approximately zero.
3. It follows from the above that the trend of the curve without categorization (we call this type of curve an "aggregated-data curve" – see Figures 8.1, 8.2, and 8.3) is identical overall to that of the curve for article "A," whereas the curve for article "P" corresponds to more stable and constant baselines of the aggregated-data curve.
4. One may conclude from this that the aggregated-data curve (aggregate approach) may closely represent the curve for article "A" (disaggregate approach "A"), and that because of the ease and compactness of obtaining the relevant data, the disaggregate approach "A" is found to be much less practical and less applicable than the aggregate approach. Therefore, it may be fair to say that the preceding hypothesis has been justified.

8.5 ANALYSIS OF THE OBSERVED WATER SAVING PHENOMENON: ANOTHER EXAMINATION OF THE WORKING HYPOTHESIS

Now the analytical viewpoint will be turned from the perceived level to the actual level of awareness during a water saving episode. It is intended to compare the SPRD, which is measured by the area of relevant articles, to the actual level of awareness. If a satisfactory coincidence is discovered between them, then our preceding working hypothesis may be supported.

The percentage of the water saved against the water used otherwise (abbreviated as the "water saving percentage") is calculated by the following formula:

$$\textit{Water Saving Percentage} = [(\textit{Normal State Demand} - \textit{Drought State Demand})/\textit{Normal State Demand}] \times 100$$

where the Drought State refers to the situation in 1994 and the Normal State represents the average state of the three consecutive nondrought years that preceded the 1994 event. Here the daily amount of water used was first aggregated into ten-day units of each month so that the influence of temperature and weather factors could be removed from the data. The aggregated data for ten-day units were used to calculate the water saving percentage as defined above. The results are shown in

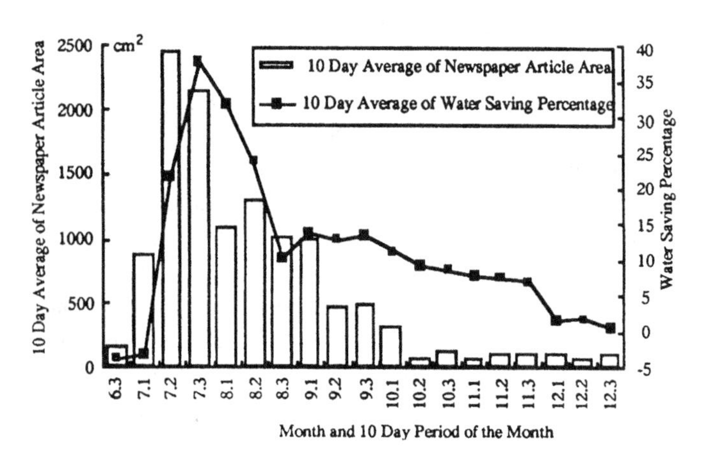

Figure 8.7. Comparison of percentage of water saved vs. SPRD (Takamatsu, 1994).

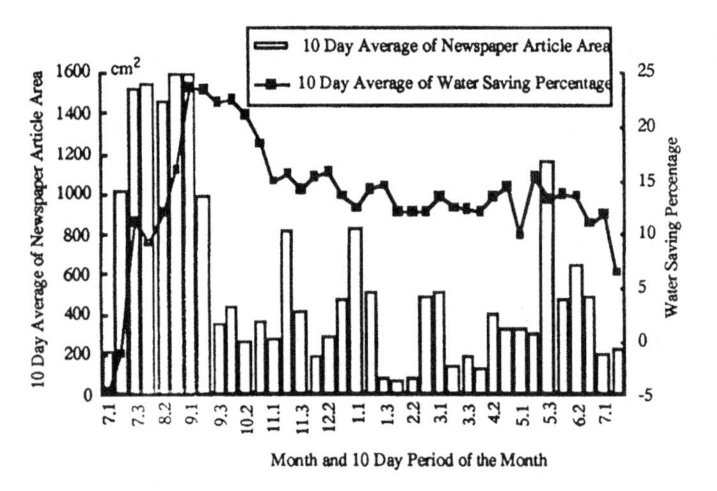

Figure 8.8. Comparison of percentage of water saved vs. SPRD (Fukuoka, 1994–95).

Figures 8.7 and 8.8. It is noted that the SPRD article area data is also aggregated in ten-day units. From these figures one can learn that:

1. As the SPRD sharply rises, indicating a sudden commencement of water supply regulation, the water saving percentage also rises, but with a time lag.
2. As compared to the SPRD, the water saving percentage shows a much more moderate decay.
3. One may interpret this as an envelope curve that is smoothly plotted to touch directly above the peaks of the SPRD curve and then lagged behind by a certain time period. This may be a good approximation of the water saving percentage.
4. This indicates that the SPRD curve is validated by the observed water saving phenomenon since there is a significant amount of coincidence between the data at both the perceived and observed levels.

8.6 MODELING THE SPRD-WSP TRANSFORMATION MECHANISM: AN ANALOGY OF THE WATER SAVING ACTION USING THE "TANK" MODEL

In order to gain further insight into the above coincidence between the data at both the perceived and observed levels, an attempt to model the transformation mechanism between the SPRD and the water saving action (represented by the water saving percentage, or WSP for short) will be made (see Suzuki et al. 1997). It is assumed that this mechanism can transform the SPRD into the WSP with a time lag. It is noted that the "tank" model proposed by Sugawara (1972), which is commonly applied to hydrological runoff mechanism models, may be extended to model society's transformation mechanism, since both phenomena exhibit a similar mathematical structure; that is, there is a storing and retarding process for water /SPRD through a reservoir/ society.

As conceptualized in Figure 8.9, one can model the transformation mechanism by use of the analogy given with a basic tank model. First it is assumed that:

$$q = \lambda h \tag{8.1}$$

holds between h and q, with h and q representing the level of the SPRD and its outflow (WSP), respectively. Then, the condition of continuity requires:

$$r(s) - q(s) = \frac{1dq}{\lambda ds} \tag{8.2}$$

to hold with r, s, and λ representing the inflow, the time, and a constant, respectively. It follows from this that:

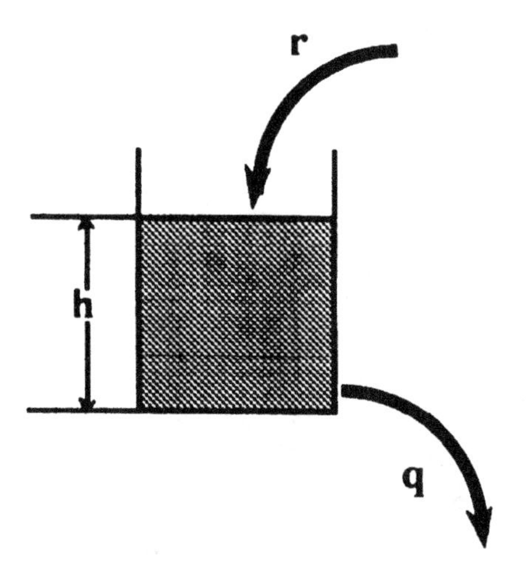

Figure 8.9. Conceptual diagram of the "tank" model.

$$q(s) = e^{-\lambda s}\{\lambda \int r(s)e^{\lambda s}ds + const\} \tag{8.3}$$

Setting $s = t - \tau$ $(0 < \tau \leq \infty)$, one can reduce the above equation to:

$$q(t) = e^{-\lambda t}\left\{\lambda \int_0^\infty r(t-\tau)e^{\lambda(t-\tau)}d\tau\right\} = \int_0^\infty \lambda r(t-\tau)e^{-\lambda\tau}d\tau \tag{8.4}$$

and from equation 8.1, one obtains

$$h(t) = \frac{1}{\lambda}q(t) = \int_0^\infty r(t-\tau)e^{-\lambda\tau}d\tau \tag{8.5}$$

It is noted that h and q are analogous to the storage level and the discharge in Sugawara's tank model respectively.

Furthermore, one may assume the following relationship among $S(t)$, $h(t)$, $R(t)$, $R'(t)$, and $R''(t)$:

$$S(t) = ah(t) + bR(t) + cR'(t) + dR''(t) + e \tag{8.6}$$

where $S(t)$ denotes the WSP at time t, $R(t)$, $R'(t)$, and $R''(t)$ denote the dummy variables (either zero or one) indicating respectively if rank 1 (the most severe), rank 2, or rank 3 (the most moderate) regulation is exercised and e denotes a constant term. More specifically, the regulation of each rank is defined as the following:

rank 1: 15- to 19-hour suspension of daily water supply
rank 2: 10- to 14-hour suspension of daily water supply
rank 3: 5- to 9-hour suspension of daily water supply

For the calculation of $h(t)$, the discrete form of equation 8.5, in units of ten days, is:

$$h(t) = \sum_{\tau=0}^{\infty} r(t-\tau)e^{-\lambda\tau} \tag{8.7}$$

Given this model, the parametric values are identified by use of regression analysis. The results are listed in Table 8.2. One may interpret parameters b, c, and d as representing the forced portion of change in WSP as the regulation increases by one or two modes (note that blanks in Table 8.2 refer to the fact that that particular regulation shift was bypassed to a higher mode). With these parametric values identified as such, the simulated WSP values are plotted against time and are compared to the observed WSP as illustrated in Figure 8.10 (Takamatsu, 1994), Figure 8.11 (Fukuoka, 1994–95), and Figure 8.12 (Fukuoka, 1978–79).

Table 8.2. *Values of coefficients obtained for each case*

	1994 Takamatsu	1994–1995 Fukuoka	1978–1979 Fukuoka
a	2.678×10^{-5}	2.405×10^{-5}	1.882×10^{-5}
b	0.0180	0.0190	0.0273
c		0.0520	0.0886
d	0.2429		0.0906
e	−0.0455	−0.01164	0.0352
λ	0.11	0.08	0.12

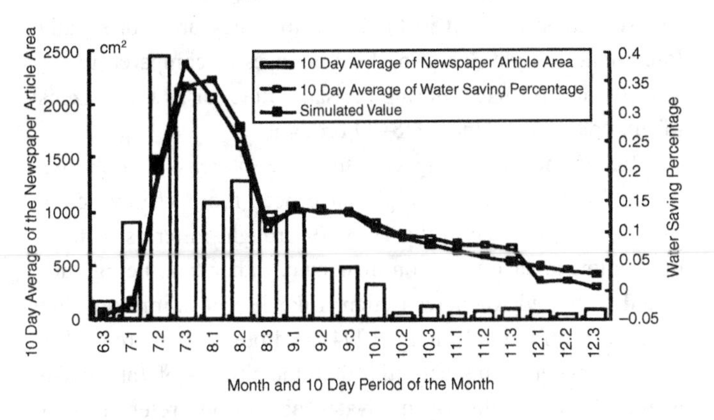

Figure 8.10. Comparison of observed SPRD, WSP, and simulated WSP (Takamatsu, 1994).

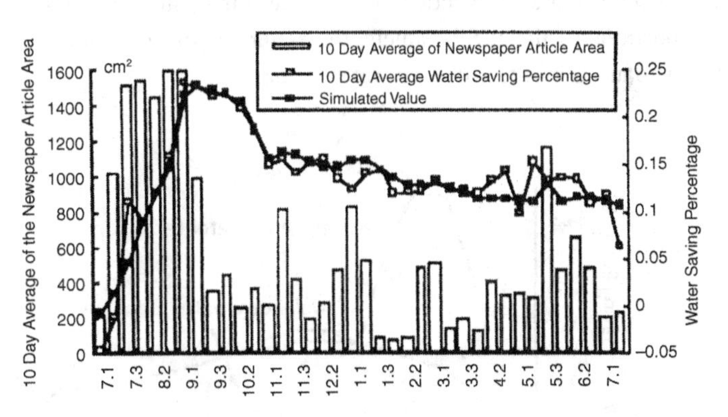

Figure 8.11. Comparison of observed SPRD, WSP, and simulated WSP (Fukuoka, 1994–95).

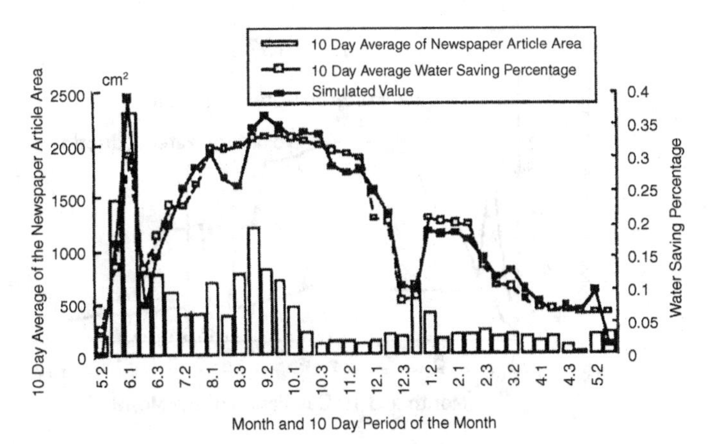

Figure 8.12. Comparison of observed PRD, WSP, and simulated WSP (Fukuoka, 1978–79).

Study of Table 8.2 shows that a, which is interpreted as the decaying rate of SPRD, is found to be smaller in value for the case of Fukuoka, 1994–95 than in the other two cases (Takamatsu, 1994; Fukuoka, 1978–79). This is indicative of the high likelihood that the citizens of Fukuoka have become more robust against drought in the sense that they are more ready to combat drought due to past experiences as compared to the citizens of Takamatsu during the same time, or as the citizens of Fukuoka during the 1978–79 drought.

This immediately suggests that the citizens of Fukuoka in the 1994–95 drought might have been more visibly active than they used to be in the 1978–79 drought in terms of their motivation to save water on their own, rather than being subjected to forced water restriction. Likewise, it is inferred that the citizens of Fukuoka in the 1994–95 drought are more visibly active in water saving activities than the citizens of Takamatsu in the 1994 drought. By motivated action, one refers to that portion of the WSP which comes about among the citizens as the SPRD is produced in proportion to the level of the SPRD. By forced water saving action, a reference is made to the residual portions of the WSP, namely, the water saving percentage that is caused by the enforced water regulation.

The above inferences may be easily verified by taking note of the parameter a, the value of which is interpreted as the degree of society's reaction which contributes to water saving as if it flows out from the "society's reservoir of SPRD." Comparison of the parametric values of a for each case shows that it is almost identical for Fukuoka, 1994–95, and Takamatsu, 1994, and it has increased for Fukuoka, 1994–95, as compared to 1978–79.

The plots in Figure 8.13 illustrate these points, where the "voluntary water reduction" portions refer to the self-motivated WSP (corresponding to the parameter a) and the "administered water reduction" portions refer to the forced WSP (corresponding to the residuals). We may conclude from this that in Takamatsu, 1994, regulation (forced saving) tended to start up more sharply at the outset but later tended to relax much earlier and mere swiftly than in Fukuoka, 1994–95. As for the self-motivated water saving, it has been found that Fukuoka, 1994–95, shows a more sustained (constant-level maintained) water saving than Takamatsu, 1994. Comparing the 1978–79 drought and the 1994–95 drought, one may find that in Fukuoka City, the self-motivated water saving is more dominant and vital for the latter case than for the former case. This

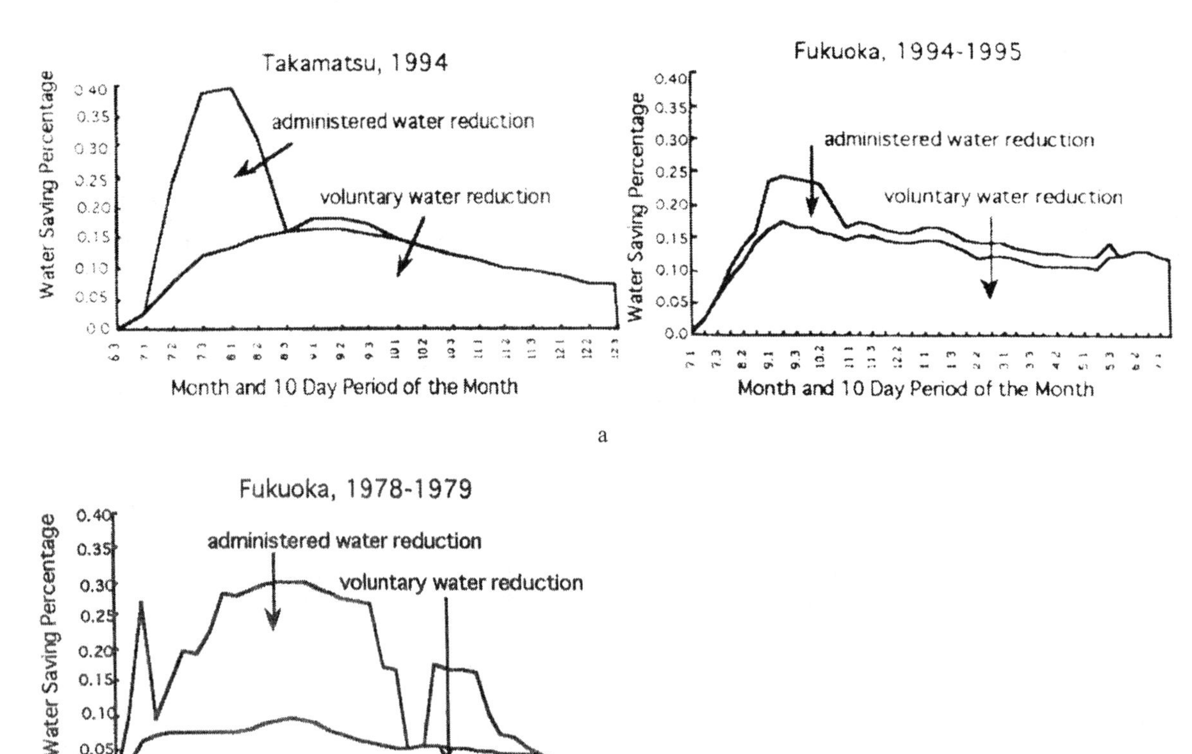

Figure 8.13. WSP based on action type.

suggests that the citizens of Fukuoka have changed their water use practices in such a way that they would more eagerly volunteer to save water during a drought situation even if there is no regulation exercised by the city.

From the above discussions it may be fair to state that the model of the SPRD-WSP transformation mechanism provides a useful analytical vehicle to gain insight into this otherwise unobservable mechanism of society's behavior.

8.7 CONCLUSIONS

The primary objective of this chapter was the development of an appropriate social indicator that represents society's robustness against drought. A sociopsychological approach based on the concept of social representation has been presented. As the working hypothesis, society's perceived level of readiness against drought has been defined and measured in terms of the social message sent out by the relevant newspaper articles by article area. With actual drought cases for the cities of Takamatsu and Fukuoka, this working hypothesis has been examined and reexamined from two different analytical viewpoints, that is, a contextual analysis and an analysis of a water-saving phenomenon. Based upon these findings, the mechanism of the SPRD-WSP transformation has successfully been modeled using an analogy with the tank model. This model has been found to be an effective analytical tool to gain further insight into this hidden transformation mechanism. On the whole, the results have been basically positive in support of the working hypothesis. Such being the case, one may further claim that based on the indicator of the SPRD, an invisible aspect of the society's robustness against drought can be assessed. Also, significant policy implications may be derived from such an assessment in reference to the manner in which the level of the invisible infrastructure and its bottlenecks are discussed. This study is a first step toward this goal.

REFERENCES

Sugawara, M. (1972) Methods of runoff analysis, *Hydrology Series* 7, Kyoritsu Shuppan, Tokyo.

Sugimori, N., Yamori, K., and Okada N. (1993) A basic analysis of community's perceived readiness against disasters and its changing process – a case study of Nagasaki. *Proceeding, Annual Conference of the Japan Society of Water Resources and Hydrology*, 14–15.

Suzuki, K., Okada, N., and Ikebuchi, S. (1997) Modeling indicators of drought and analysis of its changing process – a case study of 1994 drought. *Infrastructure Planning Review*, 13.

Suzuki, K., Yamori, K., and Okada, N. (1994) An analysis of social image of disasters and a systematic information processing approach. *Proceedings of the 1994 Annual Conference of the Japan Society of Civil Engineers*, 230–31.

9 Determination of capture zones of wells by Monte Carlo simulation

W. KINZELBACH, S. VASSOLO, AND G.-M. LI*

ABSTRACT

Effective protection of a drinking water well against pollution by persistent compounds requires the knowledge of the well's capture zone. This zone can be computed by means of groundwater flow models. However, because the accuracy and uniqueness of such models is very limited, the outcome of a deterministic modeling exercise may be unreliable. In this case stochastic modeling may present an alternative to delimit the possible extension of the capture zone. In a simplified example two methods are compared: the unconditional and the conditional Monte Carlo simulation. In each case realizations of an aquifer characterized by a recharge rate and a transmissivity value are produced. By superposition of capture zones from each realization, a probability distribution can be constructed which indicates for each point on the ground surface the probability to belong to the capture zone. The conditioning with measured heads may both shift the mean and narrow the width of this distribution. The method is applied to the more complex example of a zoned aquifer. Starting from an unconditional simulation with recharge rates and transmissivities randomly sampled from given intervals, observation data of heads are successively added. The transmissivities in zones that do not contain head data are generated stochastically within boundaries typical for the zone, while the remaining zonal transmissivities are now determined in each realization through inverse modeling. With a growing number of conditioning data the probability distribution of the capture zones is shown to narrow. The approach also allows the quantification of the value of data. Data are the more valuable the larger the decrease of uncertainty they lead to. By reducing the size of the zones of equal parameter values it is seen that the aquifer can be replaced by a homogeneous one if zones become small against the typical dimensions of the capture zone. The need for identification of large-scale features is stressed.

* Institute of Hydromechanics and Water Resources Engineering, Federal Institute of Technology, Zurich, Switzerland.

9.1 INTRODUCTION

The bacterial pollution of wells in Germany has been successfully prevented by the introduction of well head protection zone II, which restricts polluting activities in the fifty-day travel zone around the well (DVGW 1995). Protection against persistent pollutants for which the fifty-day zone is irrelevant such as chlorinated hydrocarbons or nitrate in aerobic aquifers has been much less successful. It requires measures in the complete capture zone of the well, for example, changes in fertilization practices in agricultural areas.

It may not be possible to arrive at a complete protection of a regional aquifer; therefore, measures have to focus on the actual capture zones of wells. The effectiveness of such measures depends on the accuracy with which the geographical extent of the capture zone is known. In the case of a homogeneous areal recharge N, the area A of a capture zone can be computed by continuity arguments as $A = Q/N$ where Q is the average pumping rate of the well. The shape of the zone, however, depends on the conductivity structure of the aquifer as well as the screening of wells. If recharge is spatially varying, the shape of the capture zone becomes more complex and its area in general is rather different from the value given by the above formula.

Given the travel times involved and the small number of quasi-ideal tracers available, the capture zone usually can not be determined experimentally and the use of groundwater flow models is the method of choice. From an accurate head distribution, pathlines can be constructed by particle tracking. To determine whether a point on the ground surface belongs to the capture zone or not, a pathline is started in that point. If it ends up in the well the point belongs to the capture zone. By starting a dense set of pathlines, the capture zone can be delineated.

9.2 PROBLEMS ARISING IN THE DETERMINISTIC DETERMINATION OF CAPTURE ZONES

Head distributions obtained by interpolation of observed heads at piezometers are usually not sufficiently accurate for pathline construction for several reasons: A general interpolation algorithm does not take into account continuity and Darcy's law. But even if the interpolation is based on a flow model, uncertainty remains.

First of all, a model of given structure calibrated on the basis of local heads is nonunique in the usual situation where regional fluxes are only known within rough bounds. If, for example, recharge is known to lie within an upper and lower bound, transmissivity can only be determined within corresponding bounds as well, even if accurately measured heads are available.

Second, the model structure itself may not be unique. A model usually works on the basis of zoned equivalent properties of an aquifer. Choosing the shapes and sizes of zones amounts to determining the model structure, which may not be identifiable from data.

A further problem may lie in the dimensionality of the model used. While aquifers are three-dimensional, data are usually only sufficient to build a two-dimensional model. This is allowed under certain conditions which, however, will have to be ensured (Kinzelbach, Marburger, and Chiang 1992).

Finally, the capture zone relevant for groundwater protection has to be based on the long-term average situation. If there are climate changes, however, the boundaries of a capture zone are not constant in time. For variations that are fast compared to the travel time from the center of mass of the capture zone to the well, a mean flow field can be adopted.

9.3 DETERMINATION OF A CAPTURE ZONE BY MONTE CARLO SIMULATION

The basic idea of the stochastic approach is illustrated by a greatly simplified example. An aquifer with homogeneous transmissivity is bounded by two rivers acting as discharge zones on the northeastern and southwestern sides, while the other two sides are impervious due to faults. The aquifer receives a homogeneous recharge on all areas that are not covered by thick loess. On the loess a constant fraction of that recharge rate is assumed. Two wells abstract water from the aquifer at a constant rate. The transmissivity of the fractured sandstone block is unknown, as is the recharge. From regional experience the transmissivity T should lie between a minimum value T_{min} and a maximum value T_{max}. Likewise the recharge can be estimated between bounds Nmin and Nmax from regional low-flow hydrograph analysis at the two rivers. Dividing the T- and N-ranges into ten intervals each, the plane of combinations T-N is represented by 100 values. Due to lack of information, the probability distribution of T- and N-values is assumed equally distributed between minimum and maximum values. If there was a priori knowledge about the shape of the distribution being different from the equal distribution, this could without difficulty be incorporated into the method by weighting of realizations.

For each of the realizations thus generated, a deterministic flow computation using a standard finite element model is performed and from it the capture zone is obtained by particle tracking. For particle tracking the method of Cordes and Kinzelbach (1992) is used, which makes optimal use of all information contained in Galerkin finite elements. Superimposing the 100 capture zones leads to a probability distribution from which we can read off the probability of any point in the aquifer to belong to a capture zone (Figure 9.1).

The results in this case disagree with the hydrogeologist's judgment of the situation, as for various reasons the subterraneous connection between the two rivers seems improbable. The reason the 100 realizations do not yet lead to satisfactory results may simply be that they do not take into account the fact that there is further knowledge on heads in the two wells available. Some of the unconstrained realizations may lead to heads in the wells which are unreasonable (e.g., lie above ground level). Therefore, in a second approach, only the range of recharge N is divided into a number of intervals, for example, 100. For each value of N the flow model is run in inverse mode to determine the T-value which leads to satisfactory heads in the two wells. The inverse identification procedure uses the Marquardt-Levenberg method (Marquardt 1963; Press et al.

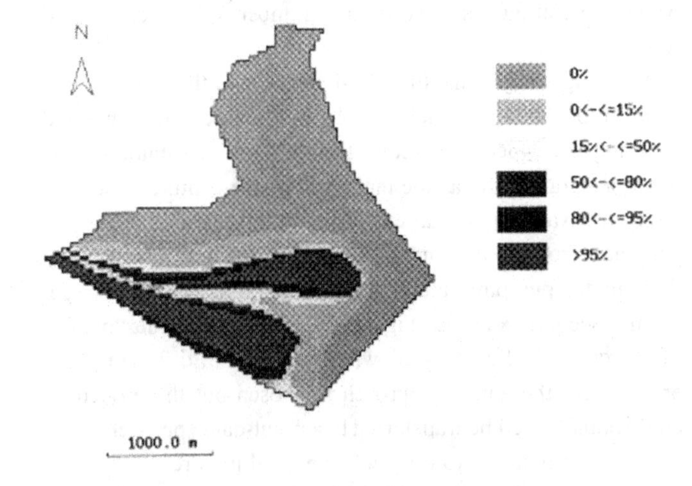

Figure 9.1. Unconditional determination of capture zone distribution.

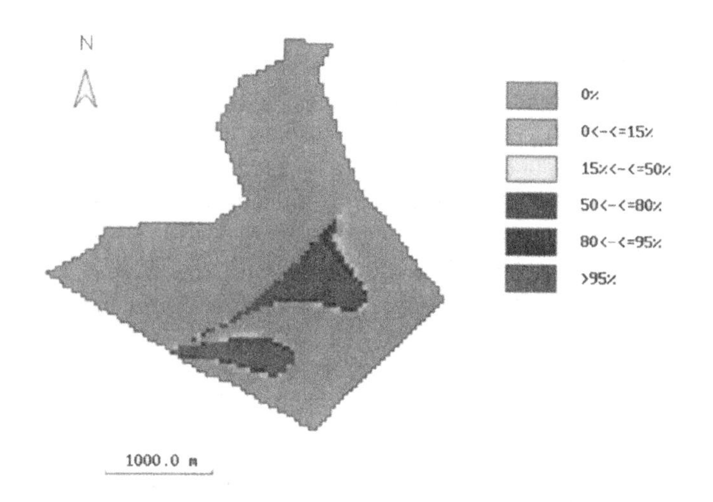

Figure 9.2. Conditional determination of capture zone distribution.

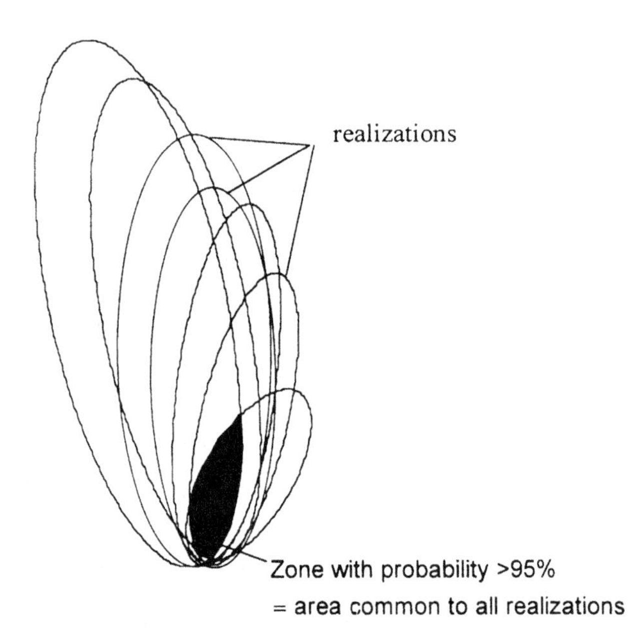

Figure 9.3. Capture zone distribution with direction of main axis changing from realization to realization.

1992). Again 100 capture zones are obtained, but now conditioned on the head data. The resulting probability distribution is narrower and obviously very different from the first one (Figure 9.2).

As a matter of fact the first, unconditioned distribution should contain all realizations of the second exercise. Obviously it doesn't. Of course in the first approach the realistic realizations obtain a different weight in the overall statistics as they are "diluted" by realizations that are unfeasible as far as the computed heads are concerned. This is an important point, as one might naively believe that the results of a stochastic modeling exercise are the more conservative the wider the distributions of the input parameters are chosen. The example shows this may not be so and the maximum of the unconditioned distribution may be far off the region in which realistic solutions lie. Realistic solutions may have a small probability in the overall distribution and therefore be missed completely when generating an insufficient number of unconstrained realizations.

One might argue that the 50%-percentile of the capture zone distribution should coincide with the capture zone obtained by using the expected value of the recharge distribution. This is not generally true, as the main axis of the capture zone may shift with varying recharge. Therefore, the area of the 50%-percentile zone is in general larger than the area obtained by dividing the pumping rate by the average recharge (Figure 9.3).

In a second example (Figure 9.4), a more complex picture of an aquifer is drawn by allowing for zones of different properties. Here the zoning approach is chosen but the procedure could equally well be transferred to a point-data approach using kriging interpolation to complete the areal picture.

The fractured basalt aquifer is divided into seven zones with different transmissivities, mostly on the basis of known faults and pumping tests at 5 of the 11 piezometers and wells available (Figure 9.5). At the eleven piezometers, time series of head values exist from which average values can be abstracted. The recharge distribution is divided into three zones, a zone with a higher rate in places with thin soil cover above the fractured basalt, a zone with a lower rate in places with a thick loess cover, and a zone with almost zero recharge within the settlements. The western boundary is formed by impervious Devonian rock, the northern boundary is a water divide, the eastern boundary is formed by the edge of the basalt aquifer and can be considered impervious as extremely little water flows in the adjacent tertiary formation, while the southern boundary is formed by the river Wetter. Together with two springs and the two wells of the local water supply, it receives all the groundwater formed on the model domain.

9.4 RESULTS

The results of unconditional modeling are shown in the upper left graph of Figure 9.6. The distribution results from the superposition of 125 realizations of the capture zone where the recharge in the three recharge zones and the transmissivities in the seven transmissivity zones are chosen at random from the ranges of these parameters. The ranges were estimated from the available pumping tests and regional low-flow discharge data. The distribution shows that a back infiltration from the river into the wells may occur in some realizations. This was never

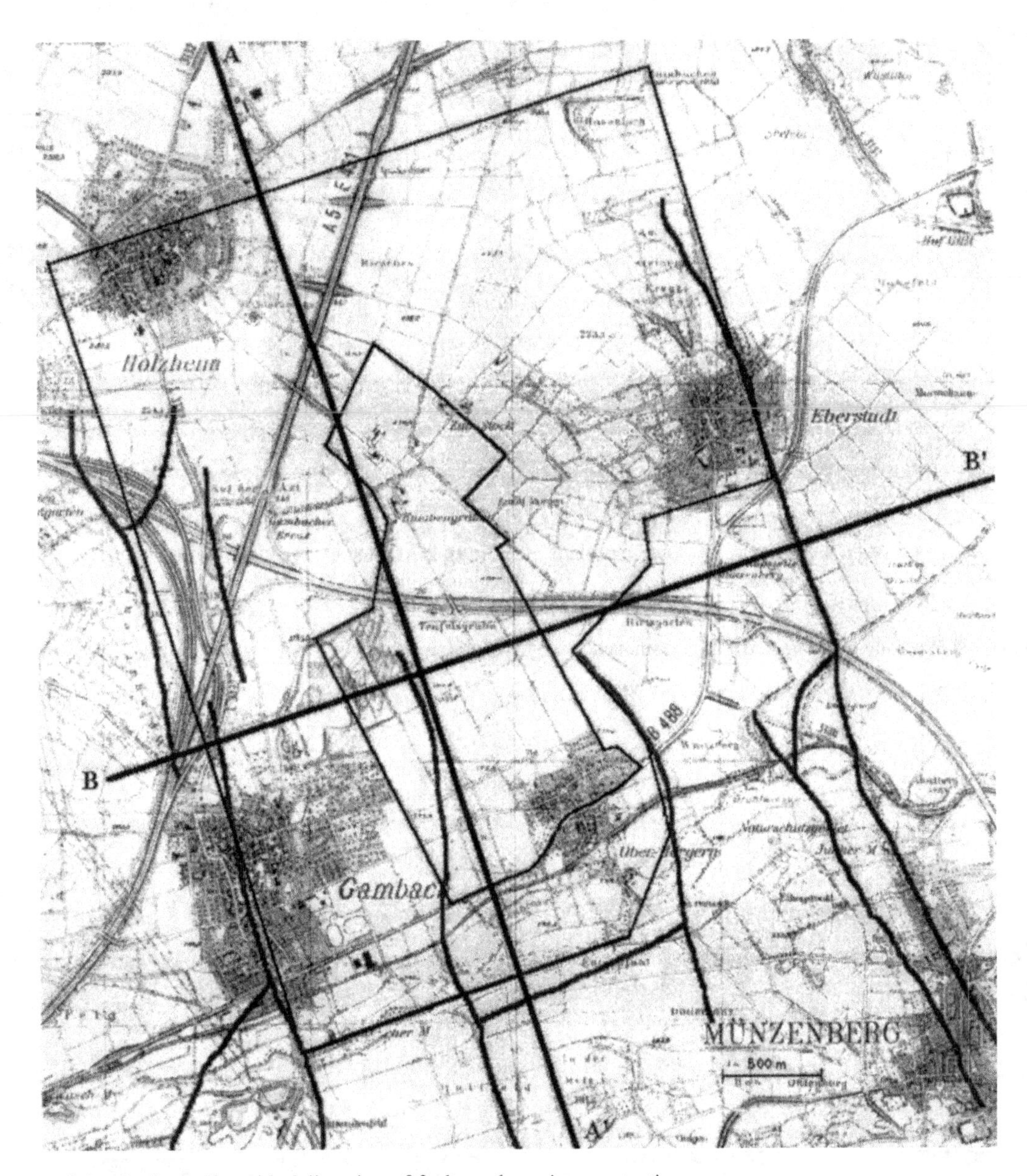

Figure 9.4. Extent of Gambach aquifer with delineation of faults and previous protection zone.

observed in the field. On the whole, the width of the distribution is large and there is a nonvanishing probability that almost the whole model domain belongs to the capture zone.

In the following, the head information was successively included for conditioning of the stochastic model. Let us consider that m heads are given with m between 1 and the total number of piezometers (here 11).

The realizations leading to the distribution of capture zones are obtained in the following manner: From the intervals of groundwater recharge in the three recharge zones defined above, three values are randomly chosen and assigned to the model nodes within the three zones. In all transmissivity zones without piezometer head information a transmissivity value is chosen likewise. Then, using the given piezometer heads, the transmissivities in the remaining transmissivity zones are determined by an inverse procedure using the Marquardt-Levenberg method. The sum of squared deviations of measured and computed heads is chosen as criterion for goodness of fit. On the basis of the optimized transmissivities a head distribution is computed which leads to a capture zone by particle tracking. The distributions for growing numbers of conditioning heads are shown in Figure 9.6, and the distribution using all available 11 heads for conditioning is shown in Figure 9.7. About 125 realizations were used in each case. It should be noted that in some transmissivity zones there is more than one head value available, while in one zone representing the

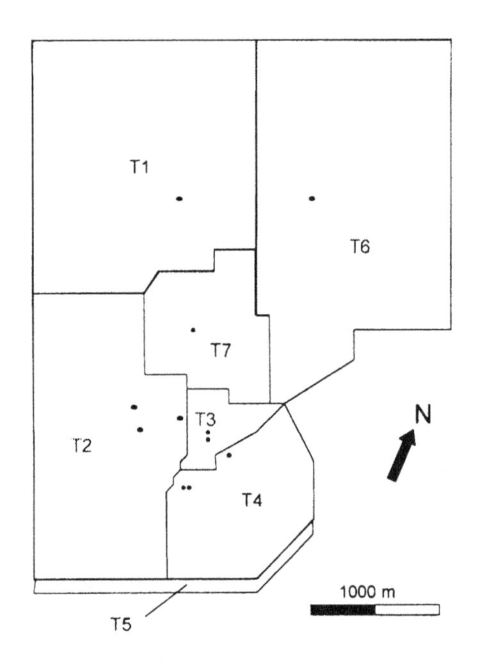

Figure 9.5. Zoning and location of piezometers in the Gambach aquifer.

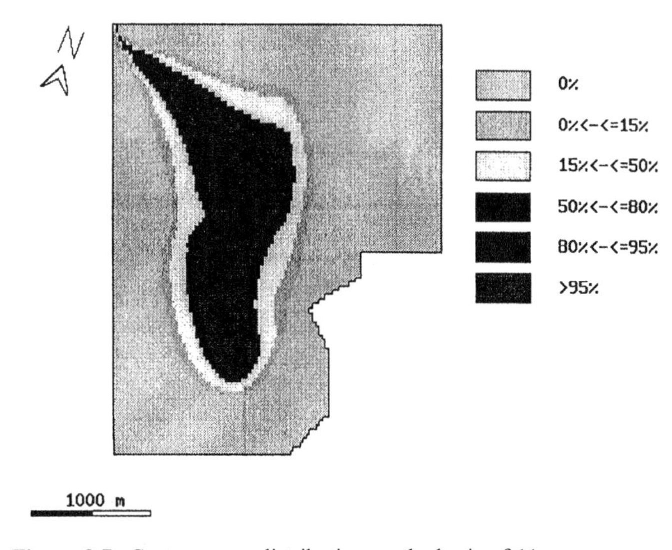

Figure 9.7. Capture zone distribution on the basis of 11 conditioning heads.

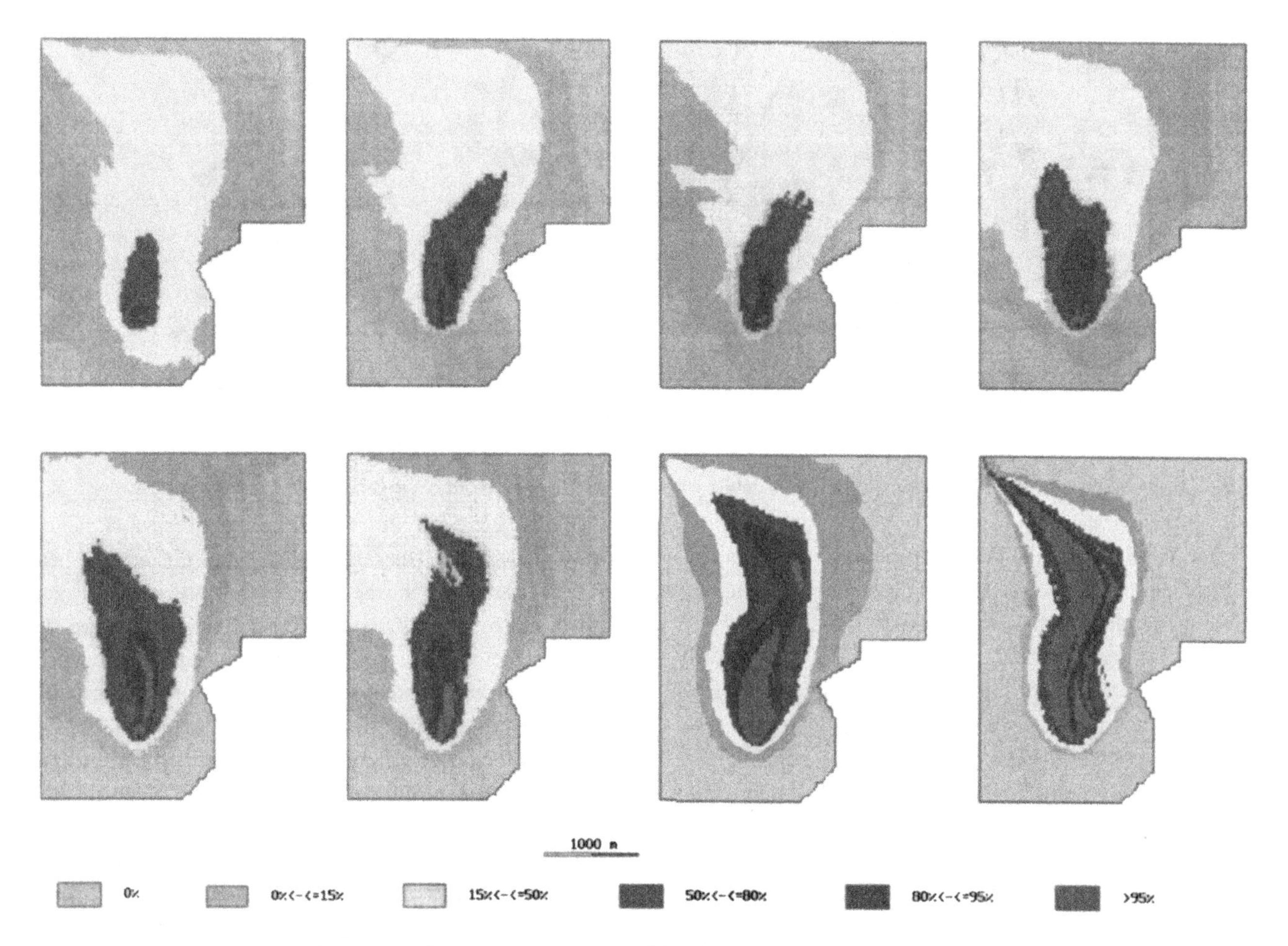

Figure 9.6. Capture zone distributions for conditioning with 0 through 7 heads.

colluvial material close to the river no head value is available. This explains the fact that there is not much difference between capture zone distributions with seven and eleven conditioning heads. It would be ideal to have at least one head per zone, preferably at a location that is centrally positioned within the zone.

The general observation that can be made is that with a growing number of heads included in the conditioning, the high probability zone is getting larger while the low probability "halo" of the distribution shrinks. Consequently, the uncertainty of the capture zone extension is decreased. The procedure also shows the incremental value of additional data. Data which lead to a significant decrease in the distribution width are valuable. Of course the order in which data are added may change the incremental decrease of uncertainty associated with it. A good strategy establishes the major longitudinal gradient first by two to three measurements and then adds points sidewise which give information on the lateral gradients (Vassolo 1995).

If the number of zones is increased without an increase in number of piezometer data points, the distribution of capture zones resembles more and more the distribution one obtains if the aquifer is assumed homogeneous, that is, with one transmissivity zone only, the value of which is given within bounds.

It is evident that only larger scale structural elements will make the capture zone distribution deviate from the one for a one-zone aquifer. The paramount task of geological surveying is therefore to identify structural elements of a size of the order of the typical extensions of the probable capture zone.

9.5 CONCLUSIONS

If data are sparse in an aquifer considered, the only possible way to say something meaningful about capture zones is a stochastic assessment. Whether this assessment is of value depends on the outcome of the exercise. If for given conditioning data the resulting width of the capture zone distribution is small enough to allow operational decisions, the exercise is successful. If the width of the resulting distribution is still large, either a large protection zone can be established to be on the safe side or further data have to be collected. In that sense the conditional simulation in principle allows to determine whether data are sufficient for the intended purpose or not.

REFERENCES

Cordes, C. and Kinzelbach, W. (1992) Continuous groundwater velocity fields and pathlines in linear, bi- and trilinear finite elements. *Water Resources Research* 28(11): 2903–11.

DVGW (1995) *Richtlinien für Trinkwasserschutzgebiete.* Technische Regel W101, Eschborn, 23 p.

Kinzelbach, W., Marburger, M., and Chiang, W.-H. (1992) Determination of groundwater catchment areas in two and three spatial dimensions. *Journal of Hydrology* 134: 221–46.

Marquardt, D. W. (1963) An algorithm for least-squares estimation of non-linear parameter. *Journal of the Society of Industrial and Applied Mathematics* 11: 431–41.

Press, W., Teukolsky, S., Vetterling, W., and Flannery, B. (1992) *Numerical Recipes*. 2nd ed. Cambridge University Press, 675–83.

Vassolo, S. (1995) *Stochastische Bestimmung von Einzugsgebieten von Brunnen*. Ph.D. Thesis, Civil Eng. Dept., Kassel University, Kassel, Germany.

10 Controlling three levels of uncertainties for ecological risk models

THIERRY FAHMY,* ERIC PARENT,** AND DOMINIQUE GATEL***

ABSTRACT

Bayesian methods have been developed to analyze three main types of uncertainties, namely: the model uncertainty, the parameter uncertainty, and the sampling errors. To illustrate these techniques on a real case study, a model has been developed to quantify the various uncertainties when predicting the global proportion of coliform positive samples (CPS) in a water distribution system where bacterial pollution indicators are weekly monitored by sanitation authorities. The data used to fit and validate the model correspond to water samples gathered in the suburb of Paris. The model uncertainty has been evaluated in the reference class of generalized linear multivariate autoregressive models. The model parameter distributions are determined using the Metropolis-Hastings algorithm, one of the Monte Carlo Markov Chain methods. Such an approach, successful when dealing with water quality control, should also be very powerful for rare events modeling in hydrology or in other fields such as ecology.

10.1 INTRODUCTION

The bacterial pollution indicators are understood here as the coliforms, a group of bacteria that is "public enemy number one" for water suppliers. Their occurrence in domestic distributed waters is a major concern for many utility companies. The coliform group includes many different species, the most famous one being *Escherichia coli*. Part of the bacteria belonging to the coliforms group are fecal bacteria and may provoke gastroenteritis or other digestive problems. The other inoffensive part is generally considered as an indicator of a possible presence of their more dangerous cousins.

The water companies have to face local sanitary regulations concerning this group. In France, the inlet into the distribution systems must by law be free of fecal coliforms, and only a 5 percent annual positive sample rate is tolerated. In this chapter we use the word "coliforms" for total coliforms, whereas fecal coliforms are only a subgroup of this family. The first results of this research are part of a larger project of water network quality control whose objectives were to:

1. Establish a relationship between the explanatory variables and the expected percentage of coliform-positive samples;
2. Calculate the uncertainty in the model parameters, in the predictive number of coliform-positive measurements for a given week, and in the model itself;
3. Re-evaluate the relevance of the variables under consideration for the phenomenon of coliform occurrence;
4. Estimate the effectiveness of the present quality control devices and, if necessary, establish more adequate experimental sampling designs.

The coliform regulation will have to be faced not only at the exit from the treatment works, but also at the inlet of the different subdivisions and interconnections. Although the presence of total coliform is unlikely to contaminate globally the whole distribution system, there is evidence that regrowth events can occur locally (Block et al. 1995; Camper et al. 1993; Lechevallier, Babcock, and Lee 1987). A model to predict coliform occurrence should therefore be able to take into account the bacteriological history and some explanatory variables linked to the distribution system working conditions.

10.2 A NEW ALERT MODEL

10.2.1 Selected variables

Occurrence of coliforms in distributed water samples has often been correlated with high temperature, high nutrients in the water, and the absence of chlorine residual (Colbourne et al. 1992; Lechevallier, Shaw, and Smith 1994), even though some authors suggest that there is no evidence that these conditions are sufficient (Goshko, Pipes, and Christian 1983; Camper, Jones, and Goodrum 1994). There is, however, a wide

* Anjou Recherche, 92982 La Défense, France.
** ENGREF, 75014 Paris, France.
*** Compagnie Générale des Eaux, 92982 La Défense, France.

consensus to consider that coliforms, including thermotolerant coliforms, are able to survive and multiply under conditions found in drinking water distribution systems (Camper et al. 1991; Block et al. 1995).

The explanatory variables that have been chosen are the temperature as a growth factor, residual of free chlorine as a growth inhibitor, and the turbidity at the end of the treatment process as an indicator of the performances of the treatment. The data that will be used in the model will consist of weekly and spatial means of the previously specified variables. A probabilistic model has been developed using autoregressive generalized linear model techniques also called generalized autoregressive models (GAM).

10.2.2 The mathematical model

The mathematical model is defined as follows:

(a) At week t the observations y_t (the number of CPS) are binomial counts with parameters π_t and N_t. The likelihood reads:

$$L = \prod_{t=1}^{T} C_{N_t}^{y_t} (\pi_t)^{y_t} (1-\pi_t)^{N_t - y_t} \tag{10.1}$$

T represents the total number of weeks with available data, N_t is the sample size in week t (which is supposed to be set by legal regulations), and π_t, expressed by equation 10.2, is the parameter of interest at time t that expresses the network water quality, that is, its contingency to yield to a coliform contamination:

$$\pi_t = \frac{E(y_t)}{N_t} \tag{10.2}$$

(b) It is desirable to link π_t, which characterizes the state of the network at week t, with the working conditions up to time t since they might have been of influence for coliform occurrence. One considers here the same variables that were previously taken into account by Volk and Joret (1994) in his four qualitative risks level model, except for the plant turbidity (Turb) that was added later, that is: free chlorine residual (FCR) as a measure of the disinfectant potential, the temperature (Temp) as a growth positive factor, total direct counts (TDC) as a measure of the general biomass quantity, and the difference of biodegradable dissolved organic carbon between sampling points on the DS and the produced water at the end of the treatment process (ΔBDOC) as a measure of the bacterial activity. The TDC were quickly removed because they include the coliforms themselves and because it has been observed that the link between coliforms and TDC is not stable at all. A model with few parameters can be designed to take into account this influence under the following recursive form:

$$\pi_{t+1} = f(\theta, X_t, \pi_t) \tag{10.3}$$

In equation 10.3, $f(\ )$ is the system function, X_t denotes the values of the explanatory variables (FCR, Temp, Turb, ΔCODB) at date t, and θ represents a set of structural parameters which are considered as random variables, as usual with the Bayesian approach.

The "plant equation" relationship $f(\ )$ can be specified by a linear form mapped onto the [0, 1] interval as follows:

$$\pi_{t+1} = f(\alpha\pi_t + \beta X_t, \lambda) \tag{10.4}$$

The generic parameter vector θ has the following components:

- α, coefficient of autoregressive dependence between two consecutive time steps,
- β, itself a vector with five components (β_{FCR}, β_{Temp}, β_{Turb}, $\beta_{\Delta CODB}$, β_m) that are a function of the linear effect of the explanatory variables plus a constant to account for the average effect.
- λ, the multistructure model parameter (see below).

The system function used in this chapter can be described a priori by the logistic or log-logistic models. This kind of model is generally used to relate a proportion or a probability to explanatory independent variables. The logistic dependence is:

$$\pi_{t+1} = f_1(\alpha\pi_t + \beta X_t) = \frac{exp(\alpha\pi_t + \beta X_t)}{1 + exp(\alpha\pi_t + \beta X_t)} \tag{10.5}$$

Whereas one possible form of the log-logistic function is given by:

$$\pi_{t+1} = f_2(\alpha\pi_t + \beta X_t) = exp(-exp(\alpha\pi_t + \beta X_t)) \tag{10.6}$$

The method looks similar to linear regression but the dependent variable has to lie in the interval [0,1].

Using a power transformation (Box and Tiao 1973), a multistructure model has been designed by simply including a structure parameter named λ, so that when λ varies between 0 and 1, the model moves continuously from the log-logistic function to the logistic function.

$$\pi_{t+1} = f(\alpha\pi_t + \beta X_t, \lambda) = \frac{1-\lambda}{exp[(1-\lambda)exp(-(\alpha\pi_t + \beta X_t))] - \lambda} \tag{10.7}$$

Figure 10.1 sketches the logistic and log-logistic curves as well as the λ-logistic curve for $\lambda = 0.5$; contrary to the logistic curve that is fully sigmoid and symmetric, the log-logistic is

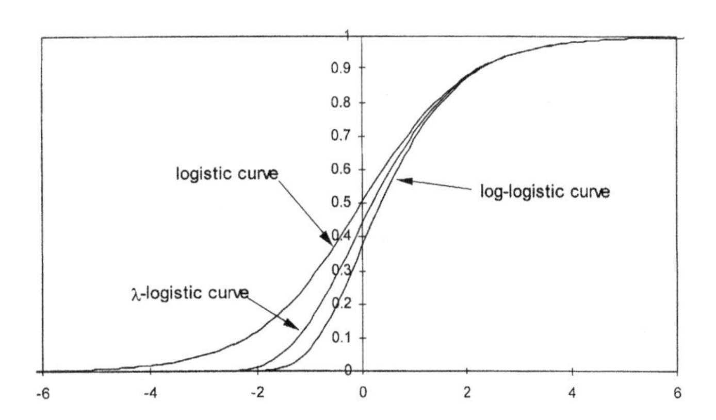

Figure 10.1. The *log-logistic, λ-logistic, and logistic* curves.

asymmetric and "shrunk" at the bottom. Since the experimental data are mostly gathered around small frequencies of coliform occurrence, the latter form seems much more suitable than the former. The λ-logistic is an intermediary solution between two well-known functions.

10.2.3 Evaluation of uncertainties

Because the sampling methods cannot be perfect, the data are only a more or less false image of reality and even if one supposes that the model describes perfectly the reality, some uncertainty will arise upon the model parameters which are determined using the data under the model structure and some probabilistic considerations as constraints. If the model does not describe perfectly the phenomenon (which is most often true), there will be more uncertainties on the parameters because part of the description is missing or false. So the uncertainties concerning the model parameters that will be described by a distribution function (see section 10.2.4) will represent both model and measurement uncertainties.

As we have built a multistructure model by adding one parameter, the uncertainties linked to this parameter will mostly show the uncertainties for the structure of the model, though they can also represent part of the measurement errors: the measurement errors could lead to uncertainties for the shape of the model.

Finally, when using the model to predict what the number of CPS, M_t would be on a week t knowing that N_t samples will be collected, if π_t is the expected proportion of CPS, one needs to take into account the sampling errors. These can be easily described because they correspond to the well-known binomial probabilistic model $B(N_t, \pi_t)$. As π_t is involved in this calculation, the uncertainties on π_t that come from the parameter's uncertainties will increase the uncertainties for M_t. Using a large sample of trajectories for π_t, which is easy with the MH algorithm (see below), it is easy to determine a range for each M_t by simulation.

10.2.4 Estimation of model parameters; the Metropolis-Hastings (MH) algorithm

Coliform pollution events are rare and therefore little information will be available to assess parameters with precision. Consequently, the variability of the observed y_t will in turn influence the model parameters and the model ability to forecast correctly global water quality in the distribution system. It is thus of utmost importance to study in detail the corresponding uncertainties. The model has been designed taking into account π_t, the actual proportion of CPS for week t as an indicator of the bacteriological pollution inertia to evaluate $\pi_t + 1$. This link between present and future CPS implies a severe technical complication: the model is now autoregressive, and the ordinary mean square algorithms commonly used to optimize general linear models cannot be used. This kind of model is usually called Generalized Autoregressive Models or Markov Models of order 1 (Fahrmeir and Tutz 1994). Two main algorithmic approaches to solve this problem and to determine parameter vector θ from y_t, the data collected weekly are now described.

The *Metropolis-Hastings (MH) algorithm* is described in Smith and Roberts (1993), Fahmy (1994), and Bois, Block, and Gatel (1997). It aims at creating some ordinary probability distributions to describe parameter uncertainty by stochastic simulations: this algorithm therefore refers to a Bayesian context. The main idea of the technique is to use prior distributions for the parameters to obtain starting drawings and then to design an ergodic π-unreducible and aperiodic Markov chain. This stochastic chain is such that its limiting states are distributed according to the posterior joint distribution of the parameters. The likelihood given in equation 1 is used to control the convergence of the Markov chain, so as to force gradually the stochastic algorithm to converge toward the equilibrium probability distribution of the parameters. The MH algorithm is a very powerful algorithm that can generate by a sampling procedure the complete joint probability distribution of the parameters.

The procedure to generate the chain is as follows:

The transition for θ from a ϕ value at step i (of the algorithmic procedure) to a ϕ' value at step $i + 1$ is given by an arbitrary matrix $q(\phi, \phi')$. Here the elements of the matrix are chosen to be $N(\phi, \Sigma^2)$, which is a common strategy and one easy to compute. The algorithm defines for θ a Markov chain with a transition probability $p(\phi, \phi')$ given by:

$$p(\phi,\phi') = \begin{cases} q(\phi/\phi')\cdot \rho(\phi,\phi') & \text{if } \phi \neq \phi' \\ 1 - \int_{\phi''} q(\phi/\phi'')\cdot \rho(\phi,\phi'')d\phi'' & \text{if } \phi' = \phi \end{cases}$$

$$\text{where } p(\phi,\phi') = \begin{cases} min\left\{\dfrac{\Pi(\phi')\cdot q(\phi'/\phi)}{\Pi(\phi)\cdot q(\phi/\phi')}, 1\right\} & \text{if } \Pi(\phi)\cdot q(\phi/\phi') > 0 \\ l & \text{if } \Pi(\phi)\cdot q(\phi/\phi') = 0 \end{cases}$$

$\Pi(\phi)$ corresponds to the posterior density function for step i, and $\Pi(\phi')$ is the posterior density function for step $i + 1$. A comparison between $a(\phi,\phi')$ and a U value sampled from a $U[0, 1]$ law, will lead to the decision whether ϕ' is accepted or not:

if $U > \rho(\phi, \phi')$, *then* ϕ' *is rejected and* $\theta_{i+1} = \theta_i$
if $U < \rho(\phi, \phi')$, *then* ϕ' *is accepted and* $\theta_{i+1} = \phi'$

The stochastic simulation results constituting the chain are stored and can be used in many ways:

(a) For instance, one can obtain parameter marginal posterior distributions and some statistics: the posterior mean θ can be taken as a point estimate of θ and its standard deviations will also be computed from the Monte Carlo simulations of θ.
(b) Equation 10.7 allows to associate a π_t trajectory to each θ and X_t (assumed to be fixed and known for each week t). It is therefore possible to evaluate 5 percent and 95 percent empirical quantiles by sampling a thousand of such trajectories.
(c) For a π_t trajectory, a second level random trial according to the binomial law (1) can generate y_t predictive values: they are used to assess a credible set for the CPS to be detected by the sampling devices. Figure 10.7 has been drawn by such a procedure from the case study data: the inner "tube" around the model sketches a 90 percent credibility zone for π_t. The larger zone gives the corresponding confidence interval for the predictive values of y_t.

10.2.5 Statistical test for selecting a model

To know whether or not all the explanatory variables are necessary to model the coliform apparition phenomenon, one must compare the model using all variables with models based on a subset of explanatory components. In this study, the classical procedure is employed to test the assumption that a model using p parameters is better (in the sense of a trade-off between parsimony and fit) than one using only p–q parameters: the logarithm of their likelihood ratio is compared to a chi-square value with q degrees of freedom at a required significance level (Agresti 1990; Wilks 1938).

10.3 CASE STUDY DESCRIPTION

10.3.1 The distribution system in the suburb of Paris

The Paris Suburb Water Authority distributes water to roughly 4 million inhabitants in 144 cities. The water is extracted from the Seine, Marne, and Oise Rivers. The raw water is heavily loaded with both organic compounds (initial DOC between 2 and 4 mg C/l) and fecal indicators (thermotolerant coliforms from 1,000 to 100,000 CFU/100 ml). The three treatment units use the following processes: ozone pretreatment if required, flocculation-settlement, biological filtration through sand, advanced oxidation to remove pesticides, bacteria, and parasites, biological filtration through activated carbon, and chlorination. The water is then distributed with an average free chlorine residual of 0.2 mg/l.

10.3.2 The database

Forty to sixty samples have been gathered by Compagnie Générale des Eaux every week throughout the whole distribution system, divided into three zones (DS1, DS2, and DS3) depending on where the water comes from, during the spring and summer periods of the years 1992, 1993, 1994, and 1995. Furthermore, sixty weekly samples collected by the public control authority in each zone during the same period have been merged with these databases. The data used to fit the model have been collected from April to September 1992 and from March to October 1994 on DS2. Two different years were necessary to take into account several different events, as each year alone seemed to be too specific. To validate the model, the data corresponding to DS2 in 1995 and DS1 in 1992 and 1994 have been used. The results presented here concern only DS2.

10.4 RESULTS

The four variables Temperature, FCR, Turb, and ΔBDOC were first tried. The ΔBDOC parameter was not significantly different from zero, which means that the corresponding variable does not carry any significant information into the model. Because of the autoregressive characteristic of the model, it has been necessary to introduce two more parameters $p0$–92 and $p0$–94, corresponding to the CPS expected rates π_t for the initial weeks $t = 0$ of years 1992 and 1994.

Table 10.1 presents the general statistics calculated for each component of the parameter vector θ from 4,000 vectors generated by the MH algorithm. Except for *FCR* and *Temp*,

Table 10.1. *General statistics for the components of parameter vector θ*

	Auto-regression α	Temperature β_{Temp}	Free chlorine residual β_{FCR}	Plant turbidity β_{Turb}	Structure λ	Start 92 *p0–92*	Start 94 *p0–94*	Constant β_m
Min	6.4989	0.0253	−2.0610	40.1683	0.9693	0.0050	0.0002	−11.675
Max	13.3438	0.1294	−0.0007	67.8025	1.0000	0.0500	0.0100	−7.639
Mean	9.4506	0.0744	−0.5759	52.6982	0.9970	0.0340	0.0059	−9.429
Std. dev.	0.9748	0.0149	0.3573	4.4920	0.0032	0.0095	0.0023	0.603

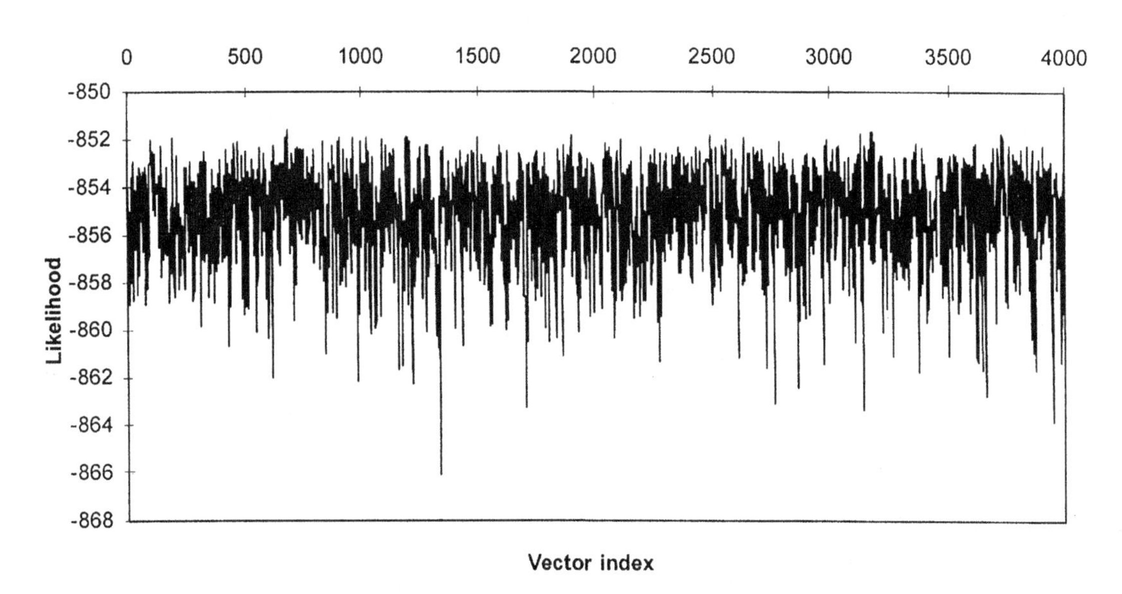

Figure 10.2. Likelihood evolution for one chain after convergence is reached.

one can see from those data that the effects of the variables are significant, as there is little uncertainty for the corresponding parameters (the standard deviations are a lot weaker than the means). The structure parameter λ is almost certain with value 1, which corresponds to the logistic model. So the effect of the variables in the low risks is more exponential than linear.

Figure 10.2 shows the likelihood values for a 4,000-vectors chain generated by the Metropolis-Hastings algorithm after convergence is reached. To test convergence, the Gelman method was used, which is a variance test with three chains running simultaneously. Figure 10.3 shows the corresponding chain for the *Turb* parameter values. For each parameter, such a chain is available, and from the 4,000-value sample it is possible to build histograms to approach the marginal distribution of the parameter. The histograms for λ and *Turb* are presented in Figures 10.4 and 10.5.

Correlations between parameters can also be studied as shown in Figure 10.6. When a correlation is detected, it is advised to write a relation between parameters to reduce the number of sampled parameters and increase the speed of the algorithm. The negative correlation observed in Figure 10.6 shows that the model hesitates between putting the effect into the constant γ and putting it into the *Turb* parameter.

In Figure 10.7 and Figure 10.8, the model and two-levels confidence range curves against the data are represented. In Figure 10.7, the data were used to fit the model, and Figure 10.8 corresponds to a validation on the data of year 1995. A third level of confidence range corresponding to the model uncertainties couldn't be represented on those plots as the value of λ is almost certain and the corresponding curves are superimposed with the first level curves, corresponding to the parameter's uncertainty (measurement and model errors). Results for the year 1992 were fine though a bit optimistic, but the plot for year 1995 shows that the model can also be pessimistic. Depending on the year, and then on the treatment conditions and on the distribution system characteristics, it seems that the model can behave in very different ways.

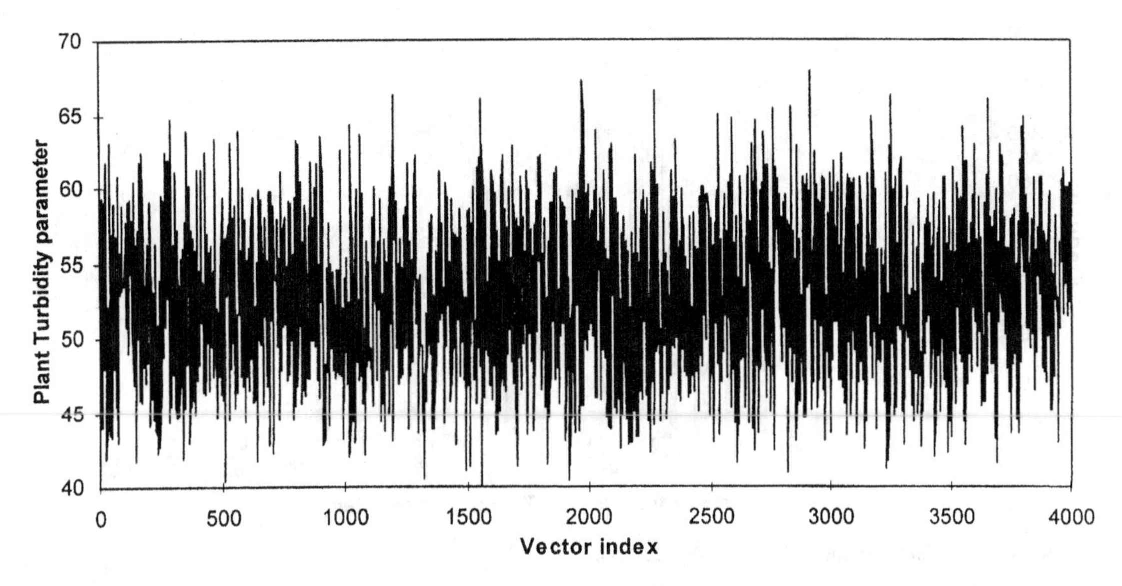

Figure 10.3. *Turb* parameter evolution for one chain after convergence is reached.

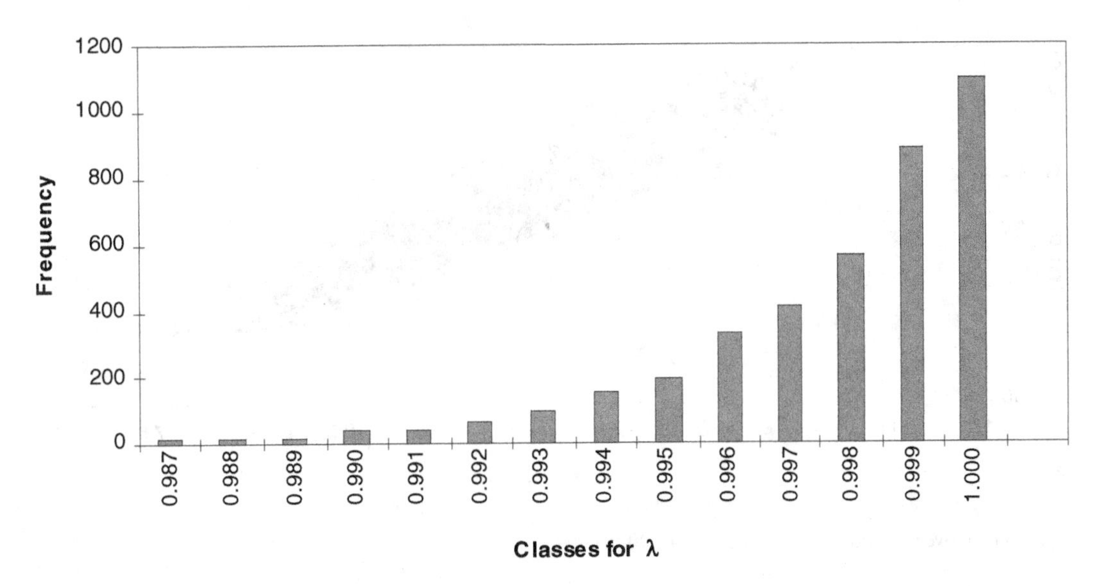

Figure 10.4. Histogram for parameter λ (from 4,000 vectors).

10.5 DISCUSSION

10.5.1 Operational results

The model is mostly influenced by the plant turbidity variable, and less by the temperature and the residual of free chlorine. Figure 10.9 shows that the proportion of CPS grows from nearly 0 percent to 10 percent, when the log of turbidity grows from 4 to 5, for usual values of FCR, temperature, and π_t. This sensitivity of the model to turbidity is related to the events chosen to design it, and therefore the model will be really efficient if the coliforms origin is an increase of turbidity at the inlet to the distribution system. If the origins are more complex, such as a filter wash, corrosion, or local contaminations, the model might be blind and miss its detection goal. This is why it is suggested that some other coliform contamination indicators should be monitored, especially before the water enters the distribution system.

10.5.2 Limits and perspectives

As for any extreme event modeling, the main problem that appeared while working on this model is the rarity of the Coliform Positive Samples, except during a short period in spring 1992. Consequently, this yields very uncertain model parame-

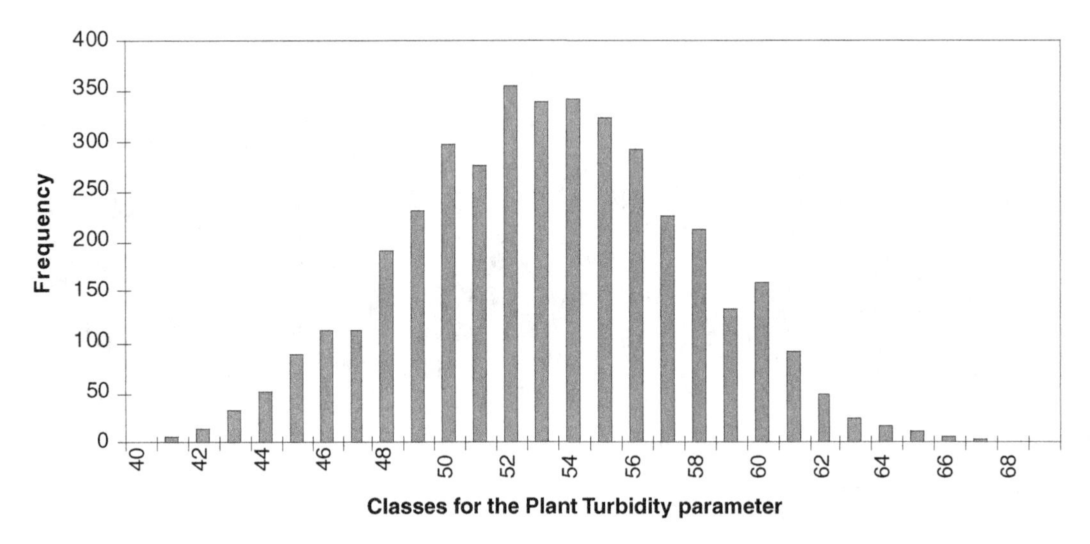

Figure 10.5. Histogram for the *Turb* parameter (from 4,000 vectors).

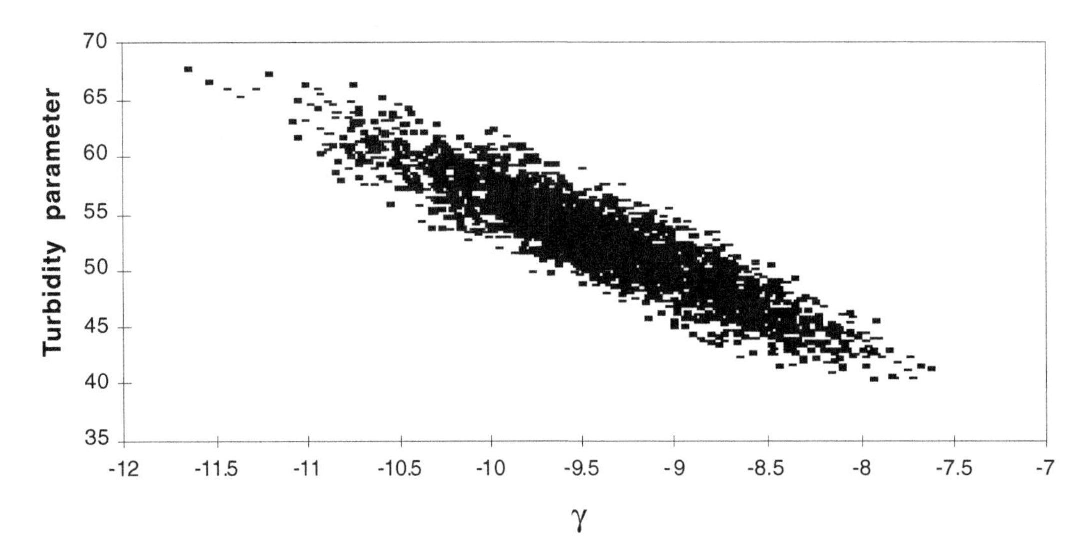

Figure 10.6. Correlations between γ and *Turb* parameters (4,000 vectors).

ters and results should be taken with care before adopting definite conclusions.

Furthermore, each of the five basic hypotheses is questionable:

(H1) The approach that consists in focusing only on a presence/absence can be criticized as bacteriological analyses give more information, namely a number of coliform colonies that developed.

(H2) For this study, it has been assumed that it is reasonable to consider the whole Distribution System (DS) as homogeneous for the measured variables. This allowed the use of spatially averaged values for each variable. This is a strong hypothesis, obviously too simplistic although it yields encouraging results to develop simple probabilistic models based on Poisson or binomial counts. Maul, Vagost, and Block (1989) used other distributions, negative binomial law for instance, allowing to take into account the distribution system heterogeneity because of their structure. Another way to get rid of the spatial homogeneity assumption is to use the much more complex hierarchical models for a regionalized approach.

(H3) Measurement errors during the biochemical analysis and experimental factors of variability of the sampling procedure (distance from reservoirs, size of pipes, for example) are likely to be important and should be modeled.

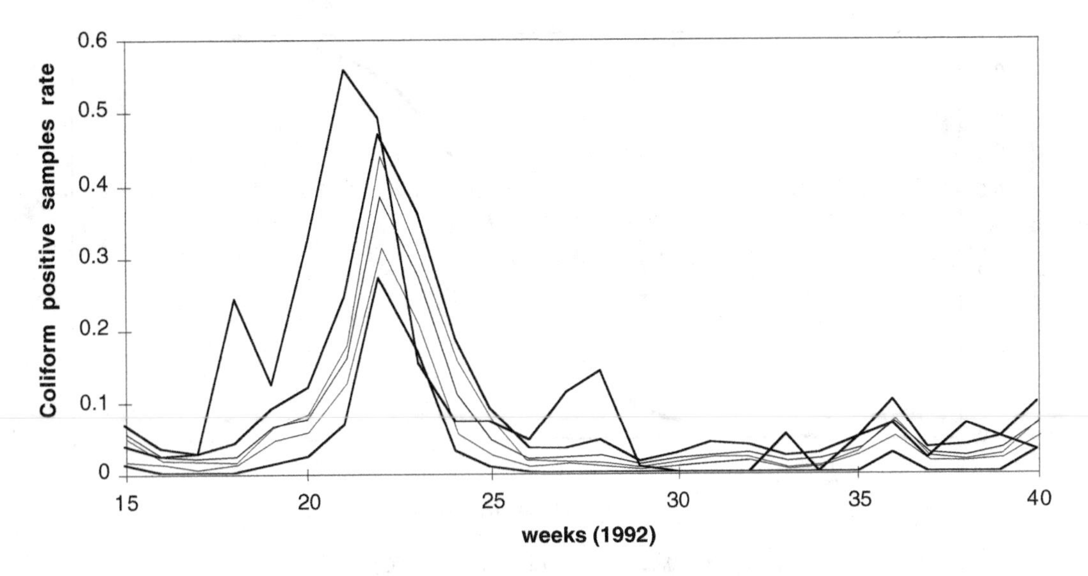

Figure 10.7. Two-level uncertainties for the model in 1992 (fitted on years 1992–94 data).

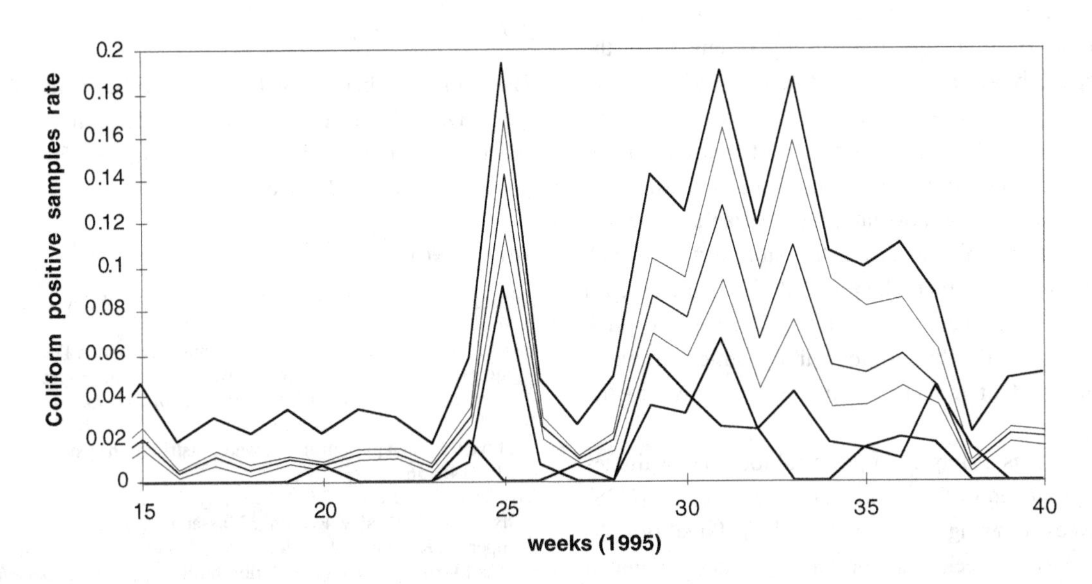

Figure 10.8. Two-level uncertainties for the model in 1995 (fitted on years 1992–94 data).

(H4) The quality manager does react to the observation of CPS by re-injecting chlorine in the DS if he or she feels it necessary. This feedback link should also be incorporated in the model, at least if one wishes to adapt the regulation.

(H5) The transition should not be deterministic and a stochastic state noise should be incorporated, at least to express that near ignorance conditions are more common in biomathematical modeling than it is generally admitted by experts.

(H6) Here it is considered that a single model can describe any situation. Maybe, as it has already been described in hydrology, several models should be included in a multimodel procedure to take into account the various possible configurations of a distribution system.

10.6 CONCLUSIONS

This prediction model has been written to mimic complex bacteriological phenomena whose dynamic is not very well known. To take into account various assumptions, and to make possible the evaluation of the uncertainties, a model mixing both GLM and autoregressive characteristics has been

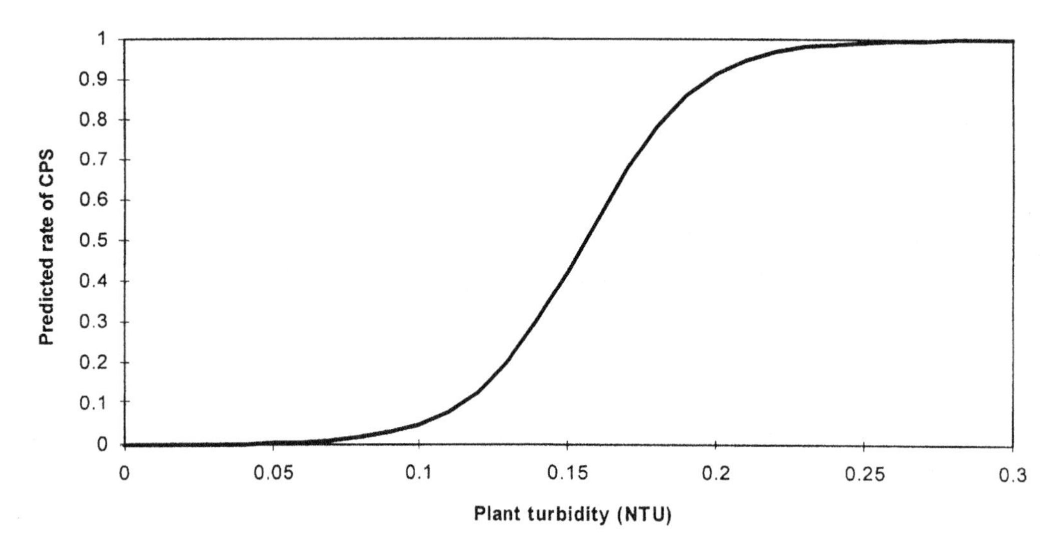

Figure 10.9. Model's behavior when turbidity varies (π_t = 0.02; FRC = 0.2 mg/l; T = 15°C).

created within a Bayesian approach. It has been optimized using Metropolis-Hastings and gradient algorithms, to fit the weekly and spatially averaged values of several variables collected in 1992 and 1994, and validated using data collected during other years. Classical statistical tests based on the model led to the conclusion that, in the specific case study with the available data, the most important variable is the turbidity measure at the inlet to the distribution system. Much uncertainty still remains due to the poor information conveyed by the few coliform occurrences in the DS and the lack of knowledge about the bacteriological dynamics in water networks. This kind of model may be very sensitive to the distribution system.

This new model is a simple but powerful tool that evaluates the expected proportion of Coliform Positive Samples (CPS) for the next week knowing a few easily collected quality indicators for the current week. The confidence ranges calculated for each prediction give very important information, especially for water quality managers who might need to readjust the treating processes or activate other control means, if the results do not correspond to legal rules.

This kind of risk model should be easy to apply to other fields, as it enables one to take into account quality of the past because of its autoregressive characteristics and several variables that have more or less a sigmoid or partially linear effect on the explained variable (here the proportion of coliform positive samples). Testing the role of new explanatory variables characterizing the working conditions of the distribution system is an easy task: without changing the structure of the model, statistical tests that have been described above can evaluate the worth of the information they convey.

Acknowledgments

This work has been possible through the financial support of the SEDIF (Syndicat des Eaux d'Ile de France, Paris, France) and the data collected by the Compagnie Générale des Eaux (Paris-La Défense, France)

REFERENCES

Agresti, A. (1990) *Categorical Data Analysis*. Wiley InterScience, New York.

Block, J. C., Mouteaux, L., Gatel, D., and Reasonner, D. J. (1995) Survival and growth of *E.coli* in drinking water distribution systems. *Leeds International Conference on E.Coli Proceedings 1995.*

Bois, F., Fahmy, T., Block, J.-C., and Gatel, D. (1997) Dynamic modeling of bacteria in a pilot drinking water distribution system, *Water Research*, 31: 3146–56.

Box, G. E. P. and Tiao, G. C. (1973) *Bayesian Inference in Statistical Analysis*. Addison Wesley, Reading, Massachusetts.

Camper, A. K., Hayes, J. T., Jones, W. L., and Zilner, N. (1993) Persistence of coliforms in mixed population biofilms. *WQTC Proceedings, Miami*, 1653–61.

Camper, A. K., Jones, L. W., and Goodrum, L. (1994) The effect of chlorine on the persistence of coliforms in mixed populations biofilms. *WQTC Proceedings, San Francisco*, 685–88.

Camper, A. K., Mc Feters, G. A., Characklis, W. G., and Jones, W. L. (1991) Growth kinetics of coliform bacteria under conditions relevant to drinking water distribution systems. *Applied Envir. Microbiol.* 57: 2233–39.

Colbourne, J. S., Dennis, P. J., Rachwal, A. J., Keevil, W., and Mackerness, C. (1992) The operational impact of growth of coliforms in London's distribution systems. *WQTC Proceedings*, 799–810.

Fahmy, T. (1994) Modélisation de l'évolution des populations bactériennes dans les réseaux de distribution d'eau potable; *rapport de DEA*, Institut National Agronomique Paris-Grignon.

Fahrmeir, L. and Tutz, G. (1994) *Multivariate Statistical Modeling Based on General Linear Models*. Springer Series in Statistics; Springer-Verlag, New York.

Gale, P. and Lacey, R. F. (1995) Predicting coliform detection in water supply. *Leeds International Conference on E.Coli Proceedings 1995.*

Goshko, M. A., Pipes, W. O., and Christian, R. R. (1983) Coliform occurrence and chlorine residual in small water distribution systems. *J. Am. Wat. Wks. Assoc.* 79: 74–80.

Lechevallier, M. W., Babcock, T. M., and Lee, R. G. (1987) Examination and characterization of distribution system biofilms. *Applied Envir. Microbiol.* 53: 2714–24.

Lechevallier, M. W., Shaw, N. J., and Smith, D. B. (1994) Factors related to regrowth of coliform bacteria. *WQTC Proceedings, San Francisco,* 657–61.

Maul, A., Vagost, D., and Block, J. C. (1989) Stratégies d'échantillonnage pour l'analyse microbiologique sur réseaux de distribution d'eau. *Coll. Technique et Documentation, Lavoisier, France.*

Smith, A. F. M. and Roberts, G. O. (1993) Bayesian computation via the Gibbs sampler and related Markov Chain Monte Carlo methods. *J. Roy. Stat. Soc. Series B* 55: 3–23.

Volk, C. and Joret, J.-C. (1994) Paramètres prédictifs de l'apparition des coliformes dans les réseaux de distribution d'eau d'alimentation. *Revue des sciences de l'eau* 7: 131–52.

Wilks, S. S. (1938) The large sample distribution of the likelihood ratio for testing composite hypotheses. *Annals of Math. Stat.* 9: 60–62.

11 Stochastic precipitation-runoff modeling for water yield from a semi-arid forested watershed

AREGAI TECLE* AND DAVID E. RUPP**

ABSTRACT

A stochastic precipitation-runoff modeling approach is used to estimate water yield from a particular forested watershed in North Central Arizona. The procedure uses selected theoretical probability distribution functions and a random number generator to describe and simulate various precipitation characteristics, such as storm depth, duration, and time between storm events. The spatial characteristics of precipitation events are described in terms of their orographic and areal distribution patterns while temporal distributions are expressed in terms of daily events in the watershed. The generated precipitation events are used as input into a precipitation-runoff model to estimate water yield from a particular forested watershed. The method uses geographic information systems (GIS) to subdivide the study watershed into cells assumed to be homogenous with respect to watershed characteristics, such as elevation, aspect, slope, overstory density, and soil type. The total water yield is the accumulated surface runoff generated at the watershed outlet. The outcome is the development of an improved model for estimating water yield which takes into consideration uncertainty, as well as temporal and spatial watershed characteristics. This method is useful not only for providing water resources managers with a good estimate of the amount of water yield, but also for determining the reliability or failure of a source to meet desired downstream water demands.

11.1 INTRODUCTION

This chapter is concerned with the development of an appropriate precipitation-runoff model for estimating water yield from a semi-arid forested watershed. This involves combining a stochastic precipitation model and a deterministic runoff model. The first one is selected to capture the inherently uncertain characteristics of precipitation, while the latter is chosen to simplify an otherwise complex surface runoff estimation method. The stochastic approach simulates the occurrences of individual hydrologic events. Since the occurrence and magnitude of surface runoff in a semi-arid upland watershed is directly dependent upon the occurrence of runoff-producing precipitation characteristics, the probability of the latter can be used to forecast the magnitude and reliability of future water yield scenarios.

In order to improve the prediction of the amount of water yield coming from a forested watershed, the effects of spatial distribution of watershed characteristics on runoff are analyzed using ARC/INFO geographic information system (GIS). GIS enables the use of high-resolution spatial data. Variables important to runoff, such as precipitation, temperature, elevation, slope, aspect, vegetation cover, and soil type, need not be grossly averaged over a catchment as has been done historically. Instead, detailed information on these characteristics is important for developing a reliable water yield model.

The precipitation model consists of a temporal and a spatial component. The temporal component is a stochastic simulator of storm event depth, duration, and interarrival time between events. Events occurring within a few days of each other are said to be part of the same storm sequence, as they are expected to be related to the same large-scale weather system. Under this condition, two additional variables are needed to adequately describe the precipitation characteristics in the area: interarrival time between storm sequences, and the number of events occurring in a sequence. Univariate theoretical probability density functions are fit to the frequency distributions of the number of events and the interarrival data, while a bivariate probability density function is fit to the depth-duration data. A random number generator is used to aid in synthetically generating future events on the basis of the selected theoretical distributions.

The spatial component of the precipitation model estimates the precipitation at any point on the watershed. Data from a network of precipitation gages in the Beaver Creek Watersheds of Arizona are used to develop a regression equation to predict the precipitation at any location on the study watershed. The

* School of Forestry, Northern Arizona University, Flagstaff, Arizona 86011-5018, USA.

** Ph.D. candidate, Oregon State University, Corvalis, OR, USA.

location is described in terms of its elevation, latitude, and aspect. A GIS is used to generate raster, or grid surfaces of both depth and duration of a precipitation event. Then, using a combination of these two, a third grid of average event intensity is created. The generated data are then used as input into the deterministic model to estimate the amount of event surface runoff. A water balance approach is used to estimate the amount of runoff coming out of a cell. The runoff from each cell is routed downstream from one cell to another in a cascading fashion to determine the total amount of event water yield at the outlet of the watershed. The sequence of runoff events is then cumulated to give an estimate of the total seasonal water yield from the entire watershed. A simple method of estimating the reliability of a watershed to produce enough water to meet downstream water demands is also presented. Mathematical equations representing the different watershed components, such as interception, snowmelt, evapotranspiration, infiltration, subsurface storage, and the simulated precipitation data are used to estimate the amount of runoff from each cell. A GIS is used to incorporate spatial watershed characteristics such as slope, aspect, elevation, vegetation cover, and soil types in the estimation process.

11.2 STUDY SITE CHARACTERISTICS

Any attempt to develop a watershed model needs to use a real world problem to ensure reliable applicability of the developed model. The development of the precipitation-runoff model in this chapter uses, as a case study, the Woods Canyon watershed. This watershed is a part of the Beaver Creek Experimental Watersheds of the Coconino National Forest in North Central Arizona, United States (Figure 11.1). The watershed is 4,814 hectares in size and rises from a low of 1,944 meters to a high of 2,355 meters elevation above sea level. The topography of the study watershed generally inclines southwesterly with an

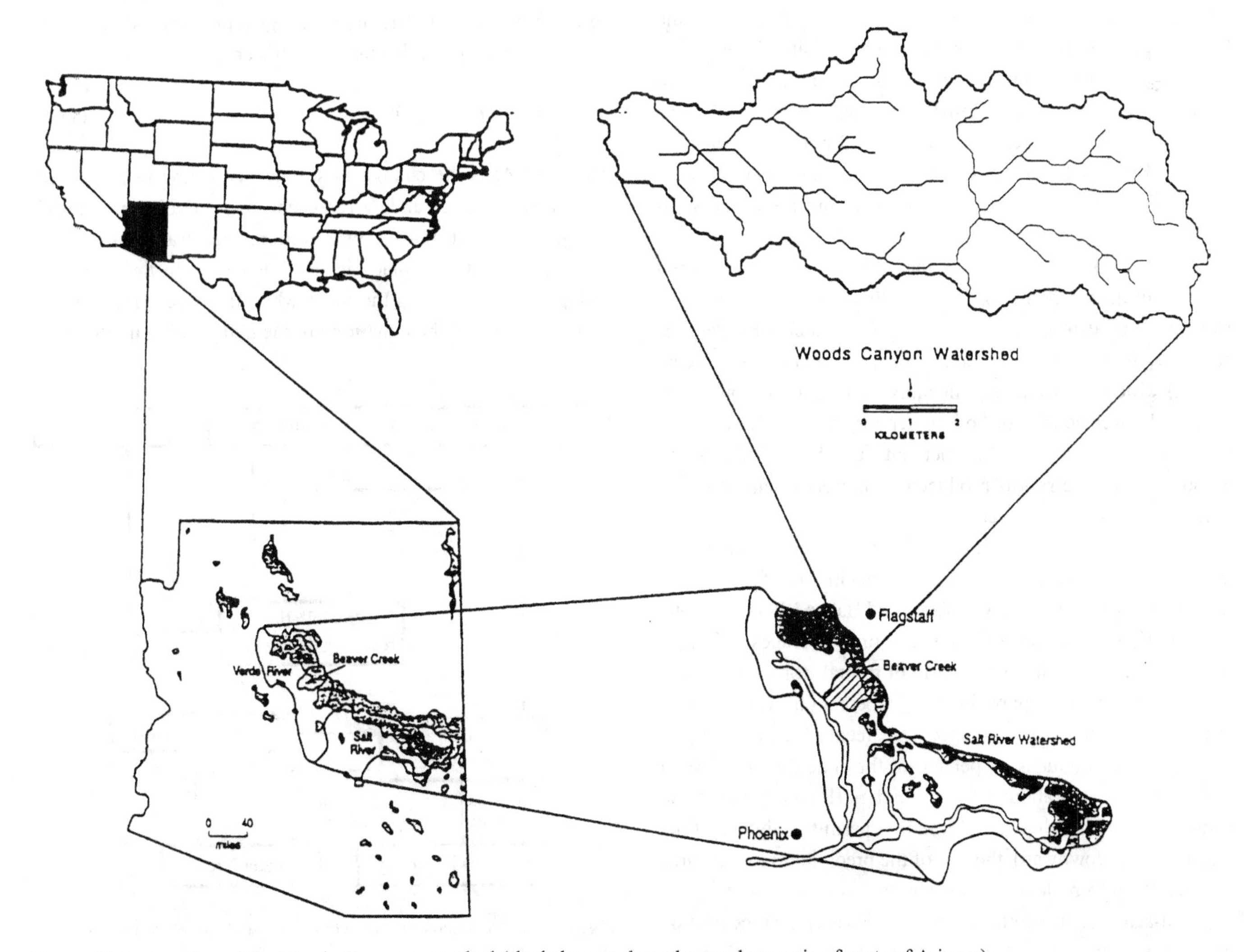

Figure 11.1. Location of the Woods Canyon watershed (shaded areas show the ponderosa pine forests of Arizona).

average slope of about 5 degrees. Other watershed characteristics that have important influence on the hydrology of the area are soil type and depth, and vegetation type and cover density. The spatial distributions of the physical and biological characteristics of the study area are described using a geographic information system (GIS). The raster-based component of the *ARC/INFO* GIS, called *GRID*, is employed to divide the watershed into over fifty thousand 30 by 30 meter cells. An individual cell is assumed to be homogeneous over its 900 m^2 area with respect to specific watershed characteristics such as soil type, elevation, slope, and aspect.

The parent material in the Woods Canyon watershed consists of a volcanic rock overlaying a sedimentary rock. The first layer has an average depth of 152 meters, and it is the result of a series of lava flow deposits (Rush and Smouse 1968). Each of these deposits has its own distinct set of vertical contraction joints which do not penetrate vertically from one layer to the other. These fractures constitute the primary passageways for downward transport of surface water through the column of the relatively impervious lava rock formation, and because they do not extend to adjacent deposits, there are no large amounts of downward water flow to the underlying sedimentary rock formation (Rush 1965). Due to such conditions, Feth and Hem (1963) estimate that only 2 percent of the precipitation that falls on the Mogollon Rim, which cuts through the Beaver Creek watershed, reaches the aquifers in the sedimentary rock formation.

The overstory vegetation on the Woods Canyon watershed is predominately ponderosa pine (*Pinus ponderosa* Laws.). But there are also other species such as Gambel oak (*Quercus gambelii* Nutt.) and alligator juniper (*Juniperus deppeana* Steud.) which contribute significantly to the canopy cover. The observed cover density of both understory and overstory types varies from about 30 to 70 percent (USFS 1992), and the amount of surface runoff produced is indirectly related to the amount of vegetation cover.

Most of the precipitation in the Beaver Creek comes during the cold season, which spans from the beginning of October to the end of April (Ffolliott, Gottfried, and Baker 1989). The total precipitation in the area during this time averages 431 millimeters, most of which is the result of frontal storms. The cold-season precipitation, approximately 56 percent of which arrives in the form of snow, accounts for 67 percent of the annual precipitation and produces 97 percent of the annual surface water yield. The latter is about 138 millimeters (Baker 1982). Now, because the underlying rock formation permits only very little water to seep downward, the rest of the precipitation in the area is considered to be lost due to evapotranspiration.

Cold-season temperatures in the ponderosa pine type portion of the Beaver Creek average 1.3°C, with a low monthly mean of −2.2°C in January (Campbell and Ryan 1982). But the average difference between the daily maximum and daily minimum temperatures during the cold season is about 17°C (Beschta 1976; Campbell and Ryan 1982). Daytime temperatures usually reach above freezing throughout the cold season. As a result, most of the snowpacks in the area, except those on the highest elevations, melt during the day.

11.3 MODEL DEVELOPMENT

Simulation of surface water yield is done within the context of the continuity equation which is restructured to represent the water balance of the study watershed. A water balance, in its simplest form, consists of a balance between the inputs, the outputs, and the change in storage of water. A schematic representation of the type of water balance model used in this study is shown in Figure 11.2. The water balance components in the flow chart represent the various parts contributing to the amount of water available in the area. The general expression for the water balance model discussed in this chapter is

$$\frac{\Delta S}{\Delta t} = P - Y - E - Tr \tag{11.1}$$

where $\Delta S/\Delta t$ is the change in storage with time in the watershed, while P, Y, E, and Tr represent, respectively, the rates of precipitation, net runoff, evaporation, and transpiration (all in cm hr^{-1}) that take place in the watershed. In the equation, precipitation represents the input while the other terms on the right hand side of the equation are the hydrologic outputs from

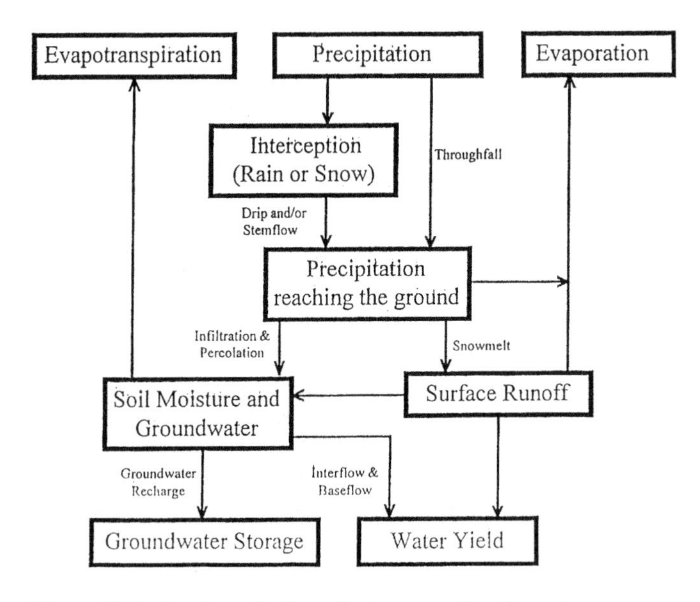

Figure 11.2. A schematic flow chart representing the water balance model in this study.

the watershed system. Implicit in equation 11.1 are the change in storage of water in the soil and the vegetation subsystems. Also implicit are snowmelt and evaposublimation from the snowpack and the overstory canopy. Any loss of water from the watershed system as groundwater flow to the distant water table below is assumed to be negligible in this model. Thus, with the obvious exception of deep seepage, the different components of the water balance model are presented below beginning with precipitation followed by the other components included in equation 11.1.

11.3.1 Modeling precipitation patterns

This section discusses the precipitation modeling approach used to generate the input into the water yield model. The developed model is designed to incorporate both the temporal and spatial characteristics of the precipitation in the study area. A stochastic event-based modeling approach is used to determine the temporal precipitation pattern, while ARC/INFO GIS is employed to describe the spatial pattern of the precipitation. The generated precipitation amount then becomes the input into a deterministic runoff model.

Temporal precipitation patterns

The temporal component of the model uses a stochastic process to describe the distribution of precipitation characteristics. The procedure is a modification of that developed by Duckstein, Fogel, and Davis (1975). The most important modification is in the definition of an event. An event, in this case, is an uninterrupted rainfall or snowfall of any duration, rather than the daily occurrence as described in Duckstein et al. (1975). An event can last from a few minutes to many hours.

The variables used to characterize event precipitation, similar to those in Duckstein et al. (1975), are:

1. time between sequences (days) (defined as ≥3.5 days),
2. number of events per sequence,
3. time between events (hours),
4. precipitation amount per event (millimeters), and
5. duration of event (hours).

The method of statistical simulation used involves fitting of known, theoretical probability distribution functions, such as gamma and exponential functions which have been used in past studies to describe the probability distributions of precipitation characteristics (Hanes, Fogel, and Duckstein 1977) to the empirical frequency distributions of the above five variables. Using the method of moments to fit probability distribution functions to the observed data requires determining values of the parameters of a distribution function (McCuen and Snyder 1986). For example, if the time between storm sequences is determined to fit exponential function, the cumulative distribution function is expressed as

$$F(\tau) = 1 - e^{-\tau/\beta} \quad for\ \tau \geq 0 \tag{11.2}$$

This expression can be used to find the probability of occurrence, $F(\tau)$, of the time between storm sequences less than or equal to τ days, and τ is obtained by rearranging equation 11.2 as follows:

$$\tau = -\ \beta ln(1 - F) \tag{11.3}$$

where F is the cumulative frequency distribution function, and β is the mean of τ, the time between sequences. The value of F is obtained using a random-number generator which gives values that lie in the interval from zero to one.

In the case of other probability distribution functions (pdf's), such as the gamma distribution function, however, analytical integration of their pdf's such as that for the exponential is not possible and numerical approximations of the integration must be implemented instead. The pdf of the gamma distribution function is:

$$f(\tau) = (\beta^{-\alpha}\tau^{\alpha-1}e^{-\tau/\beta})/\Gamma(\alpha) \quad \text{for } \tau > 0 \tag{11.4}$$

where α is the shape parameter and equals μ^2/σ^2, β is the scale parameter and equals σ^2/μ, and for a real number $\alpha > 0$, $\Gamma(\alpha)$ is the gamma function defined by the integral

$$\Gamma(\alpha) = \int_0^\infty \theta^{\alpha-1}e^{-\theta}d\theta \tag{11.5}$$

and μ and σ^2 are the population mean and variance, respectively. The test used in this study for evaluating goodness-of-fit of the theoretical pdf's to the observed data is the Kolmogorov-Smirnov test, or K-S test (Law and Kelton 1982).

Three of the precipitation model variables (time between sequences, number of events per sequence, and time between events) can be described using the techniques for the univariate simulation of random variables described above. The modeling of depth and duration, however, is more complex because the two variables are not independent of each other. Therefore, a different method of simulating two dependent random variables is used.

A procedure which involves explicitly describing the conditional probability distribution of the depth of an event given the duration of that event is used to produce a bivariate gamma pdf. This method is referred to in this study as the *explicit conditional distribution* (ECD) method because the conditional distribution is modeled directly. The method is described in Kottas and Lau (1978), but the details of the method used depend to a large extent on the data being described (Schmeiser and Lal 1982). Therefore, though the idea behind the method is that of Kottas and Lau (1978), the procedure used in this study is uniquely developed in Rupp (1995) and presented in Tecle and Rupp (1995).

To develop the bivariate model for simulating storm depth and duration, a marginal distribution for duration is first found. Assuming the marginal distribution function to be gamma, the pdf for the duration, τ_1, is

$$f_1(\tau_1) = (\beta_1^{-\alpha_1}\tau^{\alpha_1 - 1}e^{-\tau_1/\beta_1})/\Gamma(\alpha_1) \tag{11.6}$$

where α_1 and β_1 are the shape and scale parameters for the gamma distribution of duration. Then, letting τ_1 and τ_2, respectively, represent the duration time and depth of a precipitation event, the conditional pdf of depth given duration can be expressed as

$$f_2(\tau_2/\tau_1) = (\beta_2^{-\alpha_2}\tau_2^{\alpha_2 - 1}e^{-\tau_2/\beta_2})/\Gamma(\alpha_2) \tag{11.7}$$

where α_2 and β_2 are the shape and scale parameters of the gamma distribution function, respectively, for depth and are conditional upon the duration. As described previously, the shape and the scale parameters are both functions of the mean and the variance: $\alpha_2 = \mu^2{}_2/\sigma^2{}_2$ and $\beta_2 = \sigma^2{}_2/\mu_2$. To implement equation 11.7, it is necessary to know how the distribution of depth varies with duration.

Knowledge of the conditional mean and variance of depth allows estimation of the shape and scale parameters for the conditional gamma pdf equation 11.7. The shape and scale parameters for the conditional distribution of depth given duration are calculated using the following equations:

$$\alpha_2 = \mu_2^2/\sigma_2^2 = (b_0 + b_1\tau_1)^2/(c_0 + c_1\tau_1^\delta) \tag{11.8}$$

and

$$\beta_2 = \frac{\sigma_2^2}{\mu_2} = (c_0 + c_1\tau_1^\delta)/(b_0 + b_1\tau_1) \tag{11.9}$$

where b_0, b_1, c_0, and c_1 are regression coefficients, δ is constant, τ_1 is the mean of the precipitation duration data, and the rest are as explained above.

The probability density functions for each one of the five precipitation variables are determined for one gage in the study area. The remaining gages are used to analyze the spatial distribution of precipitation on the watershed.

Spatial precipitation patterns

The spatial analysis is used to define the areal distribution of precipitation event depth and duration, as well as the form of the precipitation, rain or snow, over the watershed. In general, those areas in which the air masses rise the fastest and the highest in elevation are likely to receive the greatest amount of precipitation. In mountainous regions, this rapid ascent takes place on the windward side of the orographic features (Barros and Lettenmaier 1994). For this reason, in addition to elevation, a specific set of independent variables available from the network of precipitation gages on the Beaver Creek experimental watershed were selected to predict the spatial distribution of precipitation. The variables are gage elevation, geographic location in terms of Universal Transverse Mercator (UTM) coordinates, slope, and aspect in relation to the prevailing wind direction (Oki and Musiake 1991). The dependent variables in the regression equations for depth and duration are, respectively, the ratios of precipitation depth and duration at gage i to those at gage #38. Gage #38 is located near the Woods Canyon watershed outlet, and the depth and duration of precipitation at any gage are found to be related to the corresponding variables at gage #38.

Once the prediction equations for both precipitation depth and duration are determined, GIS is employed along with the equations to map the spatial distribution of event depth and duration. The use of a GIS enables efficient determination of storm depth and duration at any cell in the watershed. For this purpose, the position, elevation, and aspect of each cell are determined first. Then, regression equations are developed in terms of these variables to simulate both the amount and duration of precipitation events at each cell. Constructing grids for the depth and duration of a precipitation event and combining them together generates a third grid that describes the spatial distribution of the storm intensity across the watershed.

11.3.2 Construction and roles of water yield model components

The characteristics of precipitation are discussed above. For this reason the discussion about the components of the water balance model used in this study is restricted to the rest of the model components. In addition to precipitation, the other components of the model described in this section include the processes of interception, evaporation, transpiration, infiltration, snowmelt, and runoff (see Figure 11.2).

Interception

In a forested environment, the overstory is assumed to intercept all precipitation until the maximum interception storage capacity is reached (Wigmosta, Vail, and Lettenmaier 1994). Accordingly, the change in interception storage, ΔS, under this condition is estimated by

$$\Delta S = P - Ep \tag{11.10}$$

where P is the precipitation rate (cm s^{-1}), and Ep is the potential evaporation rate (cm s^{-1}). During the period before maximum interception storage capacity, S_{max}, is reached, the rate of precipitation reaching the ground, P_g, is zero. After maximum interception storage capacity is reached, the precipitation rate at the ground is determined by

$$P_g = P - Ep \qquad (11.11)$$

As in Dickinson et al. (1991) and Wigmosta et al. (1994), the maximum interception storage capacity, S_{max} (cm), is estimated using

$$S_{max} = 0.01\ LAI_p F \qquad (11.12)$$

for both rain and snow. The LAI_p and F in equation 11.12 are the projected leaf area index, and overstory cover as a fraction of total surface area, respectively.

Evaporation

Evaporation can occur from either the water stored as interception on the overstory canopy, or from the water stored in the snowpack. In either case, a potential evaporation rate is estimated using the Penman-Monteith equation (Dingman 1994). This equation is expressed as

$$Ep = \frac{\Delta R_n + \rho_a c_a C_{at}(e_s - e)}{\rho_w \lambda_v (\Delta + \gamma)} \qquad (11.13)$$

where

Ep = potential evaporation rate (cm s^{-1}),
Δ = slope of saturation vapor pressure vs. air temperature (mb°C^{-1}),
R_n = net radiation (cal cm^{-2} s^{-1}),
ρ_a = density of air (g cm^{-3}),
c_a = heat capacity of air (cal g^{-1} °C^{-1}),
C_{at} = atmospheric conductance (cm s^{-1}),
e_s = saturation vapor pressure (mb),
e = actual vapor pressure (mb),
ρ_w = density of water (g cm^{-3}),
λ_v = latent heat of vaporization (cal g^{-1}), and
γ = psychrometric constant (cal g^{-1}).

The atmospheric conductance (C_{at}) in equation 11.13 is a function of the prevailing wind speed and the height of the vegetation in the area. The atmospheric conductance is computed using equation 11.14 (Dingman 1994):

$$C_{at} = \frac{v_a}{6.25(ln[(z_m - z_d)/z_0])^2} \qquad (11.14)$$

where

v_a = wind speed (cm s^{-1}),
z_d = $0.7 z_{veg}$ (m),
z_0 = $0.1 z_{veg}$ (m),
z_m = height of wind speed measurement (m), and
z_{veg} = height of vegetation (m).

The net radiation (R_n) is the sum of the short-wave and long-wave components of radiation for either at the overstory or at the ground. The net radiation and the details of the different variables used in equations 11.13 and 11.14 are described in Rupp (1995).

Transpiration

Transpiration rates are determined using the Penman-Monteith equation with the inclusion of a canopy resistance term (Dingman 1994). The canopy resistance includes expressions that account for the effects of environment as well as characteristics of the transpiring plant upon the transpiration process. The equations so constructed and a complete description of the different variables in the equations are given in Jarvis (1976), Stewart (1988), Dingman (1994), Rupp (1995), and Tecle and Rupp (1995).

Infiltration

Infiltration is determined using the Green-Ampt equation developed by Green and Ampt (1911). According to this equation, the total amount of infiltration, $I(t_f)$, at the end of rainfall and/or snowmelt event can be expressed (Rupp 1995) as

$$I(t_f) = I'(t_f) - I'(t_p) + wt_p \qquad (11.15)$$

where

w = water reaching the surface from above (cm hr^{-1})
t_p = time of ponding (*hr*), and
$I'(t_f)$ and $I'(t_p)$ are as described below.

$$I'(t) = K_{sat}\Big[\frac{(3-\sqrt{2})}{3}t + \frac{\sqrt{2}}{3}(\chi t + t^2)^{1/2} + \frac{\sqrt{2}-1}{3}\chi[ln(t+\chi) - ln(\chi)] + \frac{\sqrt{2}}{3}\chi\{ln\big[t + \chi/2 + (\chi t + t^2)^{1/2}\big] - ln(\chi/2)\}\Big] \qquad (11.16)$$

where χ equals $|\psi_f|(\phi - \theta_0)/K_{sat}$ (Salvucci and Entekhabi 1994), and $I'(t)$ is the approximate total infiltration at time t. $I'(t_p)$ is determined by substituting t by t_p in equation 11.16, and it represents the approximate total amount of infiltration at time of ponding, t_p. The other variables in equation 11.16 are K_{sat}, the saturated hydraulic conductivity of the surficial material, t, the duration of a precipitation event, φ_f, the effective tension at the wetting front (cm) (Dingman 1994), ϕ, saturated soil moisture content, and θ_0, the initial water content of the soil. Other factors such as the effect of the relationship between the amount and rate of incoming water and saturated hydraulic conductivity on infiltration rate are explained in detail in Rupp (1995).

Snowmelt

The expression for snowmelt is constructed by combining the commonly used temperature-index, or degree-day, method with a limited surface energy budget. The limited surface energy budget includes the surface radiation budget and the energy advected to the snowpack by precipitation. The expression for estimating snowmelt which also accounts for rain-on-snow events is

$$M = a_r T_{avg} + m_q(R_n + Q_r) \tag{11.17}$$

where

M = daily snowmelt depth (cm),
a_r = restricted degree-day factor ($cm\,d^{-1}\,°C^{-1}$),
T_{avg} = daily average temperature (°C),
m_q = the conversion factor for energy flux density to snowmelt depth ($cm\,d^{-1}\,(cal\,cm^{-2}\,d^{-1})^{-1}$), and
R_n = daily net radiation ($cal\,cm^{-2}\,d^{-1}$).
Q_r = the energy advected by rain (in $cal\,cm^{-2}$), and it is determined using equation 11.18

$$Q_r = \rho_w c_w T_r P_r \tag{11.18}$$

where ρ_w is the density of water ($g\,cm^{-3}$), c_w is the heat capacity of water ($cal\,g^{-1}\,°C^{-1}$), P_r is the depth of rain (cm), and T_r is the rain temperature (°C). The latter is assumed to be equal to the air temperature.

Runoff

The surface runoff consists of the rainfall reaching the ground (P_g) and/or the snowmelt (M) that neither evaporates (E_p) nor infiltrates (I) into the soil. This occurs in every cell of the watershed. The net surface runoff (Y_i) in units of centimeters from cell i is then determined using

$$Y_i = P_{gi} + M_i - I_i - E_{pi} \tag{11.19}$$

The values for the righthand side components of equation 11.19 are determined separately using the different modules discussed above.

In addition to precipitation, downstream cells also receive runoff input in the form of outflow from upstream cells. In this case, surface runoff may enter a cell from any of seven possible adjacent cells. This continues in all downgradient cells, and in the process surface runoff is routed downstream in a cascading fashion. The total runoff leaving a cell is thus an accumulation of the runoff generated from that cell plus the runoff coming from all the contributing sub-basin cells upstream. If a sub-basin is composed of n cells, then the total surface water yield leaving the nth cell or outlet cell can be expressed as

$$Y_n = P_{gn} + M_n - I_n - E_{pn} + \sum_{i=1}^{n-1} Y_i \tag{11.20}$$

All the variables in this expression are as explained in equation 11.19 above.

11.4 ANALYSIS OF RESULTS

Application and outcome of both the precipitation and the water yield models discussed above are presented in this part. Following the same sequence as above, the results of the precipitation model are discussed first, followed by those of the water yield model. In each case, the effects of both temporal and spatial analysis are given.

11.4.1 Results of precipitation modeling

Discussion on the analysis of results of precipitation modeling is in two parts. The first part looks at the temporal behavior of the various precipitation characteristics and its influence on the amounts of simulated precipitation events. The second part analyzes the influence of watershed characteristics on the spatial distribution of precipitation amounts.

Temporal analysis of precipitation events

To determine the temporal behavior of the precipitation data, theoretical probability distribution functions have been fit to the five precipitation event characteristics that represent the time between events, number of events per sequence, time between sequences, event duration, and event depth. This was determined using the Kolmogorov-Smirnov (K-S) test (Law and Kelton 1982). According to this test, the best fit theoretical probability distribution functions are the exponential and gamma distribution functions. The type of data, level of tests, their results, and the level of fitness of the theoretical distribution functions to the respective data are summarized in Table 11.1. The detailed explanation of the test procedures and the results are given in Rupp (1995).

Table 11.1. *A summary of results of fitness test of two theoretical distribution functions to precipitation characteristic data*

Precipitation characteristics	Theoretic distribution function	Level of test	Test status: Passed (Y) or Failed (N)
Time between events	Gamma	0.01	N
Events per sequence	Exponential	0.05	Y
Time between sequences	Exponential	0.15	Y
Marginal event duration	Gamma	0.01	Y
Marginal event depth	Gamma	0.01	Y

Table 11.2. *Sample and simulated parameters for duration and depth*

Parameter	Sample	ECD
m_1	4.124	4.114
m_2	7.243	7.305
s_1^2	33.821	32.818
s_2^2	154.675	168.854
r	0.792	0.780

The sample bivariate (depth/duration) density is shown in Figures 11.3a through 11.5b. Figures 11.3a,b are, respectively, the observed and simulated bivariate pdf's of the entire sample, while the pdf's in Figures 11.4a and 11.4b are for the same sample, but with the peak of the distributions removed to show the depth and duration values with low probability densities. Figures 11.5a and 11.5b, on the other hand, focus on the peak of the distribution to show only those events with both shorter durations and lower depths. The bivariate density plots of the observed data in Figures 11.3a through 11.5b reveal that the conditional distribution of depth on duration can be described using a gamma distribution function.

The simulated results show that the explicit conditional distribution (ECD) method produces a good fit to the observed data. Because of this, the ECD is selected to describe the dependence between storm depth and storm duration. The ECD simulated results compare well with the actual data, except for the short duration (less than one hour), low depth (less than two millimeter) events (see Figures 11.3a,b; 11.4a,b; 11.5a,b, and Table 11.2). This is probably because the gamma distribution overestimates the values of both duration and depth frequencies at low precipitation values. In Table 11.2, m_1, m_2, s_1^2, and s_2^2 are, respectively, the means and variances of the durations and depths of

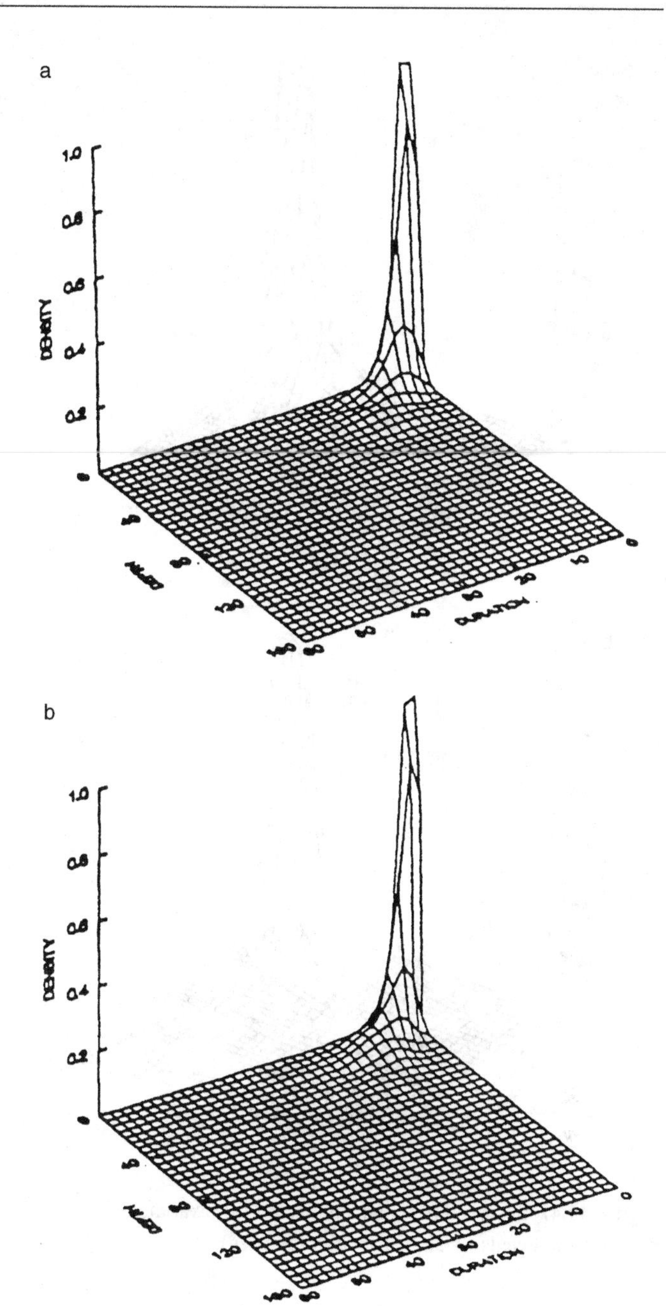

Figure 11.3. Observed (3a) and simulated (3b) bivariate density of an entire sample of precipitation events.

precipitation events; r is the correlation coefficient between depth and duration. Columns 2 and 3 are then, respectively, the sample and ECD generated values of these parameters.

Spatial analysis of precipitation events

The spatial pattern of precipitation event depth and duration in the study watershed are described in terms of five locational variables. Regression equations are developed, in terms of these

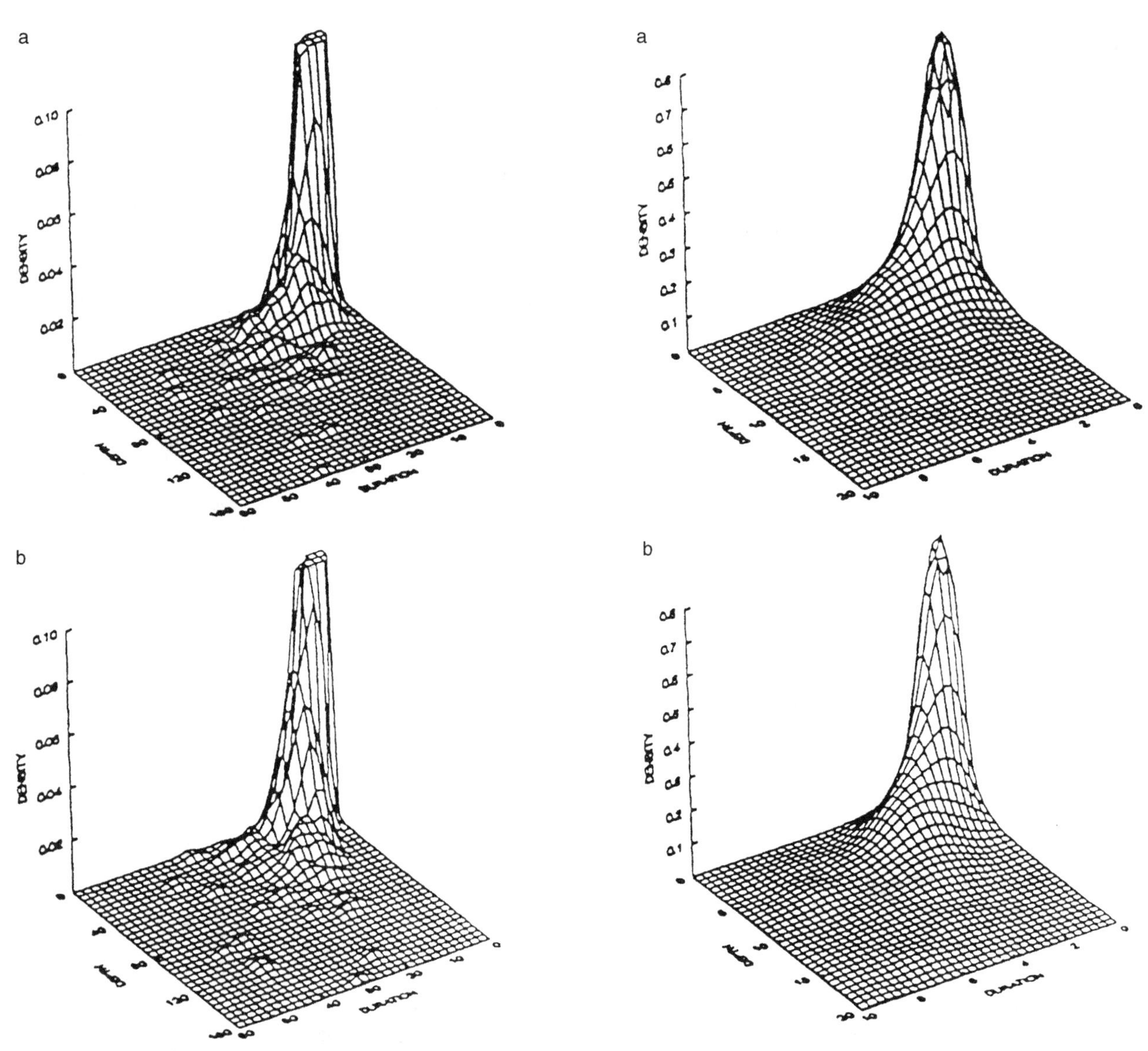

Figure 11.4. Observed (4a) and simulated (4b) bivariate density of precipitation events (peak distribution removed).

Figure 11.5. Observed (5a) and simulated (5b) bivariate density of precipitation events (focused on peak distribution).

variables, for both precipitation depth and duration. The regression model for precipitation event depth developed using a forward selection procedure is

$$P_i/P_{38} = 1.103x^{-5}[UTMY] + 4.464x10^{-4}[ELEV] - 3.340x10^{-4}[ASP] - 7.201$$

$$R^2 = 0.8121 \qquad (11.21)$$

where *UTMY* is the gage location in the *y*-direction in terms of Universal Transversal Mercator, *ELEV* is the elevation in meters, and *ASP* is the aspect in degrees from windward direction. Table 11.3 shows the correlation coefficients between any two of the variables used in equation 11.21.

As expected, equation 11.21 predicts an increase in precipitation depth with elevation and aspect in a northerly direction along the study area. There are also greater depths predicted for windward slopes than on leeward slopes, though the addition of aspect only increases the R^2 value by 0.016.

As with depth, the spatial distribution of storm duration is analyzed initially by examining the pairwise correlation coefficients among the model variables (see Table 11.4).

The highest correlation coefficients are between duration and elevation, and duration and *UTMY*, respectively.

Thus, the best regression model includes elevation and *UTMY* and has the form

Table 11.3. *Correlation coefficients for depth model variables (recording gages only)*

P_i/P_{38}	D^i/D^{38}	Elev.	UTMX	UTMY	Aspect
P_i/P_{38}	1.00000	0.57119	−0.19871	0.78391	−0.29029
Elev.		1.00000	0.60709	0.19514	−0.02089
UTMX			1.00000	−0.62291	0.21390
UTMY				1.00000	−0.22736
Aspect					1.00000

Table 11.4. *Correlation coefficients for duration model variables (recording gages only)*

	D_i/D_{38}	Elev.	UTMX	UTMY	Aspect
D_i/D_{38}	1.00000	0.73294	0.00614	0.66011	−0.15405
Elev.		1.00000	0.59329	0.17451	0.05467
UTMX			1.00000	−0.65317	0.27695
UTMY				1.00000	−0.20506

$$D_i/D_{38} = 5.517x10^{-4}[ELEV] + 7.50x10^{-6}[UTMY] - 5.025 \quad R^2 = 0.8293 \tag{11.22}$$

Any improvement due to the addition of either, or both, of the remaining variables was not significant at the 0.50 level.

11.4.2 Results of water yield modeling

The performance of the model and the efficiency of the watershed to convert precipitation into surface runoff are evaluated by comparing the model-generated results with the cold-season streamflow volumes measured at the outlet of the study watershed. Also the results from one season are analyzed by examining the daily model outputs of surface runoff, soil water, and snowpack depth, as well as the spatial distribution of total cold-season yield across the watershed. Twenty years of observed data are used for this purpose.

The model-generated results for one cold season are illustrated in Figures 11.6a through 11.6d. The histogram in Figure 11.6a shows the simulated daily precipitation depth during the entire cold season, while Figure 11.6b shows the resulting amounts of daily water yield from the study watershed. Because the amounts of precipitation and water yield during the first dozen days of the run are much higher than those produced during the rest of the season, the precipitation and water yield events are plotted again at higher resolution in Figures 11.6c and 11.6d, respectively, to highlight the smaller events. This, however, required truncating the extreme events in order to accommodate the graphs within the given space. It should be noted that in the figures, day zero corresponds to September 30 while day 212 is April 30, the final day of the cold season.

The simulated total cold-season precipitation and water yield amounts are equal to 671 mm and 59 mm, respectively. The latter amounts to about 9 percent of the former. This finding is very close to the historical findings of precipitation-runoff analysis in the study area. But the spatial distribution of the total cold-season water yield in Figure 11.7 depicts high variability in the amounts of runoff produced by each cell in the watershed. According to the figure, most of the cells contribute very little to the water yield, while a very small number of them contribute up to 500 mm of runoff per cell. Because there are many variables involved in the water balance algorithm, it is difficult to determine precisely the cause for the extreme variations in the spatial distribution of the amount of water yield shown in Figure 11.7. However, Rupp (1995) shows the amounts of surface water yield to decrease with a decrease in soil thickness and increased overstory cover density.

A third factor in the model which is not as obvious but is still evident in Figure 11.7 is that south-facing slopes seem to generate more water yield than north-facing slopes given that the overstory and soil characteristics are the same. This is because south-facing slopes receive more radiation than north-facing slopes, resulting in higher snowmelt rates which produce more runoff. In contrast, there are at least two reasons for the reduced water yield from north-facing slopes. The first one is

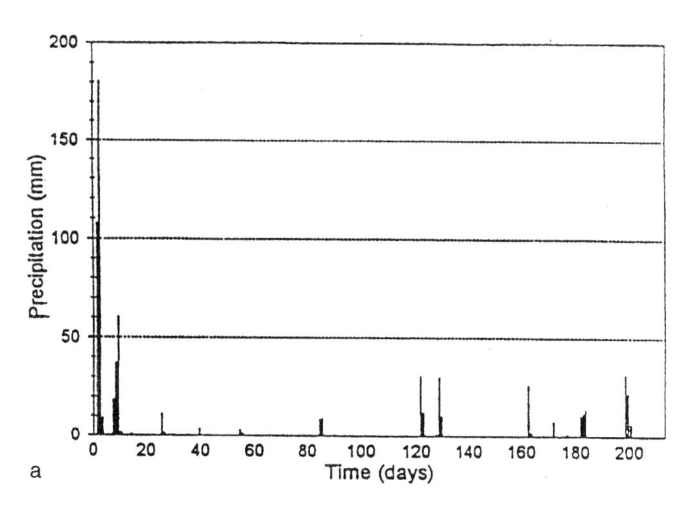

Figure 11.6a. Simulated daily precipitation depth vs. time for one cold season.

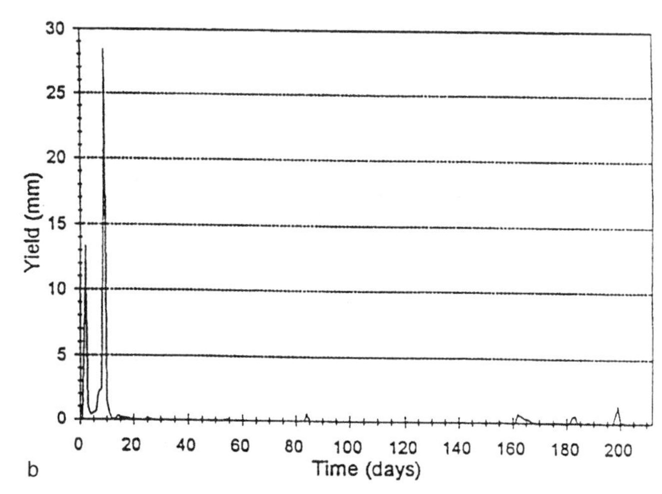

Figure 11.6b. Simulated daily water yield vs. time for one cold season.

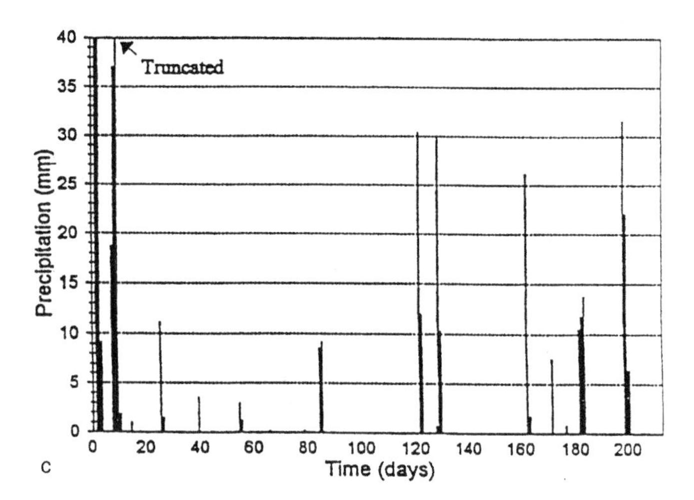

Figure 11.6c. Daily precipitation depth vs. time for one cold season simulated to show smaller values (extreme values truncated).

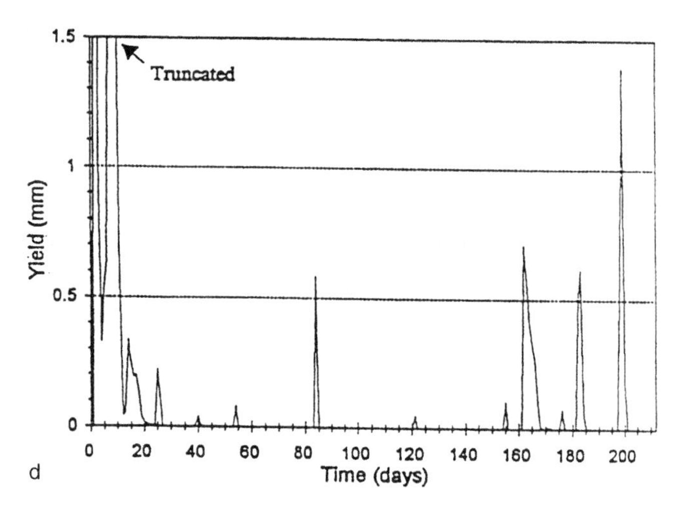

Figure 11.6d. Daily water yield vs. time for one cold season simulated to show smaller values (extreme values truncated).

that there is more possibility for water resulting from the slowly melting snow, on north facing slopes, to infiltrate into the ground leaving little amount of water for runoff. The second reason is evapo-sublimation. Because the snowpacks on the north-facing slopes tend to remain longer before melting than those on the south-facing slopes, they are more likely to evapo-sublime and not contribute as much to the runoff.

11.5 EVALUATING RISK AND RELIABILITY IN WATER YIELD

One way of applying the results of this study to a real problem is to estimate the reliability of the seasonal water yield to satisfy downstream water demands. Whether a stochastic or a deterministic process is used to evaluate the water yield problem, it is possible to assess the reliability of the watershed system to produce adequate amounts of water to meet the downstream water demands. For example, if we let d_* represent the minimum amount of water yield d, that can satisfy the downstream water demand, then the probability of failure to meet that demand can be expressed in terms of the cumulative distribution function (CDF), $F(d < d_*)$, of the amount of water yield generated. In a case where the occurrence of instant precipitation/snowmelt is the sole source of surface runoff, the CDF can be extrapolated from the bivariate cumulative distribution function of the runoff-producing precipitation events in the area. In either case, the reliability of meeting the demand,

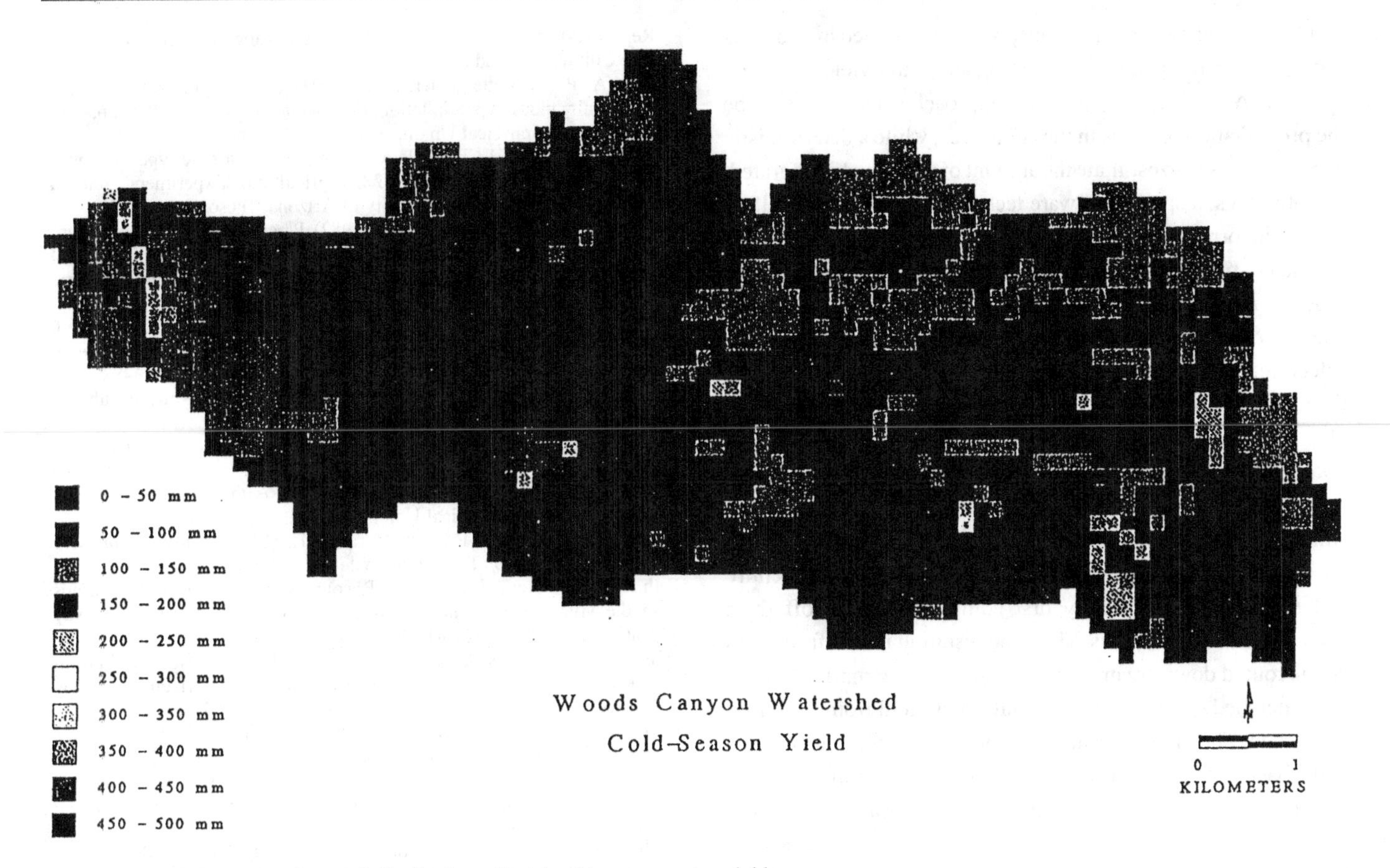

Figure 11.7. GIS generated spatial distribution of total cold-season water yield.

$R(d)$, is represented by 1-$F(d)$ (Hogg and Tanis 1983; Keeping 1995). Then, the failure rate (or hazard) to meet the demand can be estimated by dividing the probability density function of the system in failure mode, $f(d)$, by the reliability function, $R(d)$ (Wadsworth 1990).

In a deterministic problem, the relative frequency of failure to meet the demand, represented by the ratio of the number of seasons, n_s, during which the amount of water yield failed to meet the demand to the total number of seasons, n, has been recorded. This ratio, which can be represented by $P_f = n_s/n$, for the sake of brevity, is equivalent to the percentage of time the demand has not been fulfilled. Under such condition, the reliability of the watershed to produce the needed amount of runoff to meet downstream water demands is 1-P_f. Due to the semi-arid climatic conditions of the study area, there have been many occasions in which the watersheds failed to meet the amount of water needed for downstream users in Arizona. Recently, however, the problem of water shortage has, at least temporarily, been rectified in the state by importing water from the Colorado River through the Central Arizona Project (CAP).

11.6 SUMMARY AND CONCLUSIONS

The material presented in this chapter is concerned with the development of a precipitation-runoff model suitable for estimating water yield from a forested upland watershed. To ensure its applicability, the model is designed to integrate the major hydrologic processes and watershed characteristics that affect runoff generation from precipitation events. The most important processes considered are precipitation, interception, transpiration, evaporation, infiltration, and runoff. These processes have both temporal and spatial characteristics. The most important spatial watershed characteristics that can have a significant influence on the conversion of precipitation to runoff include watershed elevation, slope, aspect, soil type, and vegetation cover type and density. Integrating a number of dynamic and spatially varied processes and watershed characteristics to produce a reliable model is a complex and difficult process. However, the simplifying assumptions made and the particular steps followed in the modeling process in this study are meant to make the task easier.

Generally, the entire modeling process is handled in two parts resulting in two separate precipitation and water yield modeling schemes. A stochastic event-based approach is used to describe the precipitation patterns in the study area, while a deterministic approach is used to estimate the amount of water yield generated. In both cases, a spatial software technology ARC/INFO GIS is used to incorporate the spatial pattern of climatic and watershed characteristics into the precipitation event and water yield models. A water balance model is developed to estimate the amount of water yield. In this case, the task is made easier by selecting existing hydrologic models to estimate input values for the different components of the water yield model. The model is also designed to be flexible. Using the model, surface runoff (water yield) can be estimated for time intervals ranging from as small as 5 minutes to any length of a season, and from sizes as small as 30 m by 30 m watershed to a large stream basin.

The amount of water yield generated from the entire watershed is determined by first computing the runoff from each cell on an event basis. Then, the resulting runoff from each cell is routed downstream in a cascading fashion until it reaches the watershed outlet to give the total water yield from the entire watershed. The model can also sum up the runoff produced on a daily basis over the entire cold season to determine the total amount of seasonal water yield from the entire watershed. Using the information gained from the model, one can also determine the historical reliability or risk of depending on the forest watersheds to produce enough water for downstream users.

Even though there are some simplifying assumptions designed to make the model easy to handle, the overall performance of the model is satisfactory. For example, the model generated percent of precipitation that is converted to runoff is similar to the average historical amount of about 9 percent (Kelso, Martin and Mack 1973; Hibbert 1979). However, one simplifying assumption which should be corrected in the future is routing the runoff from cell to cell along the watershed without any loss. Other problems with the model are long CPU time and extensive data requirements. In the future, attempts will be made to make the model more efficient and perform well under small or inadequate data.

Acknowledgments

The research leading to this study was partially supported by the Bureau of Forestry Research, Northern Arizona University (ARZZ-NAU), and by a grant from the USDA McIntyre-Stennis (MS32) Program.

REFERENCES

Baker, Malchus B. Jr. (1982) Hydrologic regimes of forested areas in the Beaver Creek watershed. USDA Forest Service General Technical Report RM-90, Rocky Mountain Forest and Range Experiment Station, Fort Collins, Colorado.

Barros, A. P. and Lettenmaier, D. P. (1994) Dynamic modeling of orographically induced precipitation. *Reviews of Geophysics* 32(3): 265–84, American Geophysical Union.

Beschta, R. L. (1976) Climatology of the ponderosa pine type in central Arizona. Technical Bulletin 228, Agricultural Experiment Station, College of Agriculture, University of Arizona, Tucson.

Campbell, R. E. and Ryan, M. G. (1982) Precipitation and temperature characteristics of forested watersheds in central Arizona. USDA Forest Service General Technical Report RM-93, Rocky Mountain Forest and Range Experiment Station, Fort Collins, Colorado.

Dickinson, R. E., Henderson-Sellers, A., Rosenzweig, C., and Sellers, P. J. (1991) Evapotranspiration models with canopy resistance for use in climate models. *Agricultural and Forest Meteorology* 54: 373–88.

Dingman, S. L. (1994) *Physical Hydrology*. Macmillan College Publishing, New York.

Duckstein, L., Fogel, M. M., and Davis D. R. (1975) Mountainous winter precipitation: a stochastic event-based approach. In *Proceedings of the National Symposium on Precipitation Analysis for Hydrologic Modeling*, American Geophysical Union, 172–88.

Feddes, R. A., Kowalik, P. J., and Zaradny, H. (1978) *Simulation of Field Water Use and Crop Yield*. John Wiley, New York.

Feth, J. H. and Hem, J. D. (1963) Reconnaissance of headwater springs in the Gila River Drainage Basin, Arizona, U. S. Geologic Survey Water-Supply Paper 1619-H, Washington D. C.

Ffolliott, P. F., Gottfried, G. J., and Baker, M. B. Jr. (1989) Water yield from forest snowpack management: research findings in Arizona and New Mexico. *Water Resources Research* 25(9): 1999–2007.

Green, W. H. and Ampt G. A. (1911) Studies in soil physics. I: Flow of air and water through soils. *Journal of Agricultural Science* 4(1): 1–24.

Hanes, W. T., Fogel, M. M., and Duckstein, L. (1977) Forecasting snowmelt runoff: probabilistic model. *Journal of the Irrigation and Drainage Division*, ASCE 103(IR3): 343–55.

Hibbert, A. R. (1979) Vegetation management to increase flow in the Colorado River Basin, USDA Forest Service General Technical Report RM-66, Rocky Mountain Forest and Range Experiment Station, Fort Collins, Colorado.

Hogg, R. V. and Tanis, E. A. (1983) *Probability and Statistical Inference*. 2nd ed. Macmillan, New York.

Jarvis, P. G. (1976) The interpretation of the variations in leaf water potential and stomatal conductance found in canopies in the field. *Phil. Trans. R. Soc. Lon., Ser. B* 273: 593–610.

Keeping, E. S. (1995) *Introduction to Statistical Inference*. Dover Publications, New York.

Kelso, M. M., Martin, W. E., and Mack, L. E. (1973) *Water Supplies and Economic Growth in an Arid Environment: An Arizona Case Study*. University of Arizona Press, Tucson.

Kottas, J. F. and Lau, Hon-Shiang (1978) On handling dependent random variables in risk analysis. *Journal of the Operational Research Society* 29: 1209–17.

Law, A. M. and Kelton, W. D. (1982) *Simulation Modeling and Analysis*. McGraw-Hill, New York.

McCuen, R. H. and Snyder, W. M. (1986) *Hydrologic Modeling: Statistical Methods and Applications*. Prentice-Hall, Englewood Cliffs, New Jersey.

Oki, T. and Musiake, K. (1991) Spatial rainfall distribution at a storm event in mountainous regions, estimated by orography and wind direction. *Water Resources Research* 27(3): 359–69.

Rupp, D. E. (1995) Stochastic, event-based, and spatial modeling of upland watershed precipitation-runoff relationships. Master's thesis, Northern Arizona University, Flagstaff.

Rush, R. W. (1965) Report of geological investigations of six experimental drainage basins, unpublished report. Rocky Mountain Forest and Range Experiment Station, Flagstaff, Arizona.

Rush, R. W. and Smouse, De Forrest (1968) Geological investigations of experimental drainage basins 15–18 and Bar-M Canyon, Beaver Creek Watershed, Coconino County, Arizona, unpublished report. Cooperative project by the Rocky Mountain Forest and Range Experiment Station, Northern Arizona University, and the Museum of Northern Arizona, Flagstaff.

Salvucci, G. D. and Entekhabi D. (1994) Explicit expressions for Green-Ampt (delta function diffusivity) infiltration rate and cumulative storage. *Water Resources Research* 30(9).
Schmeiser, B. W. and Lal, R. (1982) Bivariate gamma random vectors. *Operations Research* 30: 355–74.
Stewart, J. B. (1988) Modelling surface conductance of a pine forest. *Agricultural and Forest Meteorology* 43: 19–35.
Tecle, A. and Rupp, D. E. (1995) Stochastic, event-based, and spatial modeling of cold-season precipitation. In *Mountain Hydrology: Peaks and Valleys in Research and Applications*. Canadian Society for Hydrological Sciences, Vancouver, B.C.
USFS (1992) Terrestrial Ecosystem Survey of the Coconino National Forest, Draft Report, USDA Forest Service, Region 3.
Wadsworth, H. M. (1990) *Handbook of Statistical Methods for Engineers and Scientists*. McGraw-Hill, New York.
Wigmosta, M. S., Vail, L. W., and Lettenmaier, D. P. (1994) A distributed hydrology-vegetation model for complex terrain. *Water Resources Research* 30(6): 1665–97.
Williams, J. A. and Anderson, T. C. Jr. (1967) Soil Survey of Beaver Creek area, Arizona. USDA Forest Service and Soil Conservation Service, and Arizona Agricultural Experiment Station, U.S. Government Printing Office, Washington, D.C.

APPENDIX: NOTATION OF SYMBOLS

ΔS = change in either ground surface, or interception storage (cm)
Δt = change in time (hours)
P = rate of precipitation (cm hr^{-1})
P_r = the depth of rain (cm)
P_g = the rate of precipitation reaching the ground (cm hr^{-1})
P_{gi} = the rainfall reaching the ground in cell i (cm)
Y = net runoff (cm hr^{-1})
Y_i = the net surface runoff from cell i (cm)
E = evaporation rate (cm hr^{-1})
E_p = potential evaporation rate (cm hr^{-1}, or s^{-1})
Tr = transpiration (cm hr^{-1})
$F(\tau)$ = cumulative distribution function
τ = time between storm sequences
β = mean of t, the time between sequences
$f(\tau)$ = the pdf of the gamma distribution function
α = the shape parameter of the gamma function and is equal to m^2/s^2
β = the scale parameter of the gamma function and is equal to s^2/m
$\Gamma(\alpha)$ = the gamma function
μ = the population mean of the data being considered
σ^2 = the variance of the data being considered
$f(\tau_1)$ = the gamma pdf for duration
τ_1 = duration of a precipitation event (hr)
τ_2 = the depth of a precipitation event (cm)
α_1 = the shape parameter for the gamma distribution of duration
β_1 = the scale parameter for the gamma distribution of duration
α_2 = the shape parameter of the gamma distribution for depth conditional upon the duration
β_2 = the scale parameter of the gamma distribution for depth conditional upon the duration
μ_2 = mean of precipitation depth (cm)
σ^2_2 = variance of precipitation depth (cm^2)
b_0, b_1, c_o, and c_1 = regression coefficients
δ = constant
S_{max} = maximum interception storage capacity (in cm)
LAI_p = the projected leaf area index
F = the overstory cover as a fraction of total surface area
Δ = slope of saturation vapor pressure vs. air temperature ($\text{mb}\,^\circ\text{C}^{-1}$)
ρ_a = density of air (g cm^{-3})
c_a = heat capacity of air ($\text{cal g}^{-1}\,^\circ\text{C}^{-1}$)
C_{at} = atmospheric conductance (cm s^{-1})
e_s = saturation vapor pressure (mb)
c_w = heat capacity of water (in $\text{cal g}^{-1}\,^\circ\text{C}^{-1}$)
e = actual vapor pressure (mb)
ρ_w = density of water (g cm^{-3})
λ_v = latent heat of vaporization (cal g^{-1})
γ = psychrometric constant (cal g^{-1})
v_a = wind speed (cm s^{-1})
z_d = $0.7z_{veg}$ (m)
z_0 = $0.1z_{veg}$ (m)
z_m = height of wind speed measurement (m)
z_{veg} = height of vegetation (m)
C_{can} = canopy resistance
w = water reaching the surface from above (cm hr^{-1})
t = the duration of a precipitation event (hr)
t_p = time of ponding (hr)
t_f = duration of precipitation or snowmelt (hr)
$I(t_f)$ = *Green and Ampt's* total amount of infiltration
$I'(t_f)$ = approximate total amount of infiltration at time t_f
$I'(t)$ = the approximate total infiltration at time t
$I'(t_p)$ = approximate total amount of infiltration at time t_p
x = $|\psi_f|(\phi-\theta_0)/K_{sat}$
K_{sat} = saturated hydraulic conductivity (cm hr^{-1})
ψ_f = the effective tension at the wetting front (cm)
ϕ = saturated soil moisture content (cm)
θ_0 = the initial water content of the soil (cm)
M = daily snowmelt depth (cm)
a_r = restricted degree-day factor ($\text{cm d}^{-1}\,^\circ\text{C}^{-1}$)
T_{avg} = daily average temperature (°C)
m_q = conversion factor for energy flux density to snowmelt depth (cm d^{-1} $(\text{cal cm}^{-2}\text{d}^{-1})^{-1}$)
R_n = net radiation ($\text{cal cm}^{-2}\text{d}^{-1}$, or s^{-1})
Q_r = energy advected by rain (cal cm^{-2})

T_r = rain temperature (°C)
M_i = snowmelt in cell i (cm)
n = number of cells
m_1 = mean storm precipitation duration (hr)
m_2 = mean storm precipitation depth (cm)
$s^2{}_1$ = variance of storm precipitation duration (hr^2)
$s^2{}_2$ = variance of storm precipitation depth (cm^2)
r = correlation coefficient between precipitation storm depth and duration
P_i/P_{38} = ratio of precipitation depth at gage i to that at gage 38
D^i/D^{38} = ratio of precipitation duration at gage i to that at gage 38
D^i = duration of precipitation at gage i (hr)
D^{38} = duration of precipitation at gage 38 (hr)
P_i = precipitation depth at gage i (cm)
P_{38} = precipitation depth at gage 38 (cm)
$ELEV$ = gage elevation in meters
ASP = aspect in degrees from prevailing windward direction
$UTMX$ = gage location in x-direction in terms of Universal Transverse Mercator
$UTMY$ = gage location in y-direction in terms of Universal Transverse Mercator
d = random amount of water yield
d_* = minimum amount of water yield that satisfies downstream demand
$F(d)$ = cumulative distribution function of water yield
$f(d)$ = probability distribution function of water yield
$R(d)$ = reliability function of water yield
n_t = total number of seasons
n_s = number of seasons watershed failed to meet downstream water demand
P_f = cumulative frequency distribution function of water yield

12 Regional assessment of the impact of climate change on the yield of water supply systems

RICHARD M. VOGEL,* CHRISTOPHER J. BELL,** RANJITH R. SURESH,* AND NEIL M. FENNESSEY***

ABSTRACT

Investigations of the impact of climate change on water resources systems usually involve detailed monthly hydrological, climatological, and reservoir systems models for a particular system. The conclusions derived from such studies only apply to the particular system under investigation. This study explores the potential for developing a regional hydroclimatological assessment model useful for determining the impact of changes in climate on the behavior of water supply systems over a broad geographic region. Computer experiments performed across the United States reveal that an annual streamflow model is adequate for regional assessments which seek to approximate the behavior of water supply systems. Using those results, a general methodology is introduced for evaluating the sensitivity of water supply systems to climate change in the northeastern United States. The methodology involves the development of a regional hydroclimatological model of annual streamflow which relates the first two moments of average annual streamflow to climate and drainage area at 166 gaging stations in the northeastern United States. The regional hydroclimatological streamflow model is then combined with analytic relationships among water supply system storage, reliability, resilience, and yield. The sensitivity of various water supply system performance indices such as yield, reliability, and resilience are derived as a function of climatical, hydrological, and storage conditions. These results allow us to approximate, in general, the sensitivity of water supply system behavior to changes in the climatological regime as well as to changes in the operation of water supply systems.

12.1 INTRODUCTION

Usually investigations of the relationship among climate, streamflow, and water supply combine general circulation models of the atmosphere (GCMs), rainfall-runoff models, and reservoir operations models to explore potential impacts of climate change on the behavior of a particular water supply system. Investigations of this type are too numerous to summarize here; instead, some recent comprehensive reviews may be found in Ballentine and Stakhiv (1993), Leavesley (1994), and Loaiciga et al. (1996). Much of this literature describes the nesting of several detailed simulation models including a GCM, a rainfall-runoff model, and often a reservoir operations model for the purpose of performing climate sensitivity analysis. Though validation procedures exist specifically for testing such nested hydroclimatological modeling schemes (Klemeš 1986), they are rarely applied and as a result, the complexity of the models along with associated issues of parameter and model uncertainty render most of the results questionable (Klemeš 1990). This study takes a different approach.

The behavior of many complex water supply systems is controlled primarily by year-to-year variations in hydrology and climate. Such systems, termed overyear systems, can be modeled using an annual time-scale, leading to remarkably simple modeling approaches which exploit both at-site and regional information about climate, streamflow, and storage. One objective of this study is to demonstrate that an annual time-scale provides an adequate approximation to the storage-yield behavior of a very wide class of water supply systems across the continental United States.

Hydroclimatical models with an annual time-scale are extremely simple when compared to models with monthly, weekly, or daily time-scales. The much simpler structure associated with models of annual streamflow allows us to develop more general regional models that apply to a broad geographical region rather than to a single specific river basin, as is the case for monthly, weekly, and daily models. Vogel, Bell, and Fennessey (1997) use numerous case studies to document that the methodology outlined here can be used to evaluate the sensitivity of water supply systems to climate change leading to approximately the same results as those obtained from

* Department of Civil and Environmental Engineering, Tufts University, Medford, MA 02155.

** Hughes STX Corporation, Hydrological Sciences Branch, NASA Goddard Space Flight Center, Greenbelt, MD 20771.

*** Department of Civil and Environmental Engineering, University of Massachusetts, North Dartmouth, MA 02747.

much more detailed hydroclimatological investigations that use monthly, weekly, or even daily time intervals. This study is similar to the studies by Schaake (1990) for the southeastern United States and by Rogers and Fiering (1990) which developed general graphical relations between the behavior of water supply systems and climate.

Our approach is empirical because we exploit observed relationships between streamflow and climate similar to the empirical study by Langbein et al. (1949) and the study by Revelle and Waggoner (1983) for the western United States which is based on Langbein's relations. Leavesley (1994) argues that since such empirical approaches only reflect climate and basin conditions during the time period in which they were developed, it is questionable to extend those relations to climate conditions different from those used in their development. The same criticism may be raised for any hydrological model that is not tested for geographical and climate transposability using proxy-basin and differential split-sample tests of the kinds recommended by Klemeš (1986). Our approach is to develop an empirical regional hydroclimatological model based on an extremely broad range of climate conditions, so that it would not be an extrapolation to perturb it with modest variations in climate. This chapter provides a cursory summary of our approach; more complete details are summarized in Vogel et al. (1997).

12.2 MODEL TIME-STEP FOR REGIONAL ASSESSMENT

Detailed site-specific hydroclimatological evaluations tend to employ a monthly, weekly, or even daily time-step for model simulations. Regionalization of such models would be difficult due to the complexity of the periodic (seasonal) processes. In this section we compare the behavior of water supply systems using both annual and monthly time-steps. Storage-yield curves are computed using historic annual and monthly streamflow traces at ten sites scattered across the United States. The sites are selected to reflect the complete range of variability exhibited by flow sequences as illustrated by the coefficient of variation (C_v) of the annual flow series, which ranges from 0.23 to 0.85 for these sites. Table 12.1 summarizes the name, location, record length, drainage area, annual average precipitation, runoff ratio, and coefficient of variation of the annual streamflows (C_v) for the ten U.S. Geological Survey streamflow gages. Figure 12.1 compares the storage-yield curves for these ten basins based on monthly and annual flow series. Storage-yield curves are computed using the double-cycling sequent peak algorithm which is equivalent to the use of a storage mass curve. The sequent peak algorithm estimates the minimum reservoir storage capacity that would be required to deliver the specified yield, without failure, over the historic period. Along the abscissa in Figure 12.1, we plot both level of development, α, ($\alpha = Y/\mu$) and the standardized inflow, m, which is defined as

$$m = \frac{(1-\alpha)\mu}{\sigma} = \frac{(1-\alpha)}{C_v} = \frac{\mu - Y}{\sigma} \tag{12.1}$$

where μ and σ are the mean and standard deviation of the annual inflows and α is the yield as a fraction of μ, so $Y = \alpha\,\mu$ is the constant annual yield, and C_v is the coefficient of variation of the annual inflows ($C_v = \sigma/\mu$). Both m and α are surrogates of system yield, because as yield Y increases, α increases and m decreases; however, since they are standardized, they are easier to generalize across systems. Along the ordinate in Figure 12.1 we plot the storage ratios S/μ.

As one would expect, storage-yield curves based on monthly flows are always greater than storage-yield curves based on annual flows because monthly storage-yield curves include both within-year and over-year storage requirements. Figure 12.1 illustrates that at sites with high streamflow variability ($C_v >$ 0.3), storage-yield curves based on annual flows provide an approximation to storage-yield curves based on monthly flows, with the approximation improving as C_v increases. The two sites with very low values of C_v ($C_v < 0.3$) lead to annual and monthly storage-yield curves, which are quite different. Figure 12.2 illustrates the difference between the storage ratio computed using annual flows S_a/μ and monthly flows S_m/μ as a percentage of S_m/μ. Figure 12.2 illustrates that as long as $m < 1$, the percentage difference between storage ratios is usually less than about 30 percent with that difference dropping as C_v increases. Figures 12.1 and 12.2 demonstrate that the use of an annual time-step provides a reasonable approximation (to within 30 percent) to monthly storage-yield curves as long as the standardized inflow m is less than approximately 1. This result supports the conclusions of other studies (Vogel and Hellstrom 1988; Vogel and Bolognese 1995; Vogel, Fennessey, and Bolognese 1995) which recommend the use of annual models when $0 < m < 1$. The standardized inflow m decreases as yield $Y = \alpha\,\mu$ increases and as C_v increases, hence overyear storage results from high yields, high C_v, or both.

12.3 DEVELOPMENT OF REGIONAL HYDROCLIMATOLOGICAL STREAMFLOW MODEL

The following sections describe the development of a regional hydroclimatological model of annual streamflow for the northeastern United States.

Table 12.1. *Description of ten sites used to compare annual and monthly storage-yield curves*

USGS No.	Site name	State	Length of record (years)	Drainage area km^2 (miles^2)	Annual rainfall m (inches)	Runoff rainfall	C_v
01144000	White River at West Hartford	Vermont	60	1787.1 (690)	1.04 (41)	0.55	0.23
12027500	Chehalis River near Grand Mound	Washington	59	2318 (895)	1.60 (63)	0.67	0.25
09239500	Yampa River at Steamboat Springs	Colorado	77	1564.4 (604)	0.64 (25)	0.42	0.28
07375500	Tangipoha River at Robert	Louisiana	50	1673.1 (646)	1.52 (60)	0.40	0.34
07072000	Elevenpoint River near Ravenden Springs	Arkansas	53	2937.1 (1134)	1.09 (43)	0.31	0.41
06933500	Gasconade River at Gerome	Missouri	65	7355.6 (2840)	1.07 (42)	0.29	0.48
06810000	Nishnabotna River above Hamburg	Indiana	60	7267.5 (2806)	0.76 (30)	0.18	0.65
10322500	Humboldt River at Palisade	Nevada	75	12975.9 (5010)	0.23 (9)	0.12	0.75
11152000	Arroyo Seco River near Soledad	California	87	632 (244)	0.86 (34)	0.28	0.81
08146000	San Saba River at San Saba	Texas	73	7889.1 (3046)	0.66 (26)	0.04	0.85

12.3.1 Regional hydroclimatological database

Our approach is semi-empirical and our ability to relate streamflow, climate, and water supply results from extensive use of regional streamflow, geomorphic, and climate data for the northeastern United States; hence we begin by describing those information resources. A listing of the streamflow, geomorphic, and climate data may be found in Fennessey (1994) and Bell (1995).

Streamflow database

The streamflow dataset consists of records of average daily streamflow at 166 sites located in the northeastern United States with drainage areas ranging from less than 2 mi^2 (5.18 km^2) to nearly 7,000 mi^2 (18,130 km^2). This dataset is a subset of the Hydroclimatological Data Network (HCDN) available on CD-ROM from the U.S. Geological Survey (Slack, Lumb, and Landwehr 1993). Figure 12.3 illustrates the location of the 166 sites. The streamflow records at these 166 basins are all in excess of twenty years long and contain no overt regulation due to diversion, augmentation, storage, or groundwater pumping.

Climate database

Daily precipitation and temperature data for the period-of-record, October 1, 1951 through September 30, 1981, were obtained from 323 NOAA Summary-of-the-Day climate stations. These data were then aggregated into annual values for those years with no missing data. Mean annual precipitation, P, and temperature, T, values were then estimated. The mean values of P and T were interpolated to the 166 streamgage location coordinates using an algorithm developed by Fennessey (1994). Using climate data from the nearest five climate stations, annual mean P and T values were interpolated to a given streamgage with interpolation weights proportional to the record length and inversely proportional to the square of the distance between the climate station and the streamflow gage.

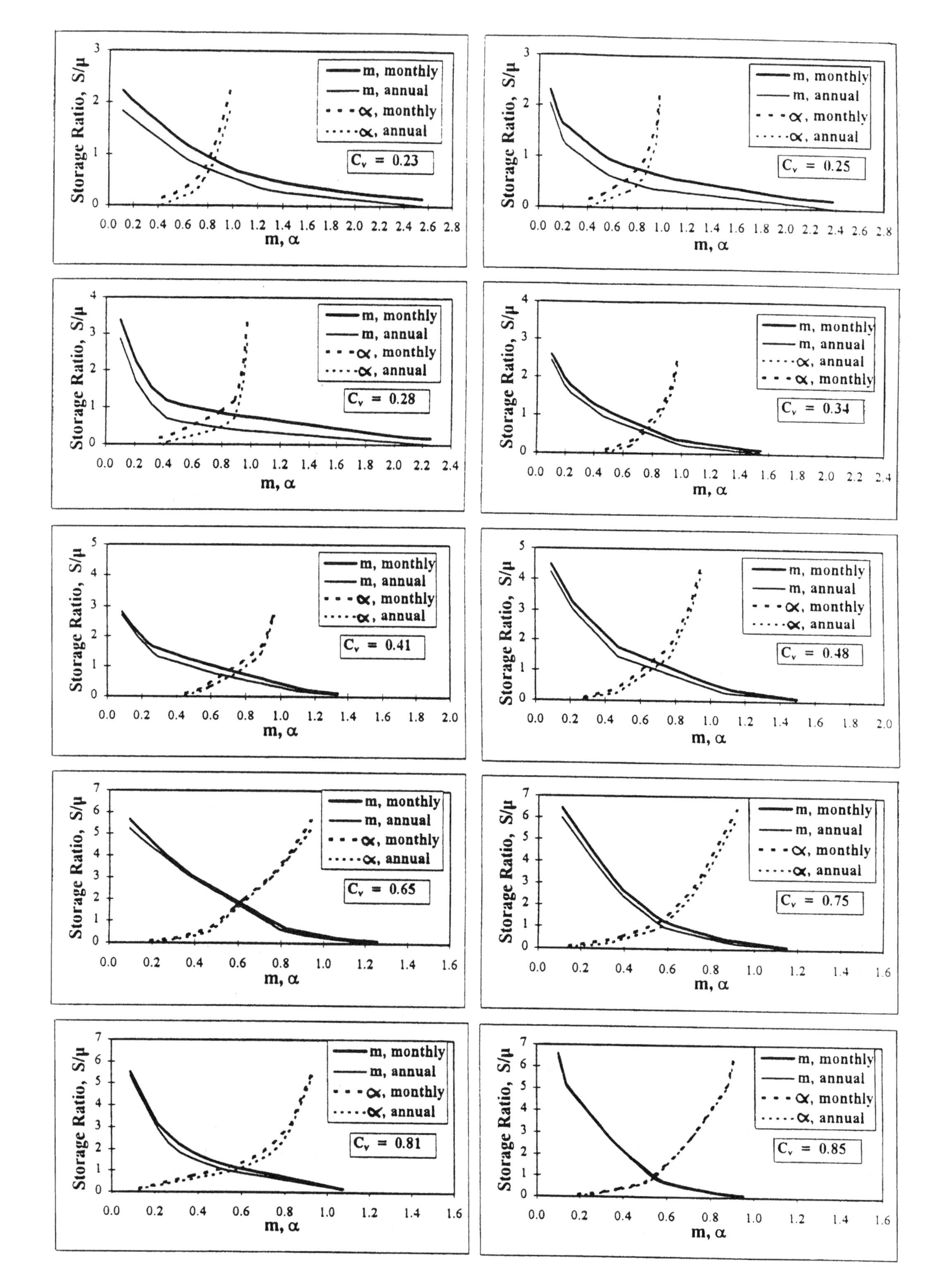

Figure 12.1. Comparison of storage-yield curves based on annual and monthly time-steps for the ten sites summarized in Table 12.1.

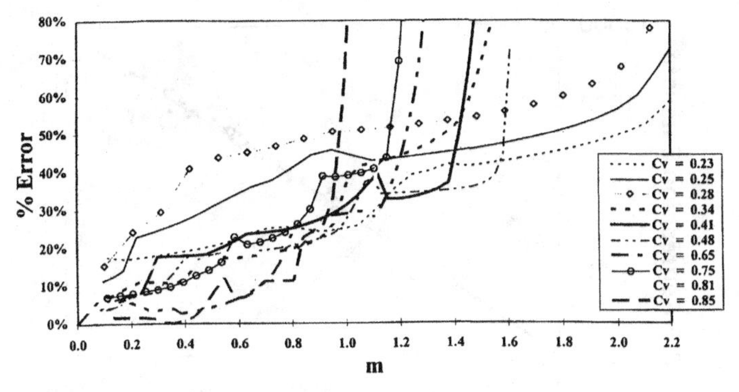

Figure 12.2. The percentage difference between estimates of storage capacity based on monthly and annual time-step as a function of the standardized inflow m.

12.3.2 Regional hydroclimatological regression models of annual streamflow

This section describes the development of regional hydroclimatological relationships for the mean, μ, and standard deviation, σ, of annual streamflow. Regional regression procedures led to multivariate relationships between μ and σ and mean annual precipitation P, mean annual temperature T, and watershed area A. Weighted least squares (WLS) multivariate linear regression procedures were used to fit models of the form

$$\theta = e^{c_0} \cdot A^{c_1} \cdot P^{c_2} \cdot T^{c_3} \cdot e^{\varepsilon}, \tag{12.2}$$

where θ represents either of the dependent variables μ or σ, the independent variables A, P, and T are drainage area, precipita-

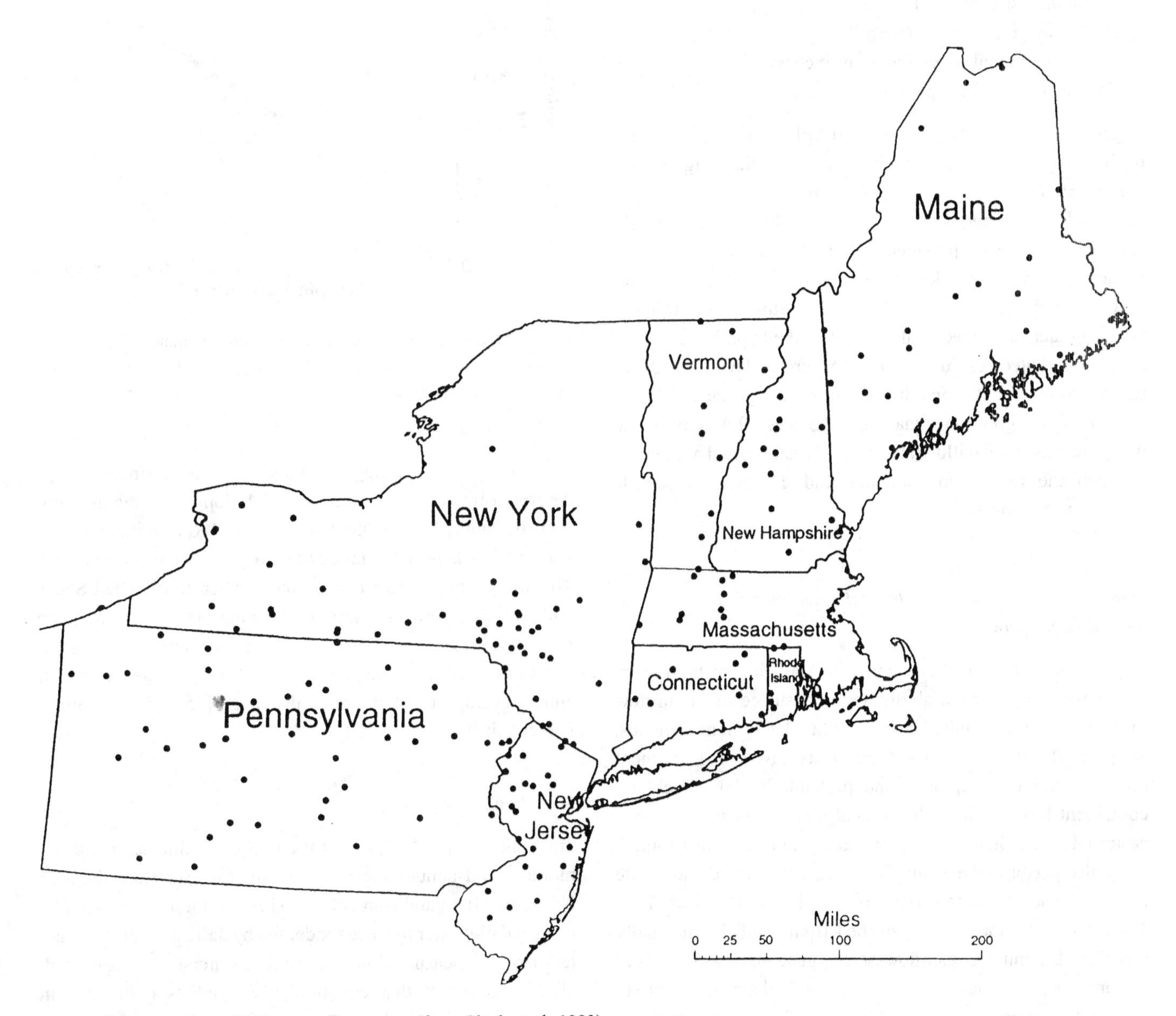

Figure 12.3. Location of 166 streamflow gages (from Slack et al. 1993).

tion, and temperature, respectively, and the ε are normally distributed errors with zero mean and variance, σ_e^2. WLS procedures account for the fact that each basin has a different streamflow record length leading to estimates of the dependent variable with varying degrees of reliability. The resulting regional regression models are:

$$\hat{\mu} = A^{1.000}P^{1.177}T^{(-1.005)} \tag{12.3}$$

$$\hat{\sigma} = 0.0124A^{0.978}P^{0.976}. \tag{12.4}$$

where

$\hat{\mu}$ = WLS regression estimate of mean annual streamflow in cfs;
$\hat{\sigma}$ = WLS regression estimate of standard deviation of annual streamflow in cfs:
A = drainage area in square miles;
P = mean annual precipitation in inches;
T = mean annual temperature in °F;

Vogel et al. (1997) provide further details on the development of these regression equations, including how the weights were computed. The t-ratios of the model parameters in equations 12.3 and 12.4 range from [7.9–129], indication of remarkably precise model parameter estimates. Bell (1995) uses influence statistics to document that none of the 166 sites used to develop equations 12.3 and 12.4 exhibit unusual leverage or influence and we could not reject the hypothesis of normally distributed residuals using a 5 percent level hypothesis test on normality. Perhaps the best summary of the goodness-of-fit of the regression equations 12.3 and 12.4 is provided in Figure 12.4, which illustrates the remarkably good agreement between the regression estimates and the original sample estimates of μ and σ.

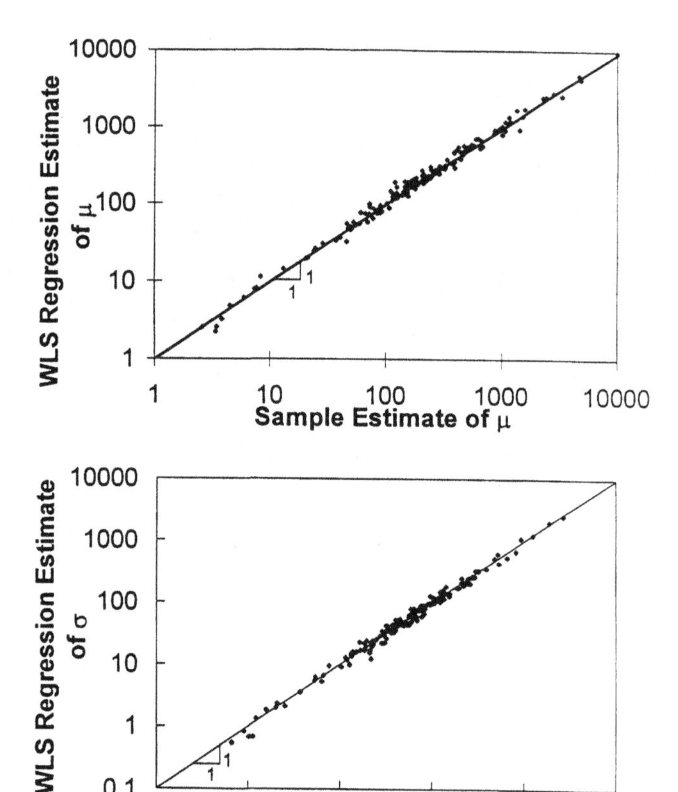

Figure 12.4. Comparison of WLS regression estimates of μ (equation 12.3) and σ (equation 12.4) with original sample estimates of μ and σ at 166 sites.

Summary of regional hydroclimatological model of annual streamflow

Vogel et al. (1995) perform numerous statistical tests to show that the year-to-year variability and persistence of streamflow in the northeastern United States is remarkably homogeneous. Vogel et al. (1995) use hypothesis tests based on L-moment ratios, L-moment diagrams, and probability plot correlation coefficient tests to show that annual streamflow in the northeastern United States is approximately normally distributed. They also performed a Monte Carlo experiment to demonstrate that the observed time-series of annual streamflow at these 166 sites could not be distinguished from synthetic, normally distributed, annual streamflow series generated with a fixed lag-one serial correlation coefficient $\rho = 0.19$ across the broad geographic region shown in Figure 12.3. In summary, the regional hydroclimatological model of annual streamflow is described by equations 12.3 and 12.4 along with the assumption that annual streamflow follows an AR(1) normal model with a fixed lag-one serial correlation coefficient of $\rho = 0.19$. This model applies to basins in the northeastern United States with values of drainage area in the range [1.5, 6,780] square miles ([3.89, 17,560] km^2), values of annual average precipitation in the range [31, 63] inches ([0.79, 1.6] m), and values of annual average temperature in the range [35.4, 54.5] degrees Fahrenheit ([1.9, 12.5] °C).

Model validation

Tung and Haith (1995) report the impact of climate change on monthly and annual watershed runoff for four watersheds in New York. Tung and Haith (1995) simulate monthly streamflow using a daily water balance model fed by daily precipitation and temperature measurements for the four watersheds. Vogel et al. (1997) document that equation 12.3 predicts reductions in annual streamflow under future climate conditions, of about the same magnitude as the procedure employed by Tung and Haith, yet our approach can be applied in a fraction of the time required by Tung and Haith.

12.4 STORAGE-RELIABILITY-RESILIENCE-YIELD RELATIONSHIPS

Vogel and Bolognese (1995) and others have introduced analytic relationships which approximate the behavior of water supply systems dominated by overyear or carryover storage requirements. Simple analytic storage-reliability-resilience-yield (SRRY) relationships are not intended to replace more detailed simulation studies; rather, they are intended to improve our understanding of the behavior of water supply systems, to allow for comparisons among systems, and for performing regional assessments of water supply.

Vogel and Bolognese (1995, Appendix A) summarize relations among reservoir system storage capacity S, planning horizon N, standardized inflow index m, lag-one serial correlation of the inflows ρ, and N-year no-failure reliability R_N, in the form

$$S/\sigma = f(m, N, \rho, R_N) \tag{12.5}$$

for systems fed by AR(1) normally distributed inflows, with m defined in equation 12.1. Equation 12.5 was developed from Monte Carlo experiments which routed AR(1) streamflows through a reservoir using the sequent peak algorithm and is too complex to reproduce here. Vogel and Bolognese (1995, equation 17) show that N-year no-failure reliability R_N can be related to annual reliability R_a using

$$R_N = R_a\left(1 - r\left(1 - R_a^{-1}\right)\right)^{N-1} \tag{12.6}$$

where r is an index of resilience which can be estimated using

$$r = \Phi\left[\frac{1}{\sqrt{1-\rho^2}}\left(m - \frac{\rho}{\Phi(-m)\cdot exp\left(m^2/2\right)\cdot\sqrt{2\pi}}\right)\right] \tag{12.7}$$

where $\Phi(arg)$ denotes the cumulative normal density function applied at arg. The index r is defined as the probability that the reservoir system will be able to meet the stated yield Y, in a year following a failure year. Here a failure year is one in which the reservoir system is unable to deliver its pre-specified yield Y. Vogel and Bolognese (1995) document that equations 12.5–12.7 reproduce theoretical storage-reliability-yield relationships introduced by other investigators for overyear systems fed by AR(1) inflows. Vogel et al. (1995) document that equations 12.5–12.7 are useful for describing the behavior of the water supply systems that service New York City; Providence, Rhode Island; Springfield, Massachusetts, and Boston, Massachusetts, metropolitan areas.

12.5 SENSITIVITY OF WATER SUPPLY SYSTEM BEHAVIOR TO CLIMATE CHANGE

One goal of this study is to document that the simple annual regional hydroclimatological model (equations 12.3 and 12.4) when linked with SRRY relations (equations 12.5–12.7) can be used to evaluate the sensitivity of complex reservoir systems and river basins to potential climate change. Another goal is to document that our simple approach is comparable to much more costly and complex operations studies which can take months or even years to complete. The following sections evaluate the ability of our approach to meet these goals. We begin by describing another case study which compares our approach with a more detailed monthly model.

12.5.1 Validation of our overall methodology

Kirshen and Fennessey (1995) describe a recent investigation of the impact of climate change on the water supply system which services the Boston metropolitan area. They used the Sacramento soil moisture accounting model, the National Weather Service river forecast system snow accumulation and ablation model, modified Penman equation estimates of reservoir evaporation, and Penman-Monteith estimates of potential evapotranspiration to generate monthly streamflows. The simulated streamflows were routed through a monthly reservoir operations model summarized by Vogel and Hellstrom (1988). Kirshen and Fennessey (1995) used the GCM model output to provide estimates of monthly temperature and precipitation as well as monthly incident solar radiation, wind speed, and specific humidity under future potential climate conditions. Those perturbed climate values were then used to determine the influence of potential climate change on water supply system yield.

Vogel et al. (1997) document the application of equations 12.3–12.7 using values of annual average precipitation and temperature obtained from the same GCM model runs used by Kirshen and Fennessey (1995). Vogel et al. (1997) found that the methodology outlined here led to approximately the same results as obtained by Kirshen and Fennessey (1995) for this particular system. This comparison serves to validate our overall methodology for determining the influence of climate change on the behavior of a water supply system that is dominated by carryover storage requirements.

12.5.2 The general sensitivity of water supply yield to changes in climate

Since the hydroclimatological and water supply behavior of river basins is described in simple analytic terms (equations 12.3–12.7), it is possible to derive expressions that describe the sensitivity of water supply system yield or resilience to various inputs such as precipitation and temperature. This is accomplished using the chain rule. For example, the sensitivity of system yield Y to changes in annual average precipitation is obtained by deriving dY/dP. Rearranging (1) to obtain $Y = \mu - m\sigma$, and applying the chain rule, yields

$$\frac{dY}{dP} = \left[\frac{\partial Y}{\partial m}\cdot\frac{\partial m}{\partial P}\right] + \left[\frac{\partial Y}{\partial \mu}\cdot\frac{\partial \mu}{\partial P}\right] + \left[\frac{\partial Y}{\partial \sigma}\cdot\frac{\partial \sigma}{\partial P}\right] \tag{12.8}$$

Noting that $\partial Y/\partial m = -\sigma$, $\partial Y/\partial \mu = 1$ and $\partial Y/\partial \sigma = -m$ leads to

$$\frac{dY}{dP} = \left[-\sigma\cdot\frac{\partial m}{\partial P}\right] + \left[\frac{\partial \mu}{\partial P}\right] - \left[m\cdot\frac{\partial \sigma}{\partial P}\right] \tag{12.9}$$

with $\partial\mu/\partial P$ and $\partial\sigma/\partial P$ easily derived from equations 12.3 and 12.4, respectively. The challenge was to derive $\partial m/\partial P$ using the fact that

$$\frac{\partial m}{\partial P} = \frac{\partial m}{\partial S}\cdot\frac{\partial S}{\partial P} = \frac{1}{\frac{\partial S}{\partial m}}\cdot\frac{\partial S}{\partial P} \tag{12.10}$$

where the terms $\partial S/\partial m$ and $\partial S/\partial P$ are obtained from chain rule calculations similar to equation 12.8 using the analytic relations for S described by equation 12.5 and given in Vogel and Bolognese (1995, Appendix A). The details of these additional derivations are given in Bell (1995).

To reduce the number of dimensions in our evaluation of yield sensitivity dY/dP, we derive the yield sensitivity, per unit area, which we denote $d(Y/A)/dP$. Since the regression equations 12.3 and 12.4 for μ and σ, respectively, are both approximately linear in drainage area A, there is no loss of generality in examining $d(Y/A)/dP$ rather than dY/dP. Figure 12.5 illustrates the sensitivity of water supply yield to changes in precipitation $d(Y/A)/dP$ for various values of precipitation and temperature. Figure 12.5 assumes an annual reliability $R_a = 0.99$ and a serial correlation of annual flows $\rho = 0.2$. As expected, dY/dP is always positive, hence yield tends to increase as the level of development α increases and as precipitation increases. However, Figure 12.5 illustrates that for a given level of development, increases in precipitation produce greater increases in yield at lower temperatures than at higher temperatures. This is due to the increased evaporation and evapotranspiration (ET) which results from higher temperatures. Each curve in Figure 12.5 represents a fixed climate or a fixed value of T and P. Figure 12.5 illustrates that for any given climate (curve), an increase in precipitation will lead to higher yields as the level of development increases because less reservoir spillage occurs as the level of development increases.

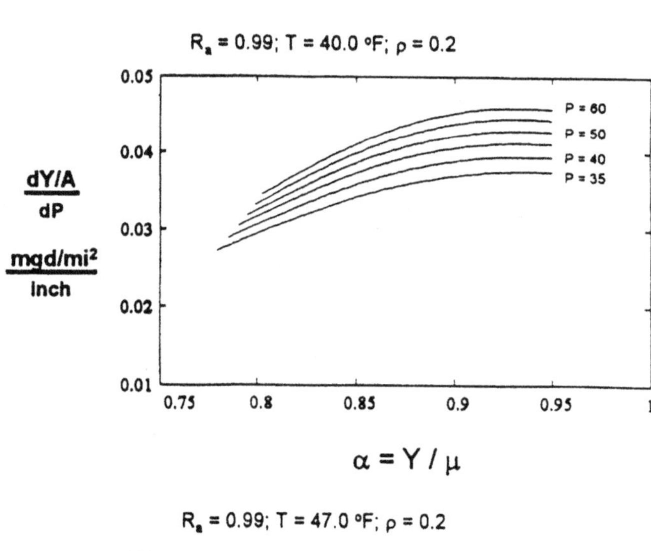

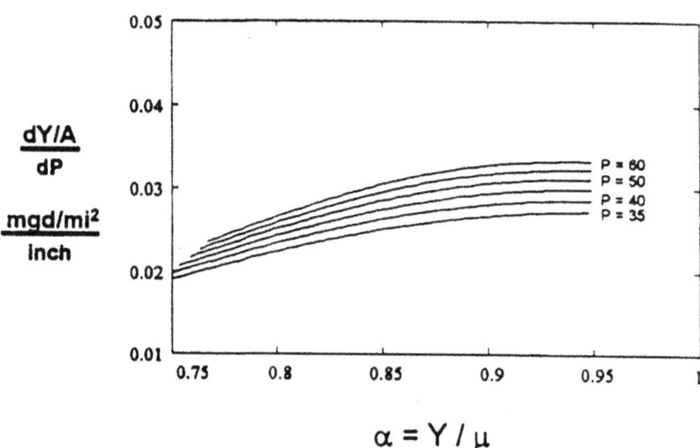

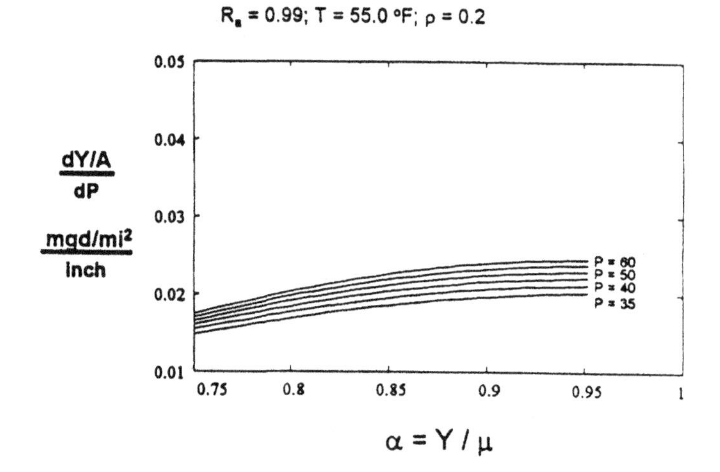

Figure 12.5. The generalized sensitivity of overyear water supply yield to changes in precipitation in the northeastern United States.

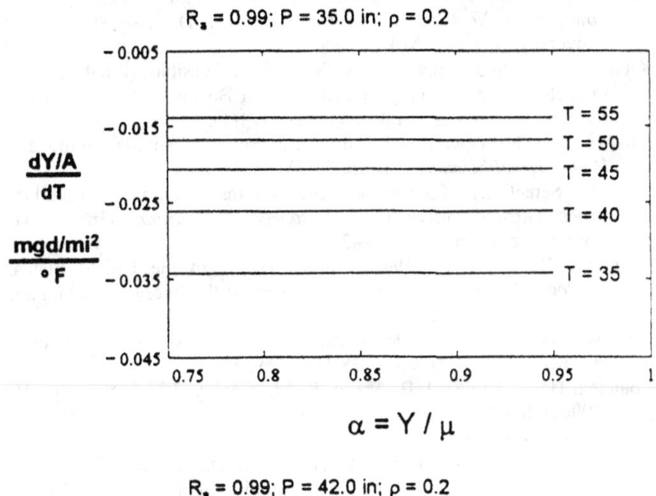

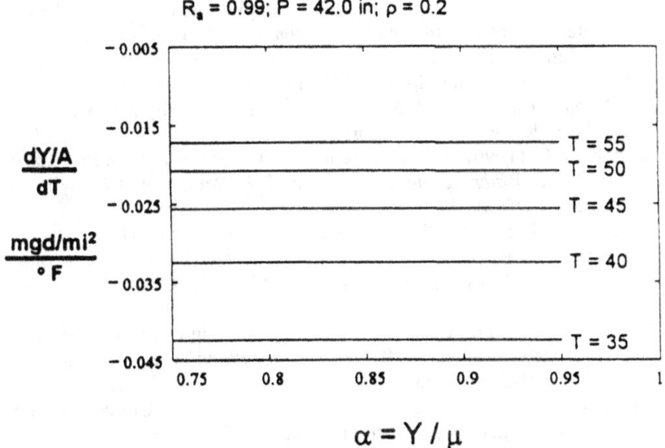

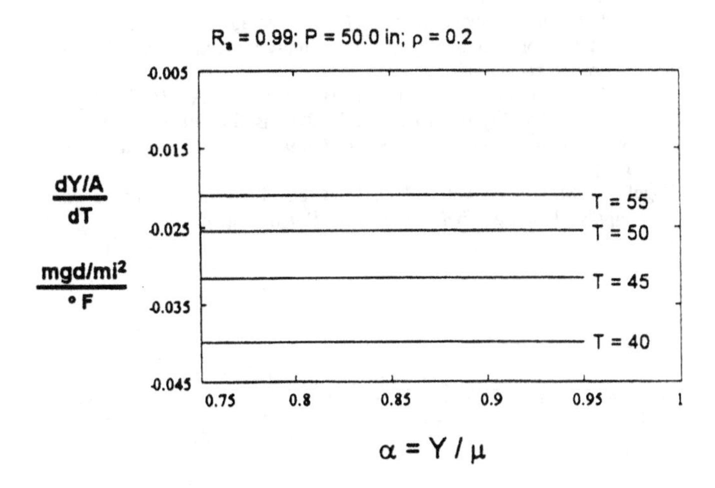

Figure 12.6. The generalized sensitivity of overyear water supply yield to changes in temperature in the northeastern United States.

The same approach was used to derive an expression for the sensitivity of yield to changes in temperature $d(Y/A)/dT$. Figure 12.6 illustrates the sensitivity of water supply yield to changes in temperature $d(Y/A)/dT$ for various values of precipitation and temperature. In this case, the yield sensitivity to temperature is always negative because increases in temperature produce higher ET leading to decreases in yield. Figure 12.6 illustrates that for a given climate, yield sensitivity to temperature is constant for all levels of development. This is because temperature only influences the mean annual inflow and not the variability of the inflows (see equations 12.3 and 12.4) in this region.

12.6 CONCLUSIONS

This study has sought to improve our understanding of the general relationships among climate, streamflow, and water supply system behavior and to develop a general methodology suitable for use in regional scale assessments of the impact of climate change on water supply. An initial experiment comparing storage-yield curves at ten basins scattered across the United States revealed that an annual time-step is adequate for modeling (approximately) the behavior of storage reservoirs for most regions of the United States. This approximation is particularly accurate for regions with high coefficient of variation of annual streamflow and/or for systems with high yields (see Figures 12.1 and 12.2). Figure 12.2 documents that as long as $m < 1$ (equation 12.1), an annual time-step provides a good approximation (to within 30 percent) to storage-yield curves. On the basis of these results, a general methodology is introduced for modeling the regional relationships among climate, streamflow, and water supply at the annual level.

Our approach involved the development of empirical regional relationships between annual streamflow, climate and drainage area using data available at 166 river basins in the northeastern United States. The resulting regional hydroclimatological model of annual streamflow was shown to be remarkably precise over this very broad geographic region. Vogel et al. (1996) document that the methodology introduced here compares favorably with much more detailed site-specific modeling studies performed by Tung and Haith (1995) for four watersheds in New York.

The idea behind developing a simple regional annual hydroclimatological streamflow model was to combine it with analytic relationships among storage, reliability, resilience, and yield (SRRY). For this purpose, we exploit the SRRY relationships recently introduced by Vogel and Bolognese (1995). Vogel et al. (1997) document that our approach to evaluating the impact of climate on streamflow and water supply system behavior compares favorably with a detailed monthly hydroclimatological modeling approach used by Kirshen and Fennessey (1995) for the water supply system that services the Boston, Massachusetts, metropolitan area.

Since our modeling approach is analytic and general, we were able to derive generalized sensitivity curves which describe the impact of changes in climate on water supply system yield and resilience. Figures 12.5 and 12.6 illustrate the impact of changes in temperature and precipitation on system yield for water supply systems in the northeastern United States which are characterized by carryover storage or overyear behavior. These curves may be used to approximate the impact of climate change on the yield of other existing systems within this broad region. It is hoped that the methodology and figures introduced here will allow municipalities and other regional authorities the opportunity to approximate the influence of potential climate change on their water supply operations. Although the results documented in Figures 12.5 and 12.6 only apply to the northeastern United States, the methodology introduced can be extended to other regions.

Acknowledgments

Although the research described in this chapter has been funded in part by the U.S. Environmental Protection Agency through grant number R 824992-01-0 to Tufts University, it has not been subjected to the Agency's required peer and policy review and therefore does not necessarily reflect the views of the Agency and no endorsement should be inferred.

REFERENCES

Ballentine, T. M. and Stakhiv E. Z. (1993) *Proceedings of the First National Conference on Climate Change and Water Resources Management*, U.S. Army Corps of Engineers, Institute for Water Resources Report 93-R-17, Fort Belvoir, VA.

Bell, C. J. (1995) Regional storage-reliability-resilience-yield-climate relations for the northeastern United States, Master of Sciences thesis, Tufts University, Medford, MA.

Fennessey, N. M. (1994) *A Hydro-Climatological Model of Daily Streamflow for the Northeastern United States*. Ph. D. dissertation. Tufts University, Medford, MA.

Kirshen, P. H. and Fennessey, N. M. (1995) Possible climate-change impacts on water supply of metropolitan Boston. *Journal of Water Resources Planning and Management* 121(1): 61–70.

Klemeš, V. (1986) Operational testing of hydrological simulation models. *Hydrological Sciences Journal* 31(1): 13–24.

(1990) Sensitivity of water resource systems to climatic variability. *Proceedings, Canadian Water Resources Association*, 43rd Annual Conference, Penticton, 233–42.

Langbein, W. B. et al. (1949) Annual runoff in the United States. *Geological Survey Circular 5*, U.S. Dept. of the Interior, Washington DC (reprinted 1959).

Leavesley, G. H. (1994) Modeling the effects of climate change on water resources. *Climatic Change* 28(1/2): 159–77.

Loaiciga, H. A., Valdes, J. B., Vogel, R. M., Garvey, J., and Schwarz, H. (1996) Global warming and the hydrological cycle. *Journal of Hydrology* 174: 83–127.

Revelle, R. R. and Waggoner, P. E. (1983) Effects of a carbon dioxide-induced climatic change on water supplies in the western United States. Chapter 7 in *Changing Climate*. National Academy Press, Washington, DC.

Rogers, P. R. and Fiering, M. B. (1990) From flow to storage, Chapter 9 in *Climate Change and U.S. Water Resources*, edited by P. E. Waggoner, John Wiley & Sons, New York, 207–21.

Schaake, J. C. (1990) From climate to flow, Chapter 9 in *Climate Change and U. S. Water Resources*, edited by P. E. Waggoner, John Wiley & Sons, New York, 177–206.

Slack, J. R., Lumb, A. M., and Landwehr, J. M. (1993) Hydro-climatic data network: streamflow data set. *U.S. Geological Survey Water Resources Investigation Report 93-4076*, U.S. Geological Survey, Reston, VA.

Tung, C.-P. and Haith, D. A. (1995) Global-warming effects on New York streamflows. *Journal of Water Resources Planning and Management*, ASCE 121(2): 216–25.

Vogel, R. M., Bell, C. J., and Fennessey, N. M. (1997) Climate, streamflow and water supply in the northeastern United States. *Journal of Hydrology* 198: 42–68.

Vogel, R. M. and Bolognese, R. A. (1995) Storage-reliability-resilience-yield relations for over year water supply systems. *Water Resources Research* 31(3): 645–54.

Vogel, R. M., Fennessey, N. M., and Bolognese, R. A. (1995) Regional storage-reliability-resilience-yield relations for northeastern United States, *Journal of Water Resources Planning and Management*, ASCE 121(5): 365–74.

Vogel, R. M. and Hellstrom, D. I. (1988) Long-range surface water supply planning. *Civil Engineering Practice* 3(1): 7–26.

13 Hydrological risk under nonstationary conditions changing hydroclimatological input

A. BÁRDOSSY* AND L. DUCKSTEIN**

ABSTRACT

Changing hydroclimatological conditions lead to changes in hydrological risk. Recent hydrological extremes such as floodings along the Rhine, Missisipi, or Oder Rivers can to a large extent be explained by the occurrence of unusual hydroclimatological extremes. Whether these extremes are part of natural variability, indicate possible climatic fluctuations, or are signals of an anthropogenically induced climate change is the first question to be answered. For this purpose, time series of different hydrological variables (atmospheric circulation patterns, rainfall, and runoff) are investigated. As hydrological risk is related to extremes, the series are investigated from that viewpoint, and not only from that of their mean behavior. Different statistical methods including nonparametric methods and bootstrap are applied to selected series to test the hypothesis of stationarity. Whenever this hypothesis is rejected, assumptions about the future have to be made. This can either be a scenario based on present trends, an assumption of stationarity at the present level, or a scenario based on a general circulation model (GCM). In the GCM case, due to the coarse resolution, a downscaling method is also needed. The next step is to assess the probabilities of extremes under these changes. It is demonstrated that areal precipitation extremes should be evaluated using not only precipitation amount but also duration and persistence of events. An example of extreme areal precipitation demonstrates this part of the methodology. On the basis of scenarios based either on present conditions or assumed trends, one may obtain a direct assessment of the flood risk. In contrast, GCM-based scenarios do not yield runoff values, thus a hydrological model has to be used to transform downscaled hydroclimatological series into runoff. Then the changed probabilities of extremes (due to the nonstationarity) have to be estimated. Corresponding to the extremes, the consequences have to be assessed using an engineering risk analysis format. In this chapter, ecological risk is assessed using a dose-response approach and economic risk is estimated using a rainfall model. A simplified but realistic example of assessment of ecological and hydrologic consequences is given so as to provide at least partial information for risk management.

* Institute for Hydraulic Engineering, University of Stuttgart, D-70550 Stuttgart, Germany.

** Systems and Industrial Engineering, The University of Arizona, Tucson, Arizona, 85721 U.S.A.

(Present address: GRESE.ENGREF 19 Ave du Maine. 75732 Paris CEDEX 15, FRANCE)

13.1 INTRODUCTION

The purpose of this chapter is to investigate the possible effect of climatic changes on hydrological and related ecological systems. As a first step, historical hydroclimatological time series of circulation patterns (CP), precipitation, flows, and temperatures are investigated. The hypothesis of stationarity of such time series under global change is tested and rejected in the case study considered (Ruhr region in Germany). Consequently, the usual engineering analysis assuming stationarity can no longer be used and a scenario of future behavior must be constructed. First, investigation of the time series of observed CP-types and linked regional precipitation is undertaken, and then scenario-based future time series are generated. Downscaling is used to condition local rainfall on daily CP-types.

The transformation of rainfall into runoff, with emphasis on extremes (high areal precipitation, hydrometeorological drought) is investigated. Economic and environmental consequences are examined, so that a trade-off between these risks may be set up. The notable feature of future risk under nonsteady conditions is that probabilities are not controllable – only consequences are so, partially at least. Various consequences of such changing input are now examined. Furthermore, a system view of global change includes the following types of consequences:

- economic, such as flood losses
- social, such as shortages or crop losses
- ecological, such as habitat loss or irreversible damage

It appears that the modeling of these variables also necessitates scenarios to estimate overall use. The input scenarios then have to be coupled with the consequence scenarios, as shown below. Management can only be done by modifying the consequences – under given coupled scenarios; furthermore, since the various types of consequences – economic, social, and ecological – are noncommensurable, a multiobjective procedure may be in order.

13.2 INVESTIGATION OF HISTORICAL SERIES

In order to assess possible future changes, time-series investigations are necessary. For Germany, Rapp and Schönwiese (1995) have performed a detailed climate trend analysis based on 100 years of data (1891–1990) covering the whole country. Their results show that significant changes in the precipitation amounts can be observed for certain areas. These changes are seasonally and regionally different. Figure 13.1 shows the relative changes for two selected areas, one in the northern and one in the southern part of Germany.

Karl, Knight, and Plummer (1995) show that for the United States, despite a zero trend in the total precipitation, there is a clear increasing trend in the intensities of the precipitation events, which corresponds to a decrease in the number of events. At selected stations in northern Germany there is an increase in frequency of wet winter days and rainfall depth per such days. Figure 13.2 shows an example of such changes in both variables for December. In contrast, Figure 13.3 shows an example (July) where the frequency of wet days does not change, but the total amount decreases.

Local climatic variables, temperature and precipitation, are strongly linked to large-scale atmospheric circulation. In Bárdossy and Plate (1992) a model was developed for the precipitation occurrence at a selected site conditioned on the actual atmospheric circulation pattern. Similar models have been developed by Wilson, Lettenmaier, and Skyllingstad (1992), and Shrestha, Duckstein, and Stakhiv (1996a, 1996b).

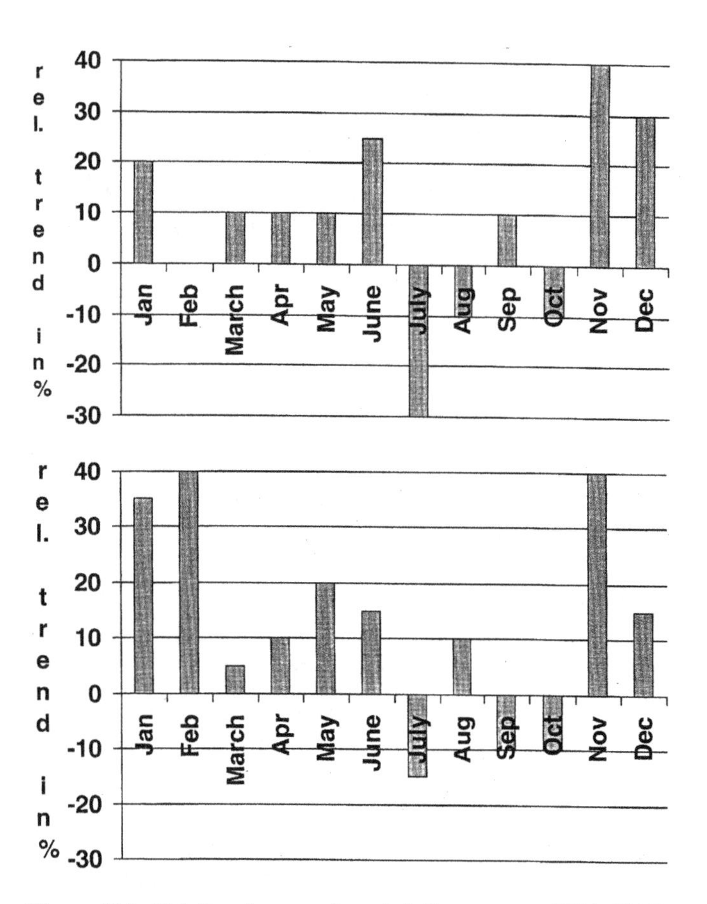

Figure 13.1. Relative changes of precipitation amounts (1891–1990) for two selected areas in Germany (Aller and Schwarzwald).

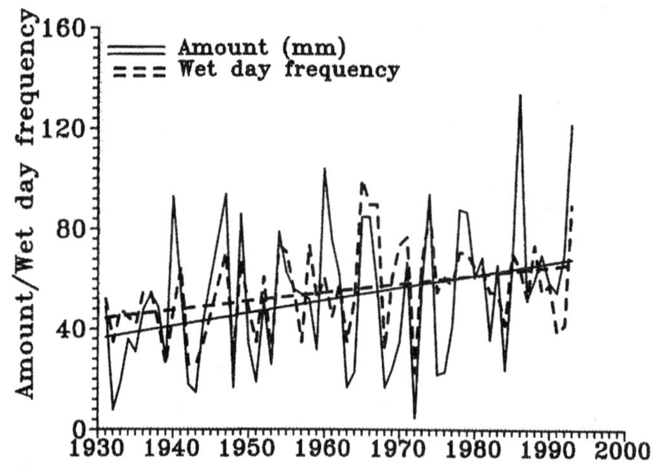

Figure 13.2. Relative changes of precipitation amounts and frequency of wet days for the Aller catchment in Germany (December).

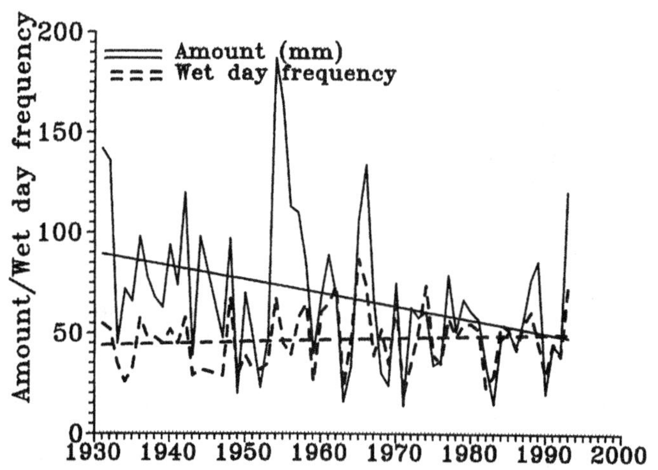

Figure 13.3. Relative changes of precipitation amounts and frequency of wet days for the Aller catchment in Germany (July).

CP types can be obtained by the classification of the surface air pressure and/or the 500 (700) hPa surfaces elevations. The classification can be done subjectively (manually), with objective (automated) methods such as k-means clustering, or using a fuzzy rule-based approach.

Possible changes in the time series of CPs have been investigated in Bárdossy (1998). Over Europe, the frequency of several types of CPs shows a nonstationary behavior. Besides the patterns' frequencies, their duration also appears to have changed during the last few years – leading to much more persistent weather. The findings based on 115 years of subjectively classified daily CP series can be supported by similar results for the objectively classified CP series for 1947–93. To investigate the persistence of the objectively defined CPs, we do not use the length of the patterns, but a different statistic based on the number of days within a given time interval. Nonparametric statistical tests based on Mann-Whitney statistics have been applied to test the significance of the frequency changes and permutation-based methods, to investigate the changes in persistence. There are several significant changes, among which the increase of the frequency and the persistence of the zonal circulations is hydrologically the most important. Figure 13.4 shows the mean duration of zonal CPs. One can observe an increase in the length, which is also confirmed by permutation tests and objectively classified series.

The persistence of precipitation is extremely important in hydrology. Persistence characteristics can be described with the wetness index w which can be defined and calculated as:

$$w(t) = \sum_{m=1}^{M} \alpha^m z(t-m) \tag{13.1}$$

where $0 < \alpha \leq 1$ is a weight and M is the number of preceding days. The higher the wetness index is, the more likely the catchment is to produce runoff.

Long time series of precipitation were investigated for possible changes of the wetness index. The annual and winter extreme values of the wetness index were considered. The mean value of $w(t)$ is proportional to the mean precipitation in the selected time period; therefore, it does not give any insight into the persistence properties. On the contrary, the high variance of $w(t)$ means that high and low values occur in the series, indicating persistent dry and wet periods. Thus, an increase of the variance of the wetness index indicates an increase of persistence.

Figure 13.5 shows the time series of the maximal values of w for the areal precipitation in the Aller catchment. The parameters used for the calculation of w in (1) were $\alpha = 0.85$ and $M = 15$ days. These series indicate a possible break point. Statistical permutation tests were used to test the stationarity of these series. At significance levels of 5 percent and 10 percent, the series indicate nonstationary behavior (for more details, see Bárdossy 1998).

All these results indicate that the assumption of stationarity cannot be applied without a reasonable doubt. As an alternative to the stationarity assumption, nonstationary scenarios may be constructed, as discussed in the next section.

13.3 NONSTATIONARY SCENARIOS

In order to calculate the future risks under nonstationary conditions, assumptions have to be made on the future changes. There are three groups of possible methods to generate hydroclimatological series for changing conditions:

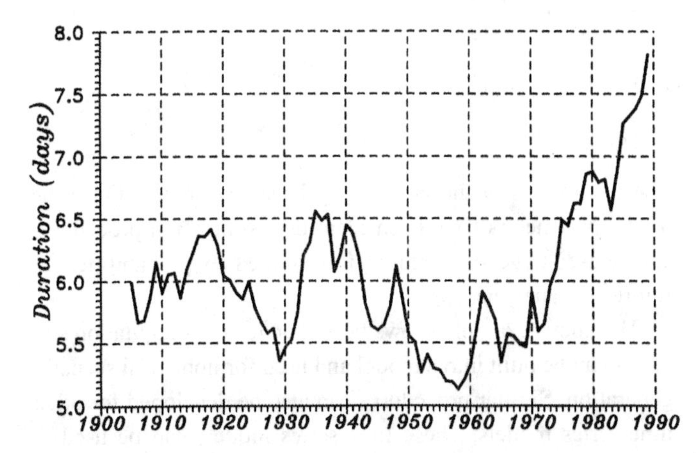

Figure 13.4. Mean duration of zonal circulation patterns (ten-year moving average, subjective classification).

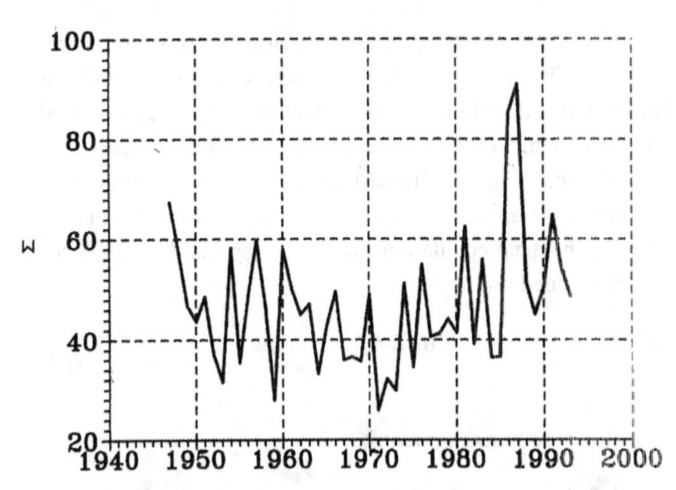

Figure 13.5. Maximal winter wetness index for the Aller catchment in winter (1947–93).

1. simple alterations of the present climate,
2. modified time-series model; and
3. GCM-based scenarios.

13.3.1 Simple alterations

As a simple alteration of the present conditions, assume an increase in temperature (typically $\Delta T = +1, +2$, and $+3$ K) and a decrease or an increase in precipitation ($\Delta Z = 0, \pm 5\%, \pm 10\%$). This change is simply applied to observed temperature series by adding ΔT and for precipitation, by multiplying the values by $(1 + \Delta Z/100)$. This approach is very simple and often claimed to give an insight to the possible range of changes and also the vulnerability of a basin. Changes in the annual cycle are usually not considered, but a monthly variation of ΔZ could be included. Both gradual changes and sudden jumps can be considered.

There are a great number of studies that use such altered time series to assess possible effects of climate change. It is claimed that climate change will remain within the investigated range, thus limits to the hydrological, ecological, or economic responses can be found. The problem with this approach is that only one of the precipitation characteristics, namely the intensity, is assumed to change. The frequency of precipitation events, their duration and spatial extent are not altered – however they might have a higher influence on the output than the rainfall depth. These parameters cannot be changed as simply as the depth, and thus a time series model is needed even for simple scenarios. The historical investigations mentioned in the previous section indicate that the duration and areal extent of precipitation events might also change.

13.3.2 Modified time-series models

Another possibility to generate scenarios is to use time-series models. One of the simplest time-series models for daily precipitation is the Markov chain model first suggested by Gabriel and Neumann (1962) for describing the occurrence of wet and dry days; an additional model for the distribution of the precipitation amounts is that of Woolhiser and Roldán (1982). Let $Z(t)$ be the precipitation amount on day t, and $Y(t)$ the occurrence defined as:

$$\tilde{Y}(t) = \begin{cases} 0 & for \quad \tilde{Z}(t) \leq Z_0 \\ 1 & for \quad \tilde{Z}(t) > Z_0 \end{cases} \tag{13.2}$$

The Markov transition matrix

$$A = [a_{ij}]$$

with

$$a_{ij} = P(\tilde{Y}(t) = j | \tilde{Y}(t-1) = i) \quad i, j = 0, 1$$

and the daily amount distribution $F(z)$ describes the model. By definition:

$$a_{i0} + a_{il} = 1 \quad i = 0, 1 \tag{13.3}$$

The frequency of wet and dry days can be calculated directly from A as:

$$\pi_0 = \frac{a_{10}}{a_{10} + a_{01}} \tag{13.4}$$

$$\pi_1 = \frac{a_{01}}{a_{10} + a_{01}} \tag{13.5}$$

The duration of wet periods L_1 has a distribution:

$$P(\tilde{L}_1 = k) = a_{11}^{k-1} a_{10} \tag{13.6}$$

The mean duration is

$$E(\tilde{L}_1) = 2\frac{a_{11}}{a_{10}} \tag{13.7}$$

To describe a scenario with a changed probability of precipitation π_1', a mean duration $E(\tilde{L}_1')$, and a relative change in the mean amount of $\Delta Z'$ one can solve equations 13.5 and 13.7 to find A':

$$a_{11}' = \frac{E(\tilde{L}_1')}{E(\tilde{L}_1') + 2} \tag{13.8}$$

and

$$a_{01}' = \frac{2\pi_1'}{(E(\tilde{L}_1') + 2)(1 - \pi_1')} \tag{13.9}$$

For the intensity distribution $F(z)$ the new intensities z' have to fulfill

$$z' = \frac{\pi_1}{\pi_1'}\left(1 + \frac{DZ}{100}\right) \tag{13.10}$$

with a distribution function:

$$G(z') = F(z)$$

The parameters of the model may have seasonality. Once the model parameters have been identified, simulation procedures can be used to generate realizations of the new probabilities for future risk analysis.

The above example shows how changes in precipitation statistics can be built into a model and used for nontrivial scenario generation. Similar procedures can also be developed for other time-series models. These time-series models can be used to generate scenarios for new stationary states and transition times with time-dependent parameters.

Historical investigations, such as the change in persistence and frequency of CPs, can be built into time-series models, by either assuming trends in their coefficients, or stationarity at a new level – for example at the level of the 1980–95 period. In the case of CP-based precipitation modeling, changes in the CP series have to be simulated. For this purpose a simulation method based on a semi-Markov process has been selected. The advantage of this representation is that the duration distributions are explicitly specified; thus, their changes can be included simply. To obtain an appropriate transition matrix which delivers the given (changed) CP frequencies is a more difficult problem. The condition to be fulfilled by the transition matrix is

$$\sum_j \frac{\pi_j}{l_j} p_{ji} = \frac{\pi_i}{l_i} \quad (13.11)$$

where (p_{ij}) is the (unknown) transition matrix of the semi-Markov chain, π' is the frequency of pattern i, and l_i is the expected duration of pattern i. The above problem usually has no unique solution. Once the transition matrix and duration distributions have been selected, a sequence of CPs can be generated. Subsequently, a precipitation series can be simulated.

The previous approaches are very flexible to generate changed precipitation series. A basic problem is that the physical background of the nonstationarity is not taken into account. The only physically based models for this purpose are the General Circulation Models (GCM). These, however, have a spatial resolution of several hundred kilometers, unsuitable for regional hydrological applications. Furthermore, the most important hydrological variable, precipitation, is not well described by GCMs. Therefore, approaches to relate the global predictions to the local climate are necessary.

13.3.3 GCM-based scenarios – downscaling

The basic problems in applying GCMs to assess hydrological consequences are the coarse spatial resolution and the inaccuracy in precipitation modeling. GCM's coarse resolution can be refined in different ways:

- using a nested model
- by stochastic downscaling
- by downscaling GCM-generated changes to local scenarios

Nested models

Nested models provide physically realistic weather data for selected regions. The boundary conditions of these models are provided by the coarse resolution models. Thus, possible errors of the coarse GCMs cannot be corrected using this approach. A realistic spatial distribution of precipitation can be achieved due to its detailed topography. More details on this approach can be found in Giorgi and Mearns (1991).

Stochastic downscaling

Another possibility to generate local scenarios is to use the previously mentioned stochastic precipitation models which are based on CPs. They provide a link between the coarse GCM resolution data and the local, hydrologically relevant precipitation. Two assumptions are related to this approach: (i) changes in the climate will mostly occur as changes in the CPs; and (ii) the linkage between CPs and precipitation will remain stable.

Paleoclimatic evidences show that past climate changes were strongly associated with CP changes (Kapsner et al. 1995). The second assumption can be supported by the stability of the linkage over the past 50–100 years. Changes in this linkage can also be assessed analyzing GCM-based large-scale precipitation. Another argument supporting this approach is that GCMs are more accurate on free atmospheric variables such as air pressure than on precipitation. Investigations show that frequencies of CPs based on 1 × CO_2 scenario (present climate) are quite similar to the observed frequencies. However, the difference between the changed climate and a control run is sometimes less then the difference between the observed and control run. Figure 13.6 shows the distributions of two different CP groups of the observed series and control run, and changed climate related GCM output.

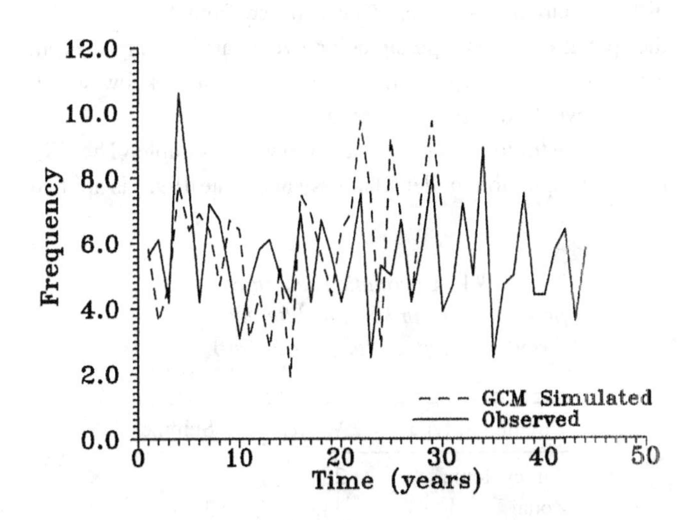

Figure 13.6. Frequencies of zonal circulation patterns for observed and GCM simulated sequences.

Downscaling and/or disaggregation of GCM generated changes

Another possibility for downscaling GCM results is to analyze their output with regard to precipitation statistics. Here, besides the usually performed comparisons between mean monthly precipitation values, an analysis of the other statistical characteristics should also be undertaken. Stochastic models with parameters reflecting these known changes can be used to generate plausible scenarios.

Due to the coarse resolution of the GCMs, a stochastic disaggregation based on local observations offers an alternative to a nested model. In this case, for each GCM time-step, the subgrid distribution is generated with a stochastic model preserving the integral values of the climatic variables for the GCM grid.

13.4 HYDROLOGICAL RISKS

The effects of changing conditions on hydrological risk are developed for a case study in the Ruhr catchment in Germany. For this hydrological risk, precipitation and runoff changes are considered.

13.4.1 Changes in precipitation

In the previous sections the time distribution of the precipitation was discussed in detail. However, for hydrological applications the spatial extent of the precipitation also plays a central role. Rainfall-runoff models require the spatial distribution of rainfall, or the areal precipitation amounts are needed. For this purpose the spatial correlation of rainfall has to be considered. Table 13.1 shows the correlation length of precipitation for different circulation types. One can see from this table that the spatial extent of a precipitation event varies strongly. From a hydrological viewpoint it is very important to know which type of events will be more frequent.

For illustration consider the following example. The CP-based precipitation model (Bárdossy and Plate 1992) is applied to the Lenne catchment (≈1200 km²) in the Ruhr basin in Germany, where forty-four rain gauges were selected. Figure 13.7 shows the subcatchments of the Lenne catchment. Daily precipitation values for fifteen years (1978–92) were used to estimate the precipitation model parameters.

The point statistics of the observed and the simulated sequences are then compared to verify the precipitation model. In addition to the sequence simulated to the observed CPs of 1978–92, two other CP sequences have been used for precipitation simulation.

1. GCM control simulation run (1 × CO_2) CP sequence obtained by classifying 10 years daily 700 hPa surfaces from the output of the ECHAM model of the Max Planck Institute (MPI) in Hamburg.
2. GCM changed climate simulation run (2 × CO_2) CP sequence obtained by classifying 10 years daily 700 hPa surfaces from the output of the ECHAM model

For each of these sequences, ten realizations (100 years) have been generated using the precipitation model at the forty-four locations. These were evaluated both for the mean and the extremes. Table 13.2 shows the observed and the GCM-based downscaled seasonal precipitation amounts. There is no change in the mean precipitation amounts. Point extremes were also considered. Table 13.3 shows the results at a selected location. Even the point extremes do not show a changing signal. In contrast, the areal precipitation extremes shown in Table 13.4 show an increase. The explanation for this is the increase of the frequency of zonal circulations in the 2 × CO_2 case to 18.6 percent from 15.6 percent for the control run. Clearly, hydrologic decisions such as reservoir operation (Shrestha et al. 1996c) would be greatly affected by such input changes.

Nonstationarity of hydroclimatological series can change, for example, flood or erosion risks. These can also be handled statistically directly by using either scenario statistics or hydrological models. Again one can begin with some simplified formulas.

Table 13.1. *Correlation length of precipitation (in km) for different circulation types (Ruhr catchment)*

Circulation type	Winter	Summer
Anticyclonic	28.6	4.0
Zonal	44.6	34.6
South + East	38.7	60.7

13.4.2 Changes in runoff

The runoff volume V for a specific rainfall event can be estimated using different empirical formulae whose general form is:

$$V = f(z, T_z, w, t) \tag{13.12}$$

where z is the precipitation amount, T_z is the duration of the event, w is the wetness index, and t denotes the time of year. For example, the U.S. SCS formula may be used. One can see that all precipitation characteristics mentioned earlier in this chapter are playing an important role.

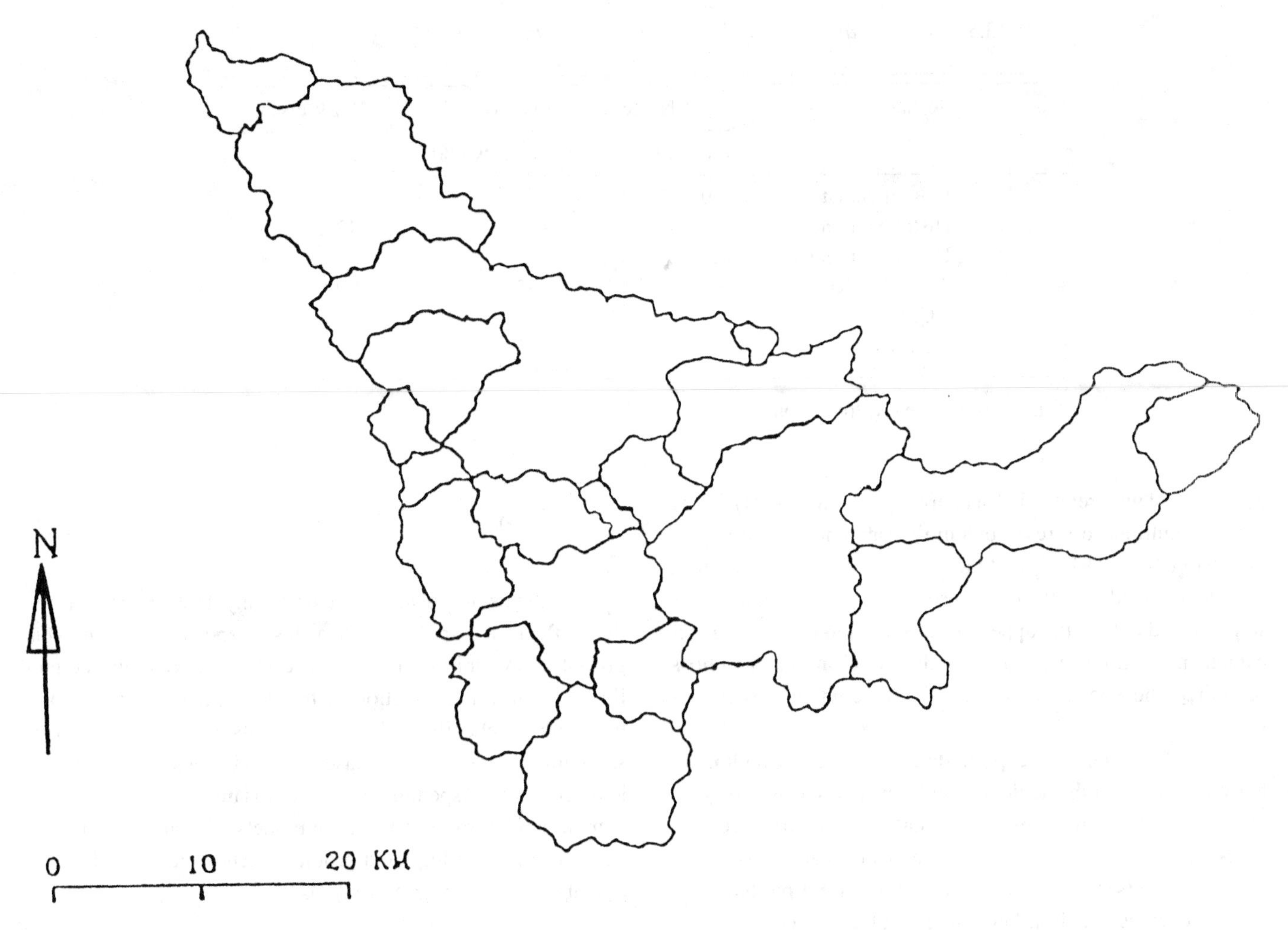

Figure 13.7. Subcatchments of the Lenne catchment.

Table 13.2. *Observed and simulated mean seasonal precipitations at the location Essen (Germany)*

	Winter	Spring	Summer	Fall	Annual
Observed	227.2	227.4	251.7	203.7	910.0
$1 \times CO_2$	213.0	208.8	217.1	210.4	849.3
$2 \times CO_2$	218.2	221.3	226.5	219.0	884.0

Table 13.3. *Extreme daily precipitation at Olpe obtained by simulation*

CP sequence	Return period			
	20	10	5	2
1981–1990	64.9	55.2	50.0	42.1
GCM $1 \times CO_2$	65.3	57.1	52.6	43.8
GCM $2 \times CO_2$	66.6	57.8	50.0	44.1

Table 13.4. *Extreme daily areal winter precipitation for the Lenne catchment*

CP sequence	Return period			
	20	10	5	2
1981–1990	43.2	39.0	50.0	30.6
GCM $1 \times CO_2$	44.5	39.1	52.6	30.6
GCM $2 \times CO_2$	53.2	46.5	50.0	30.7

In order to calculate the consequences of a changed precipitation scenario, the rainfall runoff model of Göppert (1995) was used. This distributed model based on the unit hydrograph approach was developed for the Lenne catchment (subcatchment of the Ruhr) covering 1,357 km^2 area. Figure 13.7 shows the location of the catchment and its subcatchments. It is a hilly area with the highest point at 839 m and the lowest at 98 m (above North Sea level). The area is mostly forested (58

Table 13.5. *Extremes at Bahmenohl for different climate scenarios*

Case	Scenario	Change in precipitation		Peak discharge
		amount (%)	intensity (%)	m^3/sec
1	Present climate	0	0	95.3
2	Historical trend	+25	+25	126.1
3	Historical trend	+25	+5	100.5
4	Altered series	+5	−5	90.2
5	GCM based	0 (+15)*	15	113.1
6	Intensity only	0	15	103.8

* Increase for areal extremes only.

percent) and with relatively small urban areas (8 percent). There are two multipurpose reservoirs in the catchment – the Biggetalsperre ($A_E = 289\,\text{km}^2$) and the Versetalsperre ($A_E = 24\,\text{km}^2$). In order to avoid the effects of these reservoirs, the calculations were carried out for the upper part of the catchment. The upper catchment area corresponding to the outlet at Bahmenohl is $457\,\text{km}^2$. The model was used for precipitation scenario-generated events.

Table 13.5 shows the peak discharges corresponding to ten year events. In this table the importance of the assumptions made to generate future climate scenarios is evident. Even the cases 1, 5, and 6 with unchanged annual mean deliver very different results. It may also be noted that case 4 produces a 5 percent decrease of flash flood in spite of a 5 percent increase of annual amount. For this catchment the time persistence does not play a central role; however, for larger catchments its influence could be considerable.

13.5 AN EXAMPLE OF ECOLOGICAL RISK ANALYSIS

Possible changes in the hydrological cycle clearly have an important impact on ecology. Changes in the precipitation could, for example, directly influence plant growth. In order to estimate regional ecological risks, appropriate models are necessary. Ecological effects of climate change are usually investigated at a landscape scale. This is done partly to investigate possible feedbacks. In this chapter a smaller scale example, a dose-response problem, is considered.

Dose-response functions are mostly used in the risk assessement of chemicals. The dose is the amount of material that is administered to an organism at one time (Suter 1993). The dose can also be seen as the product of concentration and time, or in a time varying case as:

$$D = \int_{t_0}^{t_1} C(t)dt \tag{13.13}$$

As organisms do not react uniformly to toxic chemicals, their effect is described with a dose-response function. This gives the probability of an adverse effect (developing certain illness, death) as a function of the dose D. In contrast to the traditional probability of adverse events $P(A)$ and the corresponding loss $L(A)$, in this case the loss is considered as a unit loss and the corresponding probability is investigated. There are a great number of dose-response models. The most commonly used ones are the logit and probit models which calculate the probability P of adverse events as:

$$P = a + b\log D \tag{13.14}$$

Here a and b are fitted constants. Figure 13.8 shows a typical dose-response curve.

The time of exposure t is usually not considered in these models. Hewlett and Plackett (1979) suggested a dose-duration-response relationship of the form:

$$P = a + b\log D + c\log t \tag{13.15}$$

Hydrometeorological changes, like increased persistence of wet and dry periods or changes in the seasonal precipitation, result in changes of the low flow conditions of a river. This changes the dose of toxic substances to which fish are exposed. Given a constant industrial or urban load of a toxic chemical L the concentration at a given time $C(t)$ is

$$C(t) = C_b + \frac{L}{Q(t)} \tag{13.16}$$

where $Q(t)$ is the discharge at time t, and C_b is the background concentration, assumed to be constant in time. Obviously changes in the runoff characteristics affect the dose and the

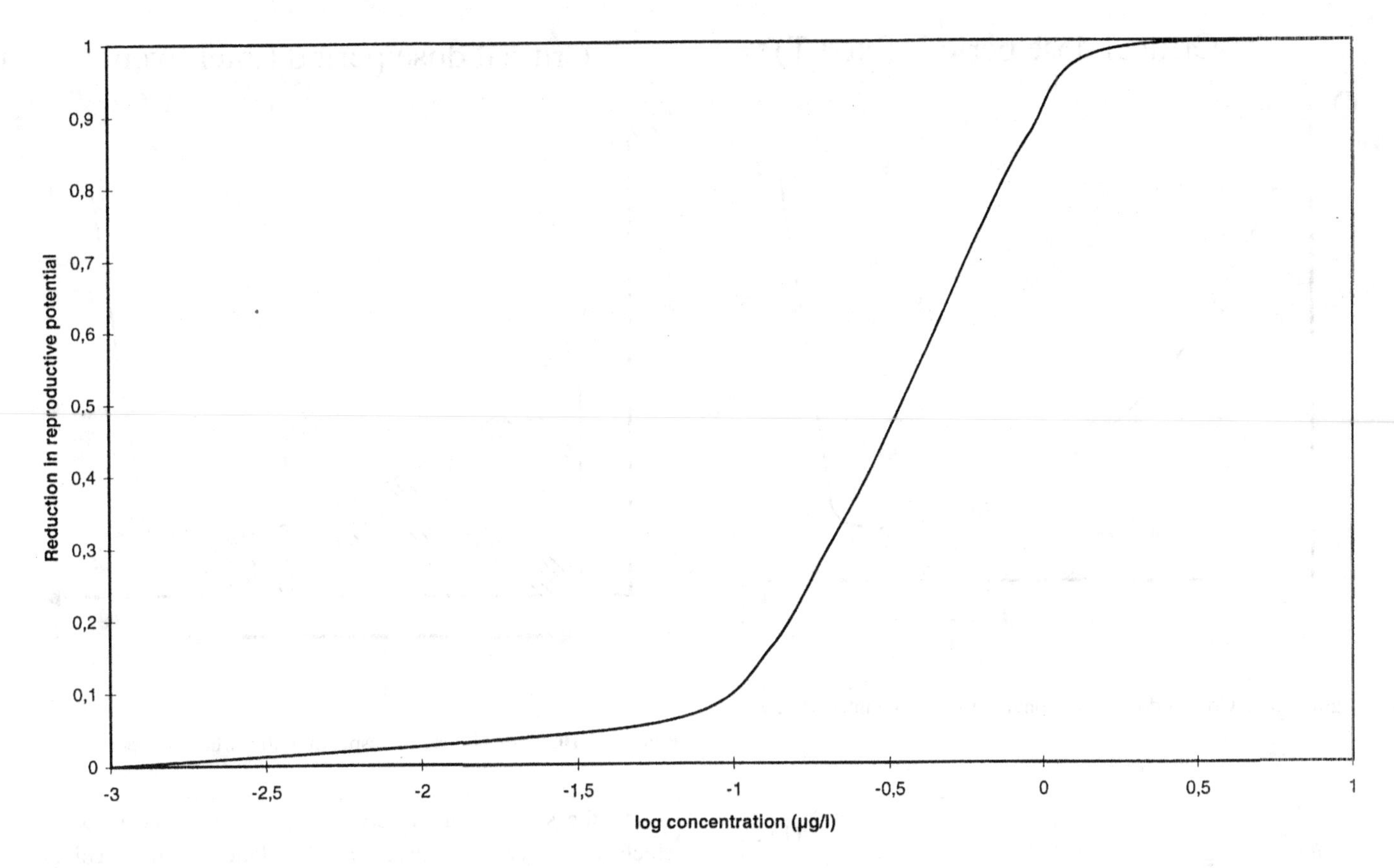

Figure 13.8. A typical dose-response curve.

exposure time. Changes of the low flow conditions can be obtained from the hydrometeorological scenarios using hydrological models or through statistics. Consider the simplest recession curve model:

$$Q(t) = Q(t_0)exp\left(\frac{t - t_0}{k}\right) \tag{13.17}$$

Inserting (13.17) into (13.16) and integrating according to (13.13) one obtains the dose for the time interval of length T beginning at time t_1 (assuming a constant load L):

$$D = C_bT + \frac{LT}{Q(t_0)}k\left(exp\left(\frac{T}{K}\right) - 1\right)exp\left(\frac{-T - t_1 + t_0}{k}\right) \tag{13.18}$$

Typical time intervals T for fish are between ninety-six hours and fourteen days (Suter 1993). The dose depends on the volume of water that dissolves the toxic load L. The highest dose occurs at the end of a dry period. Let T^* denote the length of a dry period. Taking a dose-response curve with constant time T of dose application, the maximal dose and the highest risk occurs at the end of the recession curve. Thus, $t_1 = t_0 + T^* - T$ gives the highest dose and thus the risk (Figure 13.9). The dose D and the risk P can be regarded as functions of T^*. The dose can also increase by decreasing $Q(t_0)$. Neglecting the background concentration $C_b = 0$ and substituting equation 13.18 into equation 13.14 leads to

$$\frac{dP}{dQ(t_0)} = -\frac{b}{Q(t_0)} \tag{13.19}$$

and

$$\frac{dP}{dT^*} = -\frac{b}{k} \tag{13.20}$$

These two derivatives show the sensitivity of the adverse event probability P to the change of the duration of a dry period and to the flow $Q(t_0)$. Depending on the catchment characteristics expressed by k, these two quantities can be ascertained. It is interesting to note that the sensitivities are not directly dependent on the actual duration of the dry period.

In the case when the dose is not considered over a fixed time interval T but for the whole recession period T^* one gets the dose:

$$D = C_bT^* + \frac{LT^*}{Q(t_0)}k\left(exp\left(\frac{T^*}{k}\right) - 1\right)exp\left(\frac{T^*}{k}\right) \tag{13.21}$$

This case is sketched in Figure 13.10. Neglecting C_b and substituting equation 13.21 into equation 13.14 leads to:

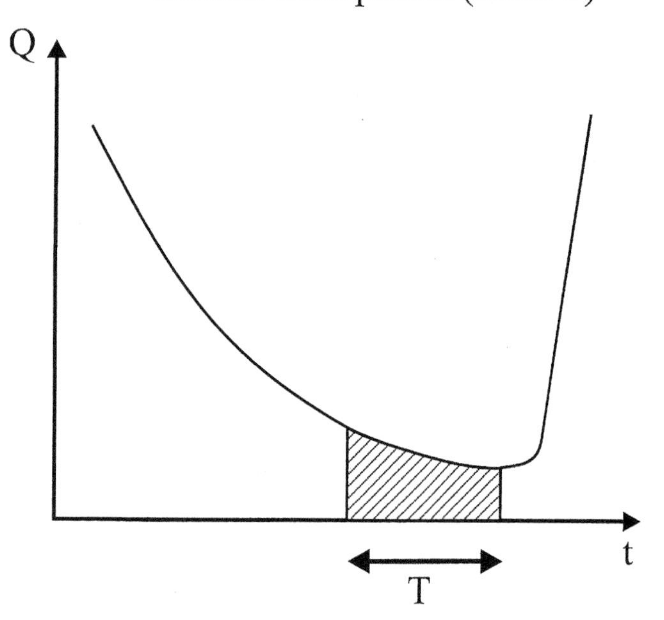

Figure 13.9. Critical dose corresponding to the time interval T.

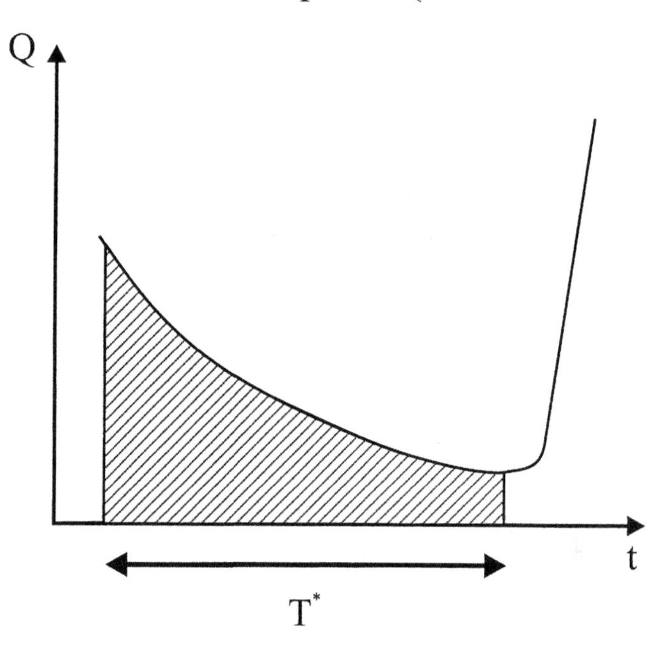

Figure 13.10. Critical dose corresponding to a whole dry period.

$$\frac{dP}{dQ(t_0)} = -\frac{b}{Q(t_0)} \tag{13.22}$$

and

$$\frac{dP}{dT^*} = -\frac{b}{k}\left(2 + \frac{exp(T^*/k)+1}{exp(T^*/k)-1}\right) + \frac{b}{T^*} \tag{13.23}$$

This latter equation shows that the adverse event probability in the cumulative case is more sensitive to the duration T^* than in the previous case. The sensitivity with respect to the initial flow does not change.

This example shows that a dose-response type of ecological risk assessment under changing conditions can be based on two related variables: the discharge $Q(t_0)$ and the time-span between two events T^*. These quantities have to be assessed from the selected scenarios. This example also shows that for an ecological risk the dry period duration plays an important role.

For known changes of the hydrometeorological input, the corresponding adverse event probability can be calculated. In this case regulations affecting the load L may be used to mitigate the risk.

In order to perform risk management, both probabilities and adverse consequences must be taken into account. The problem with changing hydroclimatological conditions is that we do not have a direct influence on the probabilities of the adverse events. Consequently, to avoid or minimize risk(s), the severity of the consequences has to be reduced, which to a certain degree can be done. In the ecological example, a reduction or a controlled timing of the load L can considerably reduce the risks. In the flood example, protection against higher discharges can be built, the economics of which can be calculated using standard engineering techniques.

13.6 CONCLUSIONS

The following conclusions can be drawn:

1. To assess the hydrological and ecological effects of changing hydroclimatological conditions, scenarios for temperature and precipitation series are required.
2. Scenarios may be obtained using assumed disturbances, historical series, or GCM output.
3. The possibility of changes in precipitation intensity, event duration, and persistence can be considered in time-series models.
4. Changes observed in historical records can be used to provide scenarios.
5. GCM-simulated changes can seldom be used directly.
6. Stochastic semi-empirical downscaling techniques provide a linkage between large-scale features and regional climate variables.

7. The spatial extent of precipitation has to be taken into account. Scenarios based only on point observations can be misleading.
8. Ecological models may be used to assess the ecological risk under hydroclimatological conditions.
9. For simple models like dose-response relationships, the changes can be assessed analytically. It is demonstrated that both precipitation amount and persistence of dry and wet periods have an important influence on ecological risk.
10. Flood risks can be assessed using hydrological models; however, every precipitation characteristic (amount, duration, space-time persistence) has to be taken into account when dealing with future scenarios.

Acknowledgments

This research was supported in part by US National Science Foundation grants CMS-9613654 and CMS-9614017.

REFERENCES

Bárdossy, A. (1998) Statistical persistence in hydroclimatological series. In *Statistical and Bayesian Methods in Hydrological Sciences*, edited by E. Parent, P. Hubert, B. Bobee, and J. Miquel, IHP-V, Technical Documents in Hydrology, No. 20, UNESCO, Paris.

Bárdossy, A. and Henze, N. (1996) Statistical investigation of time series of circulation patterns. Working paper, Institute of Hydraulic Engineering, University of Stuttgart.

Bárdossy, A. and Plate, E. J. (1992) Space-time model for daily rainfall using atmospheric circulation patterns. *Water Resources Research* 28: 1247–59.

Duckstein, L. and Parent, E. (1994) Systems engineering of natural resources under changing physical conditions: a framework for reliability and risk. In *Engineering Risk in Natural Resources Management*, edited by L. Duckstein and E. Parent. Kluwer, Dordrecht, 5–19.

Gabriel, K. R. and Neumann, J. (1962) A Markov chain model for daily rainfall occurrences at Tel Aviv. *Quarterly Journal of the Royal Meteorological Society* 88: 90–95.

Giorgi, F. and Mearns, L. O. (1991) Approaches to the simulation of regional climate change: A review. *Reviews of Geophysics* 29: 191–216.

Göppert, H. (1995) *Operationelle Hochwasservorhersage zur Steuerung von Talsperren*. Unpublished Ph.D. dissertation, IHW, Karlsruhe.

Hewlett, P. S. and Plackett, R. L. (1979) *The Interpretation of Quantal Responses in Biology*. Edward Arnold, London.

Kapsner, W. R., Alley, R. B., Shuman, C. A., Anandakrishnan, S. and Grootes, P. M. (1995) Dominant influence of atmospheric circulation on snow accumulation in Greenland over the past 18,000 years. *Nature* 373: 52–54.

Karl, T. R., Knight, R. W., and Plummer, N. (1995) Trends in high frequency climate variability in the twentieth century. *Nature* 377: 217–20.

Rapp, J. and Schönwiese, C.-D. (1995) *Atlas der Niederschlags- und Temperaturtrends in Deutschland 1891–1990*. Frankfurter Geowissenschaftliche Arbeiten, Frankfurt.

Shrestha, B. P., Duckstein, L., and Bogardi, I. (1996a) Downscaling algorithm for precipitation in semi-arid regions under climate change. Amer. Geophys. Union Spring Meeting, Baltimore.

Shrestha, B. P., Duckstein, L., and Stakhiv, E. Z. (1996b) Forcing functions under climate change. *Proceedings, ASCE Water Resources 2000 International Conference*, Anaheim, CA.

(1996c) Fuzzy rule-based modeling of reservoir operation, *ASCE J. Water Resources Planning and Management* 122(4): 262–69.

Suter, G. W. (1993) *Ecological Risk Assessement*. Lewis Publishers, Boca Raton.

Wilson, L., Lettenmaier, D. P., and Skyllingstad, E. (1992) A hierarchical stochastic model of large scale atmospheric circulation patterns and multiple station daily rainfall. *Journal of Geophysical Research* 97: 2791–2809.

Woolhiser, D. A. and Roldán, J. (1982) Stochastic daily precipitation models, 2. A comparison of distribution of amounts. *Water Resources Research* 18: 1461–68.

14 Fuzzy compromise approach to water resources systems planning under uncertainty

MICHAEL J. BENDER*

ABSTRACT

A fuzzy compromise approach is applied to two water resources systems planning examples, for the purpose of allowing various sources of uncertainty and facilitating a flexible form of group decision support. The examples compare the ELECTRE method, and Compromise Programming, with the fuzzy approach. The fuzzy compromise approach allows a family of possible conditions to be reviewed, and supports group decisions through fuzzy sets designed to reflect collective opinions and conflicting judgments. Evaluating alternatives to produce rank orderings are accomplished with two ranking measures for fuzzy sets. The ranking measures are also shown to indicate the impact of different levels of decision-maker risk aversion. Two distinct ranking measures are used – a centroid measure and a fuzzy comparison measure based on a fuzzy goal.

14.1 INTRODUCTION

Multicriteria decision analysis (MCDA) has been moving from optimization methods to more interactive decision support tools. Some areas of interest have been identified by Dyer et al. (1992) as:

> Sensitivity analysis and the incorporation of vague or imprecise judgements of preferences. Development of improved interactive software for multicriterion decision support systems.

Uncertainty is a source of complexity in decision making that can be found in many forms. Typical ones include uncertainty in model assumptions and uncertainty in data or parameter values. There may also be uncertainty in the interpretation of results. While some uncertainties can be modeled as stochastic variables in a Monte Carlo simulation, for example, other forms of uncertainty may simply be vague or imprecise.

Traditional techniques for evaluating discrete alternatives such as ELECTRE (Benayoun, Roy, and Sussmann 1966), AHP (Saaty 1980), Compromise Programming (Zeleny 1973, 1982), and others normally do not consider uncertainties involved in procuring criteria values. Sensitivity analysis can be used to express decision-maker uncertainty (such as uncertain preferences and ignorance), but this form of sensitivity analysis can be inadequate at expressing decision complexity. There have been efforts to extend traditional techniques, such as PROTRADE (Goicoechea, Hansen, and Duckstein 1982), which could be described as a stochastic compromise programming technique. A remaining problem is that not all uncertainties easily fit the probabilistic classification.

Fuzzy decision analysis techniques have addressed some uncertainties, such as the vagueness and conflict of preferences common in group decisions (Blin 1974; Siskos 1982; Seo and Sakawa 1985; Felix 1994; and others), and at least one effort has been made to combine decision problems with both stochastic and fuzzy components (Munda, Nijkamp, and Rietveld 1995). Application, however, demands some level of intuitiveness for decision makers, and encourages interaction or experimentation such as that found in Nishizaki and Seo (1994). Leung (1982) and many others have explored fuzzy decision-making environments. This is not always so intuitive to many people involved in practical decisions because the decision space may be some abstract measure of fuzziness, as opposed to a more tangible measure of alternative performance. The alternatives to be evaluated are rarely fuzzy. However, their perceived performance may be fuzzy.

An intuitive, and relatively interactive, decision tool for discrete alternative selection, under various forms of uncertainty, would be a valuable tool in decision analysis – especially for applications with groups of decision makers. This chapter explores the application of fuzzy sets in conjunction with a standard MCDA technique, compromise programming.

14.1.1 Displaced ideals

Multicriteria decision analysis techniques can approach the analysis of multiobjective problems in a number of ways. They

* Lyonnaise South East Asia (ASTRAN), Kuala Lumpur, Malaysia.

are generally based on outranking relationships, distance metrics, and utility theory. The concept of the displaced ideal was used by Zeleny (1973, 1982) to form compromise programming, a multicriteria technique that resolves criteria into a commensurable, unitless, distance metric measured from an ideal point (for each alternative). The result is a direct ranking (strong ordering) of alternatives, valid for the selected weights and the chosen form of distance measurement. The following can be used to calculate a discrete compromise programming distance metric (L), otherwise known as the Minkowski distance:

$$L = \left[\sum_i \left\{ w_i^p \left(\frac{f_i^* - f_i}{f_i^* - f_i^-} \right)^p \right\} \right]^{\frac{1}{p}} \quad (14.1)$$

where f_i is the value for criteria i, f_i^*, f_i^- are the positive and negative ideal values for criteria i, respectively; w_i is a weight, and indicates relative importance of a criteria; L is the distance from an ideal solution; and p is the distance metric exponent.

Typically, the Euclidean distance ($p = 2$) is used to penalize large deviations from the ideal. However, the exponent can also carry an economic interpretation. The Hamming distance ($p = 1$) results in a case of perfect compensation between criteria. For the Chebyshev distance ($p = \infty$) there is no compensation among criteria – the largest deviation from the ideal dominates the assessment.

Many of the traditional MCDA techniques, including compromise programming, attempt to preserve some level of transparency to problems. This is a valuable property in decision analysis tools. However, only a limited amount of information is typically utilized. Extensive sensitivity analysis is necessary to recommend a robust alternative. The marriage of a transparent technique such as compromise programming with fuzzy sets is an example of a hybrid decision-making tool available to planners.

14.1.2 Existing applications using fuzzy ideals

The concept of a fuzzy displaced ideal was probably born with the comment by Carlsson (1982): "Zeleny's theory of the displaced ideal would . . . be very useful in a fuzzy adaptation." Lai et al. (1994) used distance metrics to reduce a multiobjective problem to a two-objective problem. They are to (i) minimize the distance to an ideal solution, and (ii) maximize the distance to the worst solution.

Membership functions are assigned to the ideal and worst solutions to fuzzify the problem, weights are used to resolve the two remaining objectives. Decisions are reached by formulating the problem as a fuzzy linear programming problem, and solved using the standard Bellman and Zadeh (1970) approach.

An example of decision analysis with fuzzy composite programming can be found in Bárdossy and Duckstein (1992), where a MCDA problem is evaluated using a distance metric with one of the criteria being qualitative and subjective. A codebook, a set of membership functions used to describe categories of subjective information, is established which translates a cardinal scale selection of the subjective criteria into a fuzzy set. Application of the extension principle to combine the single fuzzy criterion with other, quantitative, criteria is demonstrated graphically. Bárdossy and Duckstein (1992) and a similar paper by Lee, Dahab, and Bogardi (1994) provide examples of using a fuzzy displaced ideal.

14.1.3 Fuzzy arithmetic operations

The theory of fuzzy sets, initiated by Zadeh (1965), defines a fuzzy set, A, by degree of membership, $\mu(x)$, over a universe of discourse, X, as:

$$\mu_A(x): X \rightarrow [0, 1] \quad (14.2)$$

Many operations on fuzzy sets use connectives called triangular norms: t-norms; and s-norms. The t models the intersection operator in set theory. Likewise, s models the union operator. The **min** and **max** operators are commonly used for t and s respectively, although the family of valid triangular norms is very large. Composition operators are also used to connect fuzzy sets. They include **sup** and **inf**. The **sup** operation is the supremum or maximum of its membership function over a universe of discourse. Likewise, **inf** refers to the minimum. The combination of composition operators and connectives produces a powerful framework for many operations. **sup-*t*** compositions (**max-min**), and **inf-*s*** compositions (**min-max**) are examples used in fuzzy operations. There are many texts on fuzzy sets, including Dubois and Prade (1978), Zimmerman (1987), Mares (1994), Sakawa (1994), and Pedrycz (1995).

Fuzzy arithmetic is made possible by the extension principle, which states that for $Y = f(X)$, $X(x)$ and $Y(y)$ are membership functions (equivalent to $\mu_x(x)$ and $\mu_y(y)$ respectively), there is:

$$Y(y) = sup_{x \in X; y = f(x)} X(x) \quad \text{where} \quad f: X \rightarrow Y, y \in Y \quad (14.3)$$

From this extension principle, fuzzy arithmetic operations such as addition, subtraction, multiplication, division, and exponentiation can be described.

14.2 FUZZY COMPROMISE APPROACH

14.2.1 Fuzzy distance metrics

By changing all inputs, for the calculation of distance metrics, from crisp to fuzzy – the distance metric is transformed to a fuzzy set. Measurement of distances between a fuzzy ideal and the fuzzy performance of alternatives can no longer be given a single value, because many distances are at least somewhat valid. Choosing the shortest distance to the ideal is no longer a straightforward ordering of distance metrics, because of overlaps and varying degrees of possibility. The resulting fuzzy distance metric, as the following approach will attempt to demonstrate, contains a great amount of additional information about the consequences of a decision.

The process of generating input fuzzy sets is not trivial. Certainly, arbitrary assignment is simple and may cover the range of possibility, but it is possible to encode a lot of information and knowledge in a fuzzy set. The process of generating appropriate fuzzy sets, accommodating available data, heuristic knowledge, or conflicting opinions, should be capable of presenting information accurately. This topic is not addressed in any great detail in this chapter. Appropriate techniques for fuzzy set generation should be considered to be specific to the type of problem being addressed, the availability of different types of information, and the presence of different decision makers (remaining discussions on properties of fuzzy distance metrics are for maximization problems; in other words, larger values for criteria are assumed to be better than smaller values, and the ideal solution tends to have larger values than the alternatives).

Fuzzification of criteria values is probably the most obvious use of fuzzy sets. There is a long history of published articles demonstrating decision problems with qualitative or subjective criteria. Fuzzy sets are able to capture many qualities of relative differences in perceived value of criteria among alternatives. Placement of modal values, along with curvature and skew of membership functions, can allow decision makers to retain what they consider degree of possibility for subjective criteria values.

Selection of criteria weights is an aspect that is typically subjective, usually with a rating on an interval scale. As a subjective value, criteria weights may be more accurately represented by fuzzy sets. Generating these fuzzy sets is also a subjective element. It may be difficult to get honest opinions about degree of fuzziness from a decision maker. It might actually be more straightforward to generate fuzzy sets for weights when multiple decision makers are involved! Then, at least, voting methods and other techniques are available for producing a composite, collective opinion. Regardless, more information can be provided about valid weights from fuzzy sets than from crisp weights.

Fuzzy sets for criteria values and criteria weights can both be expressed in three distinct forms (Figure 14.1). They are (a) uncertain (where: known with certainty is a special case with a small degree of fuzziness), (b) unknown, and (c) conflicting. Both (b) and (c) produce a somewhat conflicting interpretation of valid behavior.

Incorporation of vagueness in the ideal criteria values is an element that impacts rankings of alternatives. For example, if profit is a criteria, then what is the ideal amount of profit? Figure 14.2 shows how positive and negative ideals can be expressed as one-sided fuzzy sets. The three choices are (a) certain, (b) uncertain, and (c) unknown. The uncertain (c) case can also be considered as a fuzzy goal. Improvement in criteria value may not improve the level of satisfaction because the goal has already been completely achieved. The degree of certainty in which the ideals are known is expressed by the range of valid values.

The distance metric exponent, p, is likely the most imprecise or vague element of distance metric calculation. There is no single acceptable value of p for every problem, and it can be easily misunderstood.

Also, it is not related to problem information in any way except by providing parametric control over interpretation of distance. Fuzzification of the distance metric exponent, p, can take many forms but in a practical way it might be defined by one of the choices shown in Figure 14.3. (a) suggests the common practice of using $p = 2$. However, in (a), it is acknowledged that the distance metric exponent has a possibility of being as small as 1. (b) is the $p = 1$ equivalent. Larger values of p may also be valid but fuzzy exponential operations for large exponents result in an unmanageable degree of fuzziness

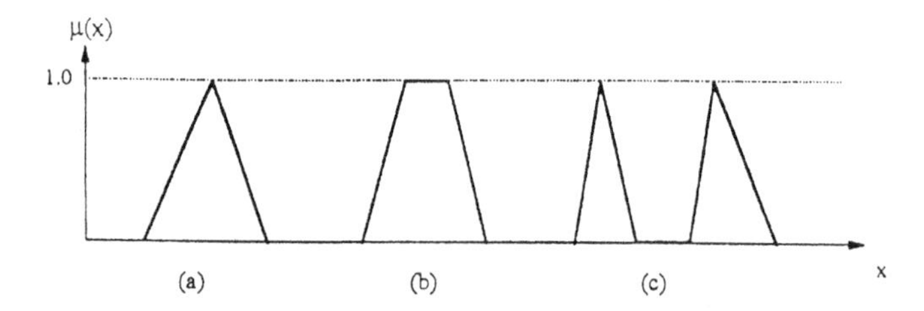

Figure 14.1. Fuzzy criteria values and weights.

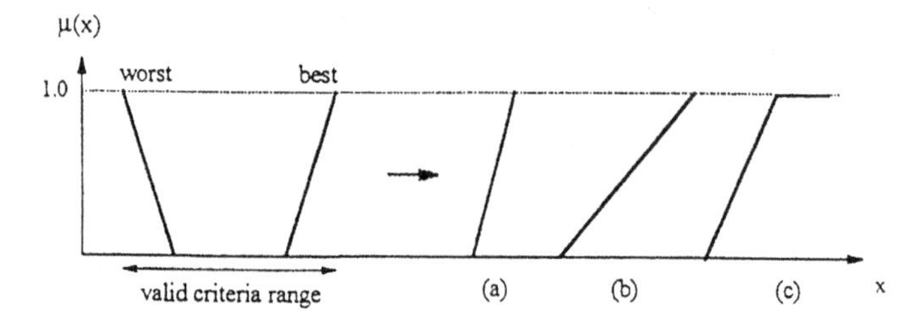

Figure 14.2. Range of valid criteria values as defined by fuzzy positive and negative ideals.

(range of possible values), making interpretation of the distance metric difficult.

The impact of fuzzy inputs on the shape of the resulting fuzzy distance metric is shown in Figures 14.4 and 14.5. Figure 14.4 shows typical shapes for L given triangular weights and criteria values, using different interpretations for p. Figure 14.5 shows the impact of unknown and uncertain weights or criteria values. For linear membership functions, areas about the modal value are impacted. The mode may be spread out, split, or both.

The benefits of adopting the general fuzzy approach to compromise programming are many. Probably the most obvious is the overall examination of decision uncertainty. Expressing possibility values with fuzzy inputs allows experience to play a significant role in the expression of input information. The shape of a fuzzy set expresses the experience or the interpretation of a decision maker. Conflicting data or preferences can also be easily expressed using multimodal fuzzy sets, making fuzzy compromise very flexible in adapting to group decision making.

14.2.2 Selecting acceptable alternatives

Nonfuzzy distance-based techniques measure the distance from an ideal point, where the ideal alternative would result in a distance metric, $L: X \rightarrow \{0\}$. In a fuzzy compromise approach, the

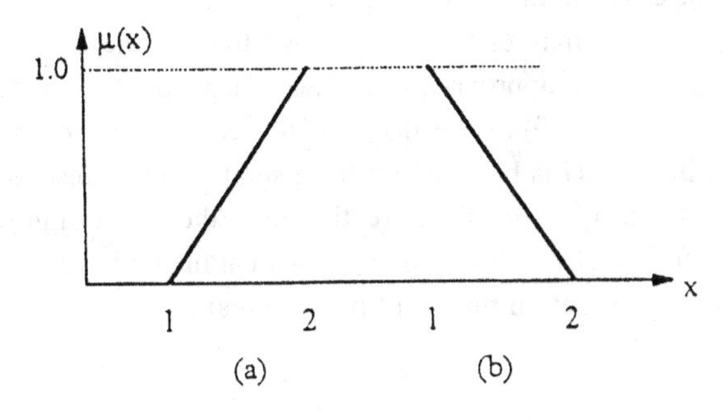

Figure 14.3. Fuzzy distance metric exponent.

distance is fuzzy, such that it represents all of the possible valid evaluations, indicated by the degree of possibility or membership value. Alternatives that tend to be closer to the ideal may be selected. This fuzzified distance metric is analogous to a sensitivity analysis for the nonfuzzy case.

As an attempt to standardize a procedure for judging which L is best among a set of alternatives, desirable properties can be defined. The most important properties are:

1. Possibility values tend to be close to the ideal, $x = 0$, distance.
2. Possibility values have a relatively small degree of fuzziness.

Some other performance indicators might favor modal values close to the ideal, or possibility values which tend to be far from poor solutions.

An aspect of comparing fuzzy distance metrics is the possible occurrence of points of indifference between fuzzy sets. If the rising limb of L_1 were to intersect the rising limb of L_2 (i.e., equal membership values at some distance from the ideal), a point of indifference would exist. This concept of indifference may vary. Interpretation of "best" depends on which side of the indifference point is considered to be interesting in the evaluation of comparative best. In the special case where the modes are equal, while the rising and falling limbs vary drastically, selection of the mode as the point of interest in ranking the sets will result in ranking the two fuzzy sets equally. Awareness of these indifference points may not be directly evident when ranking alternatives, but indifference points (depending on their location) may cause ranks to alter when fuzzy sets are examined under different "lighting" conditions.

Relative performance may be visually intuitive, but in cases where many alternatives display similar characteristics, it may be impractical or even undesirable to make a selection visually. A method for ranking alternatives may automate many of the visual interpretations, and create reproducible results. A ranking measure may also be useful in supplying additional

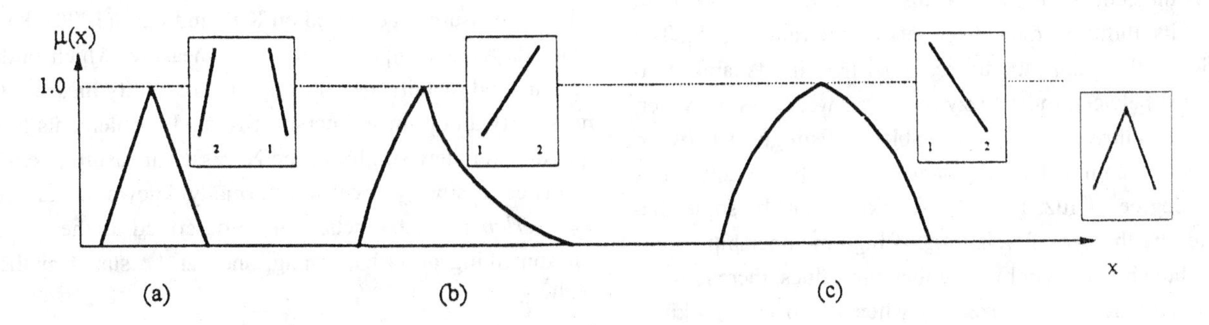

Figure 14.4. Fuzzy distance metrics for different fuzzy definitions of p.

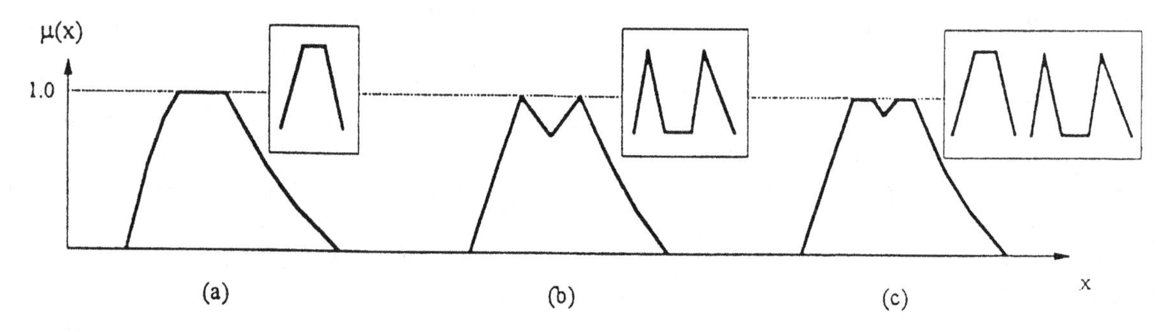

Figure 14.5. Fuzzy distance metrics for different forms of multimodal criteria values and weights.

insight into decision-maker preferences, such as (1) distinguishing between relative risk aversion and optimism in rank selection by decision makers, and (2) allowing adjustments in decision-maker emphasis for relatively extreme possibilities.

Selection of a method is subjective and specific to the form of problem and the fuzzy set characteristics that are desirable. A taxonomic examination of existing methods can be found in Bortolan and Degani (1985). There exists an assortment of methods ranging from horizontal and vertical evaluation of fuzzy sets to comparative methods. Some of these methods may independently evaluate fuzzy sets, while others use competition to choose among a selection list. Horizontal methods are related to the practice of defuzzifying a fuzzy set by testing for a range of validity at a threshold membership value. Vertical methods tend to use the area under a membership function as the basis for evaluation, such as center of gravity. The comparative methods introduce other artificial criteria for judging the performance of a fuzzy set, such as a fuzzy goal. Horizontal methods are not explored in this chapter. The following two methods are vertical and comparative, respectively.

14.2.3 Weighted center of gravity measure

Given the desirable properties of a ranking method for fuzzy compromise analysis, one technique that may qualify as a candidate is the centroid method, as discussed by Yager (1981) in terms of its ability to rank fuzzy sets on the range [0, 1]. The centroid method appears to be consistent in its ability to distinguish between most fuzzy sets. One weakness, however, is that the centroid method is unable to distinguish between fuzzy sets which may have the same centroid, but greatly differ in their degree of fuzziness. The weakness can be somewhat alleviated by the use of weighting. If high membership values are weighted higher than low membership values, there is some indication of degree of fuzziness when comparing rankings from different weighting schemes. However, in the case of symmetrical fuzzy sets, weighting schemes will not distinguish relative fuzziness.

A weighted centroid ranking measure (*WCoG*) can be defined as follows:

$$WCoG = \frac{\int g(x)\mu(x)^q dx}{\int \mu(x)^q dx} \tag{14.4}$$

where $g(x)$ is the horizontal component of the area under scrutiny, and $\mu(x)$ are membership function values. In practice, *WCoG* can be calculated in discrete intervals across the valid universe of discourse for L. *WCoG* allows parametric control in the form of the exponent, q. This control mechanism allows ranking for cases ranging from the modal value ($q = \infty$), which is analogous to an expected case or most likely scenario, to the center of gravity ($q = 1$), which signifies some concern over extreme cases. In this way, there exists a family of valid ranking values (which may or may not change too significantly). The final selection of appropriate rankings is dependent on the level of risk aversion from the decision maker. Ranking of fuzzy sets with *WCoG* is by ordering from smallest to largest value. The smaller the *WCoG* measure, the closer the center of gravity of the fuzzy set to the origin. As a vertical method of ranking, *WCoG* values act on the set of positive real numbers.

14.2.4 Fuzzy acceptability measure

Another ranking method that shows promise is a fuzzy acceptability measure, *Acc*, based on Kim and Park (1990). Kim and Park derive a comparative ranking measure, which builds on the method of Jain (1976) using the Possibility measure (*Poss*) to signify an optimistic perspective, and supplements it with a pessimistic view similar to the Necessity measure (*Nec*).

The Possibility measure, formally known as the *degree of overlap* for fuzzy sets, can be described as the possibility of something good happening, and can be stated mathematically as:

$$Poss(G, L) = sup_{x \in R}[\mu_G(x) t \mu_L(x)] \tag{14.5}$$

where L is the fuzzy set defined by $L: X \rightarrow [0, 1]$ and G is a fuzzy goal, defined by $G: X \rightarrow [0, 1]$.

The Necessity measure gives a pessimistic view, formally known as the *degree of containment*; it can be described as the necessity for ensuring something bad does not happen. The usefulness of *Nec* can be expressed as:

$$Nec(G, L) = inf_{x \in R}[\mu_G(x) t \bar{\mu}_L(x)] \quad (14.6)$$

where $\bar{\mu}_L$ is the complement $(1 - \mu_L)$ membership value.

The above two measures can be combined to form an acceptability measure (*Acc*):

$$Acc = \alpha Poss(G, L) + (1 - \alpha) Nec(G, L) \quad (14.7)$$

Parametric control with the acceptability measure (*Acc*) is accomplished with the α weight and the choice of fuzzy goal, G. The α weight controls the degree of optimism and degree of pessimism, and indicates (an overall) level of risk aversion. The choice of a fuzzy goal is not so intuitive. It should normally include the entire range of L, but it can be adjusted to a smaller range for the purpose of either exploring shape characteristics of L or to provide an indication of necessary stringency. By decreasing the range of G, the decision maker becomes more stringent in that the method rewards higher membership values closer to the ideal. At the extreme degree of stringency, G becomes a nonfuzzy number that demands the alternatives be ideal. As a function, G may be linear, but can also adapt to place more emphasis or less emphasis near the best value ($x = 0$ for distance metrics).

Ranking of fuzzy sets using *Acc* is accomplished by ordering values from largest to smallest. That is, the fuzzy set with the greatest *Acc* is most acceptable. *Acc* values are restricted on the range [0, 1] since both the *Poss* and *Nec* measures act on [0, 1], and α reduces the range of possible values by a factor of 2.

14.2.5 Comparison of ranking methods

In comparing ranking methods *WCoG* and *Acc* with those reviewed by Bortolan and Degani (1985), given the desirable properties of L, both proved to be superior to the methods given in the review. The problem with many available methods is that, although most are able to correctly identify the best fuzzy set, they may not be capable of both distinguishing degree of dominance and providing an ordinal ranking for more than two fuzzy sets. Many methods supplied ranking values, for example, as {1, 0, 0} for three fuzzy sets. Very little decision information is returned by those methods. Relative dominance among fuzzy sets is an important aspect for distinguishing between fuzzy distance metrics. Information of this type is provided by both *WCoG* and *Acc*.

WCoG is conceptually simple and visually intuitive. Its weakness in discerning between fuzzy sets with the same shape and modal value, yet with different degrees of fuzziness, is somewhat offset by the unlikely event of having distance metrics with those properties. Fuzzy distance metrics may have very similar shapes considering that all alternatives are evaluated for the same fuzzy definition of p. They may also have similar modes, depending on criteria values. Degree of fuzziness, or at least some discrepancy in shape, provides the means by which the weighting parameter, q, is able to distinguish indifference points. In general, though, interpretation of indifference points is not usually very sensitive to the choice in q.

Acc provides more comprehensive, and possibly more relevant, parametric control over the interpretation of results. *Acc* is able to explore the "surface" of fuzzy distance metrics with a meaningful interpretation of the variables used for parametric control (α, G). However, the parameters for the *Acc* measure are difficult to justify if some combination is used to recommend an alternative. The appropriate use of *Acc* is strictly to determine sensitivity, if any, of alternative rankings to different attitudes displayed by a decision maker.

Regardless of the combination of characteristics for fuzzy distance metrics, both the *WCoG* and *Acc* methods produced similar results that corresponded with visual interpretation of fuzzy distance metrics. Both methods satisfy the desirable properties for ranking fuzzy distance metrics. Both may prove to be useful in a decision-making problem with multiple alternatives. Choosing just one of these methods, or a completely different method (of which there are many), should be dependent on the desirable ranking properties of the given problem. In some cases, it may be advantageous to use more than one method as a form of verification.

14.3 EXAMPLES

The following examples are taken from the field of water resources planning. They are multicriteria decision problems that have been addressed using standard MCDA techniques to select a most desirable water management system alternative, either as a best compromise or as a robust choice. Each example below is redefined in fuzzy terms to demonstrate the fuzzy compromise approach.

14.3.1 Tisza River example

The Tisza River basin in Hungary was studied by David and Duckstein (1976) for the purpose of comparing alternative water resources systems for long-range goals. They attempt to follow a cost-effectiveness methodology to choose from five alternatives, but many of the twelve criteria are subjective. Eight criteria are subjective, and have linguistic evaluations assigned to them. Six of these subjective criteria are considered on a scale with five linguistic options {*excellent, very good, good, fair, bad*}. Two criteria are judged by different linguistic scales {*very easy, easy, fairly difficult, difficult*}, and {*very sensitive, sensitive, fairly sensitive, not sensitive*}. David and Duckstein (1976) provide numeric differences along an interval scale so that a discordance index can be calculated for the ELECTRE method.

David and Duckstein (1976) provide criteria weights to calculate the concordance index of ELECTRE. Weights were supplied from the set of {1, 2}. The technique used by ELECTRE somewhat alters the weighting issues in its use of a concordance index, and weights are not needed to calculate a discordance index, but it is not known what effect uncertainty in the weights has on assessing alternative trade-offs.

As a conclusion, David and Duckstein (1976) suggest that a mix of systems I and II would be appropriate, since they appear to somewhat dominate the other alternatives and show no overall domination over each other. Duckstein and Opricovic (1980) reached similar conclusions for the same system, using a different artificial scaling for subjective criteria. A useful improvement to evaluating water resources systems such as the Tisza River may be to treat uncertainties as fuzzy.

Fuzzy definitions may be used to form a "codebook" of linguistic terms used in assessing subjective criteria. Quantitative criteria can also be fuzzified, but are generally less fuzzy. Other fuzzy inputs include the expected ranges of criteria values, and the form of distance metric (degree of compensation) among criteria for different alternatives. Criteria weights are fuzzified on a range of [0, 1] by simple scaling of the weights as {1, 2} $\rightarrow$ {$.\tilde{33}$, $.\tilde{66}$}. All fuzzy inputs are treated in a simple form, exclusively normal and unimodal, with either triangular or one-sided membership functions.

Assuming a fuzzy definition for the distance metric exponent (p), and knowing the form of criteria values and weights to be triangular, the resulting fuzzy distance metrics (L_i) possess the characteristic shape (Figure 14.6) of near linearity below the mode, and a somewhat quadratic polynomial curvature above the mode. Although the degree of fuzziness is similar for all five alternatives, some of the alternatives are clearly inferior.

Ranking alternatives is reasonably straightforward because of the simplicity of the shapes, and similarity in degree of fuzziness. Both *WCoG* and *Acc* measures produced expected results (Table 14.1). Rankings are insensitive to changes in levels of risk aversion, as would be expected from visual inspection. The resulting ranks confirm the findings of David and Duckstein (1976), that alternatives I and II are dominant.

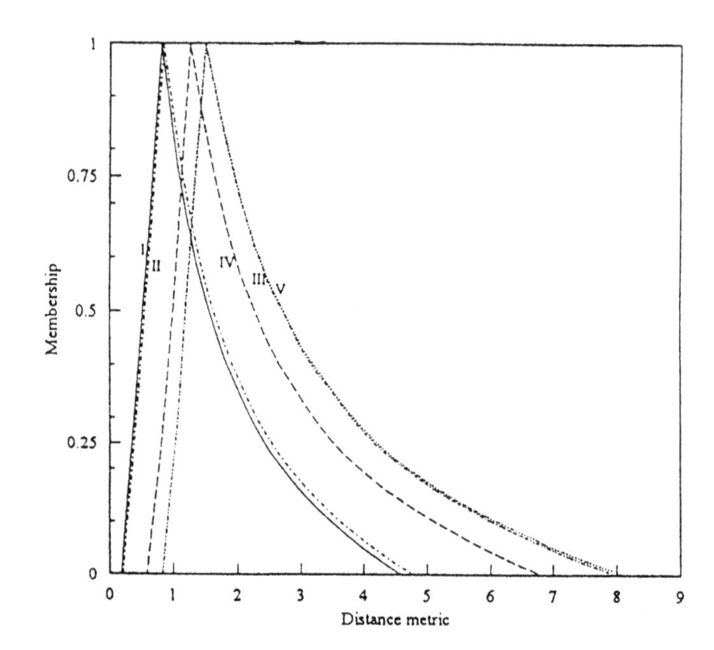

Figure 14.6. Distance metrics for David and Duckstein (1976).

Table 14.1. *Tisza River alternative rankings from WCoG and Acc measures*

Rank	Alt	WCoG (p = 1)	Acc (G:[0,8], α = 0.5)
1	1	1.49	0.81
2	2	1.59	0.80
3	4	2.38	0.75
4	3	2.83	0.72
5	5	2.85	0.71

In a live case study with multiple decision makers, there are opportunities for a group emphasis to collectively adjust fuzzy input to the Tisza River problem. The rankings may change considerably because the values defined for this experiment are predominately simple triangular membership functions, given the form of nonfuzzy input data. Adjustments in relative fuzziness, and the emergence of conflicting opinions about valid criteria values or weights, may produce an entirely new outlook, one that may be sensitive to the level of risk aversion characterized by the decision maker.

14.3.2 Yugoslavia (system S2) example

A MCDA problem for choosing a water resources system management alternative has been documented by Simonovic (1989). Simonovic explored two systems, each with the same eight criteria. Two criteria are quantitative, while six are subjective. System $S2$ has eight alternatives to choose from, and system $S1$ has six alternatives. Alternatives for both systems are evaluated with discrete compromise programming. Subjective criteria are judged on a scale {1, 2, 3, 4, 5} interpreted as *{bad, fair, good, very good, excellent}*. Weights are provided by six decision makers, for each criteria, each representing a different emphasis in decision making. Any or all of the weight sets may be valid. Criteria input came from Simonovic (1989), system $S2$.

The fuzzy compromise approach was applied to this problem. Fuzzy definitions for subjective criteria values are given triangular fuzzy set definitions on the range [0, 1]. In the Tisza River example, subjective criteria values were interpreted using normal unimodal triangular membership functions that do not overlap. For system $S2$, the range of valid x values is more continuous, there is less separation of fuzzy terms. For example, at $x = 0.4$, the fuzzy terms *fair* (2) and *good* (3) might both have a membership, $\mu(x) = 0.5$. The collection of fuzzy criteria weights (fuzzified using input from the six decision makers) can be defined as a collective opinion. Simple voting techniques or other fuzzy set generation methods can be used to generate the collective weights from a group of decision makers. Upon definition of p as linearly increasing on the range [1, 2] from a boundary {0} to the modal value {1}, the input necessary for the fuzzy compromise approach is complete. Fuzzy distance metrics can be calculated with fuzzy arithmetic operations.

Selected fuzzy distance metrics are shown in Figure 14.7. Notice, by visual inspection, that alternative {6} is dominated by {1, 3}. The alternatives {1, 3} are very similar, with slightly different modal values – {1} slightly smaller than {3}. Upon inspection of {1, 3}, there are two indifference points – on the rising limb (L-side) at $x = 0.45$, and on the falling limb (R-side) at $x = 0.8$. They suggest that ranking {1, 3} is subjective. Also notice that {3} has a lesser degree of fuzziness.

Although fuzzy weights have been defined in a collective fashion, it may be interesting to see the impact of alterations on the fuzzy inputs. For instance, there are many valid interpretations for specifying membership functions of fuzzy weights. The base case described above consists of predominately triangular input fuzzy sets, and a distance metric exponent defined as linearly increasing over [1, 2]. Using the six sets of nonfuzzy weights as the basis for generating different scenarios, the following test patterns for collective interpretation of criteria weighting are explored:

1. Conflicting opinions (multimodal weights).
2. Unknown or very uncertain preferences (trapezoidal weights).
3. Both conflicting and unknown opinions.

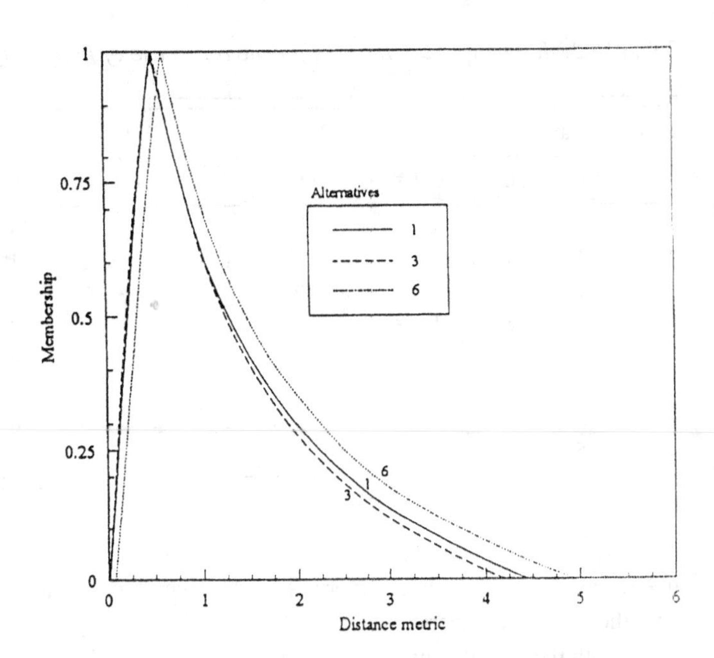

Figure 14.7. Fuzzy distance metrics: selected alternatives for $S2$.

4. Both conflicting and unknown opinions, with the distance metric exponent defined as linearly decreasing on the range [1, 2] with a modal value at $x = 1$.
5. Both conflicting and unknown opinions, with the distance metric exponent defined as nearly certain for $p = 2$, linearly increasing on the range [1.9, 2] with a modal value at $x = 2$.

For Test 1, the weights for criteria {5, 6} are treated as conflicting, where two distinct opinions emerge. The conflicting opinions expressed in the criteria 6 weight have modal values at $x = \{0.5, 0.9\}$, with valid membership values over the range [0.5, 0.95] and a single point at $x = 0.7$ where $\mu(x) = 0$. Resulting ordinal rankings from the *Acc* measure are shown in Table 14.2 for $\alpha = \{0.1, 0.5, 0.9\}$ and G linearly decreasing from $\mu(x) = 1$ to $\mu(x) = 0$, over the range [0, 4]. Although the two highest ranked alternatives {1, 3} do not change order over the sensitivity analysis of both the base case weights and the Test 1 case weights, there are several differences in ranking alternatives {2, 4, 6, 7, 8}.

Instead of interpreting the weights as conflicting, criteria {6, 7} may be interpreted to be very uncertain or unknown over part of the valid range (Test 2). The trapezoidal membership function chosen for the criteria 6 weight includes modal values on the range [0.6, 0.9], and membership values restricted on the range [0.4, 1.0]. The change in interpretation affects the rankings shown in Table 14.2 for the *Acc* measure.

Test 3 on system $S2$ uses conflicting weight definitions for criteria {5, 6} and a trapezoidal membership function to signify

Table 14.2. *Ranking of alternatives for the base case of fuzzy weights, and Tests 1 to 5*

Rank	Base case			Test 1			Test 2			Test 3			Test 4			Test 5		
	0.1	0.5	0.9	0.1	0.5	0.9	0.1	0.5	0.9	0.1	0.5	0.9	0.1	0.5	0.9	0.1	0.5	0.9
1	3	3	3	3	3	3	3	3	3	3	3	3	3	3	3	3	3	3
2	1	1	1	1	1	1	1	1	1	1	1	1	8	1	1	1	1	1
3	7	7	7	8	7	7	7	7	7	8	7	7	1	8	7	7	7	7
4	8	8	4	7	8	8	8	8	8	7	8	4	7	7	8	8	8	4
5	4	4	8	4	4	4	4	4	4	4	4	8	4	4	4	4	4	8
6	6	2	2	6	6	6	6	2	2	6	2	2	6	6	2	2	2	2
7	2	6	6	2	2	2	2	6	6	2	6	6	2	2	6	6	6	6
8	5	5	5	5	5	5	5	5	5	5	5	5	5	5	5	5	5	5

unknown preferences about criteria {7}. This combined test uses the default membership function for p varying linearly on [1, 2] with the modal value at {2}. Test 4 alters p by switching the modal point to {1}. Finally, Test 5 uses p defined on [1.9, 2] with modal value {2}.

Rankings using the *Acc* measure (Table 14.2) indicate some change in alternative ranking under the different distance metric definitions, and for varying degrees of risk aversion. Alternative 8, for example, is ranked as high as second and as low as fifth.

14.3.3 Yugoslavia (system S1) example

Another system examined by Simonovic (1989), *S1*, consists of six alternatives which are evaluated by the eight criteria given for *S2*. Simonovic considered (nonfuzzy) input from three decision makers in a compromise programming sensitivity analysis on subjective criteria values, p definition, and the sets of weights provided for system *S2*. The sensitivity analysis suggested that alternatives {3, 5} are reasonably robust in that they are consistently ranked at the top. Other alternatives that ranked high were {4, 6}. For system *S2*, criteria weights and the distance metric exponent were adjusted to reflect different interpretations of decision-maker preferences in Tests 1, 2, 3, 4, 5. Another form of uncertainty or vagueness arises when several decision makers all cast judgment on subjective criteria values. Each judgment may be, in itself, vague or imprecise. Combining the opinions of several people may serve to strengthen the impression of a subjective criterion, or it may contribute additional vagueness. This additional vagueness is demonstrated in a fuzzy test case for system *S1*.

Simonovic (1989) provides (nonfuzzy) selection of subjective criteria values by three decision makers on the range of {1, 2, 3, 4, 5}. They agree completely on a few criteria values for certain alternatives, but they generally (at least partially) disagree on most subjective criteria values.

The fuzzy interpretation of these somewhat different opinions is based on the fuzzy definitions of subjective criteria values used earlier. For instance, the three opinions for alternative {3}, criteria {4} are {4, 4, 4}. One can assume that this criteria value is quite certain about {4}. There are many combinations of partial agreement/disagreement such as {3, 3, 4} which result in more vague membership functions. There is also an instance where all three opinions differ {1, 2, 3}. This difference of opinion has been interpreted as conflicting in nature. All three examples can be found in Figure 14.8.

After redefining the fuzzy interpretation of the subjective criteria values to represent the combined opinions of the three decision makers, fuzzy distance metrics can be found. The difference in rankings between the original fuzzy interpretation, given by decision maker *A*, and the rankings realized from the collective opinion (using the *WCoG* measure to rank alternatives) is quite apparent (Table 14.3). All of the ranks vary except for alternatives {1, 2} which are consistently poor, comparatively. Amazingly, by adding the element of collaboration between decision makers, alternative 4 changed from a rank of 4 (when rated by decision maker *A* only) to the highest rating among alternatives for the collective opinion case!

14.4 CONCLUSIONS

With a technique to calculate the result of fuzzy arithmetic operations, and at least one measure to enable ranking of alternatives, the above fuzzy compromise approach may prove to be very useful to water resources systems planners. The idea of a displaced ideal is a relatively simple and intuitive concept used

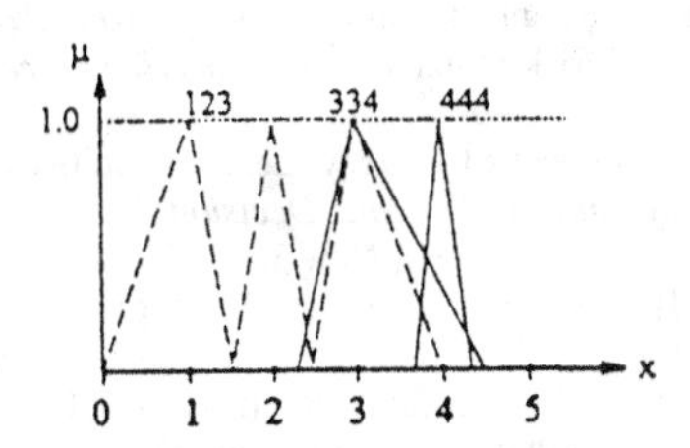

Figure 14.8. Collective criteria value definitions.

Table 14.3. *WCoG ranks for* S1 *base and test cases,* q = *{1, 2, 3}*

Alt	A			A, B, C,		
	1	2	3	1	2	3
1	5	5	5	3	4	4
2	6	3	3	5	5	5
3	3	6	6	4	3	6
4	4	4	4	6	6	3
5	2	2	2	2	2	2
6	1	1	1	1	1	1

in MCDA, and it easily adapts to fuzzy inputs. Intuitiveness and simplicity are two properties that are considered extremely important if a decision support tool is to be realized for multiple decision makers.

Fuzzy distance metrics can be inspected visually with a great degree of accuracy and minimal loss of information. This allows decision makers to make a qualitative evaluation of alternatives. Each decision maker may interpret the results differently, though, as a function of their relative degree of risk aversion. Subjective interpretation depending on risk aversion can be modeled in a ranking measure. The centroid measure (*WCoG*) discussed in this chapter is easily understood, but the acceptability measure (*Acc*) allows parametric control more specifically designed to model level of risk aversion from decision makers.

The examples demonstrate many characteristics of a fuzzy compromise approach as a multicriteria decision analysis technique. The Tisza River example showed consistency of results, compared to the ELECTRE method, without need for sensitivity analysis. The fuzzy compromise approach suggested a degree of dominance in the ordering, with both the *WCoG* and *Acc* ranking measures. The Yugoslavia examples were used to explore different interpretations in creating membership functions for the inputs. System *S2* was used to demonstrate sensitivity of rankings due to changes in the interpretation of distance metric exponent and criteria weight membership functions. System *S1* follows by showing the effects on rankings from considering collective opinions about subjective criteria values.

A fuzzy compromise approach has a number of comparative advantages over traditional (nonfuzzy) MCDA techniques. The most important is the direct and often intuitive incorporation of vague and imprecise forms of uncertainty to the decision-making process. In real decisions, many of the criteria are subjective in nature. By their very nature, subjective criteria are fuzzy. By allowing a degree of fuzziness, more realism is added to the evaluation without compromising on the technique's ability to disseminate alternative preferences. Similar observations can be made about criteria weights and the decision maker's interpretation of degree of compensation between criteria (p), all of which possess sufficient vagueness and imprecision to warrant skepticism when using traditional MCDA techniques.

By applying a fuzzy compromise approach, MCDA feedback changes from "most robust to uncertainties in perception" to "most robust to uncertainties in perception and performance." The assessment of sensitivity to degree of risk aversion allows alternatives to be chosen which are robust to the context of a given problem, in addition to the more traditional robustness which is robustness to both criteria emphasis and criteria measurement.

At a time when group decision making is becoming more common, a fuzzy compromise approach facilitates collaborative exploration of available alternatives and their associated risks. Collective opinions are incorporated by increasing (decreasing) the fuzziness of the inputs, and by locating ranges or multiple points of opinion. Fuzzy sets are able to process this kind of information, and are also able to present it effectively and intuitively.

Acknowledgments

The authors would like to thank Manitoba Hydro for continued support of this project in the form of a graduate scholarship. This work was also made possible by a University of Manitoba Graduate Fellowship, and a Natural Sciences and Engineering Research Council grant. Special thanks go to Dr. Witold Pedrycz from the Department of Electrical and Computer Engineering at the University of Manitoba, who eagerly followed our progress and participated in numerous discussions.

REFERENCES

Bárdossy, A. and Duckstein, L. (1992) Analysis of a karstic aquifer management problem by fuzzy composite programming. *Water Resources Bulletin* 28(1): 63–73.

Bellman, R. and Zadeh, L. (1970) Decision making in a fuzzy environment. *Management Science* 17: 141–64.

Benayoun, R., Roy, B., and Sussmann, B. (1966) ELECTRE: Une méthode pour quider le choix en présence de points de vue multiples. Note Trav. 49, Dir. Sci., *Soc. Econ. Math. Appl.*, Paris.

Bender, M. J. and Simonovic, S. P. (2000) A fuzzy compromise approach to water resource systems planning under uncertainty. *Fuzzy Sets and Systems* 115: 35–44.

Blin, J. (1974) Fuzzy relations in group decision theory. *J. Cybernetics* 4(2): 17–22.

Bortolan, G. and Degani, R. (1985) A review of some methods for ranking fuzzy subsets. *Fuzzy Sets and Systems* 15: 1–19.

Carlsson, C. (1982) Tackling an MCDM problem with the help of some results from fuzzy set theory. *European J. Operational Research* 10: 270–81.

David, L. and Duckstein, L. (1976) Multi-criterion ranking of alternative long-range water resource systems. *Water Resources Bulletin* 12(4): 731–54.

Dubois, D. and Prade, H. (1978) Operations on fuzzy numbers. *Int. J. Systems Science* 9(6): 613–26.

Duckstein, L. and Opricovic, S. (1980) Multiobjective optimization in river basin development. *Water Resources Research* 16(1): 14–20.

Dyer J., Fishburn, P., Steuer, R., Wallenius, J., and Zionts, S. (1992) Multiple criteria decision making multiattribute utility theory: the next 10 years. *Management Science* 38(5): 645–54.

Felix, R. (1994) Relationships between goals in multiple attribute decision making. *Fuzzy Sets and Systems* 67: 47–52.

Gershon, M., Duckstein, L., and McAniff, R. (1982) Multiobjective river basin planning with qualitative criteria. *Water Resources Research* 18(2): 193–202.

Goicoichea, A., Hansen, D., and Duckstein, L. (1982) *Multiobjective Decision Analysis with Engineering and Business Applications.* Wiley and Sons, New York.

Jain, R. (1976) Decision-making in the presence of fuzzy variables. *IEEE Trans. Systems, Man, and Cybernetics* 6: 698–703.

Kim, K. and Park, K. (1990) Ranking fuzzy numbers with index of optimism. *Fuzzy Sets and Systems* 35: 143–50.

Lai, Y.-J., Liu, T.-Y., and Hwang, C.-L. (1994) TOPSIS for MODM. *European J. Operational Research* 76: 486–500.

Lee, Y. W., Dahab, M. F., and Bogardi, I. (1994) Fuzzy decision making in ground water nitrate risk management. *Water Resources Bulletin* 30(1): 135–48.

Leung, Y. (1982) A concept of a fuzzy ideal for multicriteria conflict resolution. In *Fuzzy Information and Decision Processes*, edited by M. Gupta and E. Sanchez, Elsevier, New York.

Mares, M. (1994) *Computation over Fuzzy Quantities*. CRC Press, London.

Munda, G., Nijkamp, P., and Rietveld, P. (1995) Qualitative multicriteria methods for fuzzy evaluation problems: an illustration of economic-ecological evaluation. *European J. Operational Research* 82: 79–97.

Nishizaki, I. and Seo, F. (1994) Interactive support for fuzzy trade-off evaluation in group decision making. *Fuzzy Sets and Systems* 68: 309–25.

Pedrycz, W. (1995) *Fuzzy Sets Engineering*. CRC Press, London.

Saaty, T. (1980) *The Analytic Hierarchy Process*. McGraw-Hill, New York.

Sakawa, M. (1993) *Fuzzy Sets and Interactive Multiobjective Optimization.* Plenum Press, New York.

Seo, F. and Sakawa, M. (1985) Fuzzy multiattribute utility analysis for collective choice. IEEE *Transactions on Systems, Man and Cybernetics* 15(1): 45–53.

Simonovic, S. (1989) Application of water resources systems concepts to the formulation of a water master plan. *Water International* 14(1): 37–51.

Siskos, J. (1982) A way to deal with fuzzy preferences in multicriteria decision problems. *European J. Operational Research* 10: 314–24.

Slowinski, R. and Teghem J., (ed.) (1990) *Stochastic versus Fuzzy Approaches to Multi-objective Mathematical Programming Problems under Uncertainty*. Kluwer Academic, Dordrecht.

Yager, R. (1981) A procedure for ordering fuzzy subsets of the unit interval. *Information Science* 24: 143–61.

Zadeh, L. (1965) Fuzzy Sets. *Information and Control* 8: 338–53.

Zeleny, M. (1973) Compromise Programming. In *Multiple criteria decision making*, edited by J. Cochrane and M. Zeleny. University of South Carolina Press, Columbia.

(1982) *Multiple Criteria Decision Making*. McGraw-Hill, New York.

Zimmerman, H. (1987) *Fuzzy Sets, Decision-making, and Expert Systems.* Kluwer Academic, Dordrecht.

15 System and component uncertainties in water resources

BEN CHIE YEN*

ABSTRACT
Management of water resources is inherently subject to uncertainties due to data inadequacy and errors, modeling inaccuracy, randomness of natural phenomena, and operational variability. Uncertainties are associated with each of the contributing factors or components of a water resources system and with the system as a whole. Uncertainty can be measured in terms of the probability density function, confidence interval, or statistical moments such as standard deviation or coefficient of variation. In this chapter, available methods for uncertainty analysis are briefly reviewed.

15.1 INTRODUCTION

The planning, design, and operation of water resources systems usually involve many components or contributing factors. Each of the components or factors individually and the system as a whole are always subject to uncertainties. For example, the reliability of flood forecast depends not only on the uncertainty of the prediction model itself but also on the uncertainties on the input data. The design of a storm drain is subject to the uncertainty on the runoff simulation model used, uncertainties on the design storm determination, as well as uncertainties on the materials, construction, and maintenance used. Water supply is always subject to uncertainties on the demand, availability of the sources of water, and the performance of the distribution network. Safety of a dam depends not only on the magnitude of the flood but also on the waves, earthquake, conditions of the foundation, and maintenance and appropriateness of the operational procedure. Numerous other examples can be cited. In short, decisions on water resources are always subject to uncertainties.

Knowledge of uncertainties is useful for rational decision making, for cost-effective design, for safe operation, and for improved public awareness of water resources risks and reliability. Uncertainty of the system as a whole is a weighted probabilistic combination of the component uncertainties. Formal, quantitative uncertainty analysis rarely has been done in water resources management, but it should be for the reasons just mentioned. Comparison of relative contributions of the component uncertainties to the system uncertainty provides information on identifying the weak components whose strengthening could be most effective in uncertainty reduction of the system.

In this chapter existing methods for uncertainty analysis are briefly reviewed, including the approximate methods for first-order variance estimation (FOVE), point estimate (PE) techniques, Monte Carlo simulation, and the analytical technique of integral transformations. Availability of these methods allows quantitative determination of uncertainties of all kinds of water resources problems.

15.2 SOURCES OF UNCERTAINTIES

Uncertainties reflect our lack of perfect understanding with regard to the phenomena and processes involved in addition to the random nature of the occurrence of the events. In water resources analysis, design, and management, uncertainties could arise from the following sources (Yen, Cheng, and Melching 1986):

1. Natural uncertainties associated with the inherent randomness of natural processes;
2. Model uncertainty reflecting the inability of the simulation model or design technique to represent precisely the system's true physical behavior;
3. Model parameter uncertainties resulting from inability to quantify accurately the model input parameters;
4. Data uncertainties including (a) measurement errors, (b) inconsistency and nonhomogeneity of data, (c) data handling and transcription errors, and (d) inadequate representativeness of data sample due to time and space limitations;

* University of Illinois at Urbana-Champaign, 205 N. Mathews Ave., Urbana, IL 61801 USA.

5. Operational uncertainties including those associated with construction, manufacture, deterioration, maintenance, and other human factors that are not accounted for in the modeling or design procedure.

Yen and Ang (1971) classified uncertainties into two types: objective uncertainties associated with any random process or deductible from statistical samples and subjective uncertainties for which no quantitative factual information is available. Yevjevich (1972) distinguished the basic risk due to inherent randomness of the process and uncertainty due to the various other sources; the overall risk is the combined effect of the basic risk and other uncertainties. Burges and Lettenmaier (1975) categorized two types of uncertainties associated with mathematical modeling. Type I error results from the use of an inadequate model with correct parameter values. Type II error assumes the use of a perfect model with parameters subject to uncertainty.

The existence of various uncertainties, including inherent randomness of natural processes, is a major contributor to potential failure of water resources projects. Determination of component uncertainties is a necessary basis for estimation of the overall reliability of a water resources system (Yen and Tung 1993). The main objective of uncertainty analysis is to assess quantitatively the statistical properties, such as the probability density function (pdf) or statistical moments, of a system subject to uncertainty. In water resources design and modeling, the design quantity and model output are functions of a number of model parameters, not all of which can be quantified with absolute accuracy. The task of uncertainty analysis is to determine the uncertainty features of the model output as a function of uncertainties in the model itself and in the stochastic input parameters involved. Uncertainty analysis provides a formal and systematic framework to quantify the uncertainty associated with the model output. Furthermore, it offers the designer or modeler an insight regarding the contribution of each stochastic input parameter to the overall uncertainty of the model output. Such knowledge is essential to identify the important parameters to which more attention should be given in order to have a better assessment of their values and, accordingly, to reduce the overall uncertainty of the output.

15.3 DIFFERENT MEASURES OF UNCERTAINTIES

Several expressions have been proposed to describe the degree of uncertainty of a parameter, a function, a model, or a system. The latter three usually depend on a number of parameters; therefore, their uncertainty is a weighted combination of the uncertainties of the contributing parameters.

Traditionally an indirect way to account for uncertainties is through the use of an ignorance factor such as the safety factor or freeboard. The drawback of this approach is that the ignorance factor is determined arbitrarily and it sheds no light on the relative importance of the component uncertainties.

The most complete and ideal description of uncertainty is the pdf of the quantity subject to uncertainty. However, in most practical problems such a probability function cannot be derived or found precisely.

Another method to express the uncertainty of a quantity is to express it in terms of a reliability domain such as the confidence interval. The methods to evaluate the confidence interval of a parameter on the basis of data samples are well known and can be found in standard statistics and probability reference books. Nevertheless, this method of confidence interval has a few drawbacks, including: (a) the parameter population may not be normally distributed as assumed in the conventional procedures to determine the confidence interval. This problem is particularly important when the sample size is small; (b) there exists no means to combine directly the confidence intervals of individual contributing random components to give the confidence interval of the system as a whole.

A useful alternative to quantify the level of uncertainty is to use the statistical moments of the random variable. In particular, the second moment is a measure of the dispersion of a random variable. Obviously, either the variance or standard deviation can be used; usually they can easily be estimated if sample data of the variable are available. The coefficient of variation, which is the ratio of standard deviation to the mean, offers a normalized measure that is useful and convenient for comparison and for combining uncertainties of different variables.

15.4 METHODS FOR UNCERTAINTY ANALYSIS

Methods for performing uncertainty analysis vary in the level of sophistication. They are also dictated by the information available regarding the stochastic input parameters. In principle, it would be ideal to derive the exact probability distribution of the model output as a function of those of the stochastic input parameters. However, most of the models or design procedures used in water resources are nonlinear and highly complex, preventing the probability distribution of model output to be derived analytically. As a practical alternative, engineers frequently resort to methods that yield approximations to the statistical properties of uncertain model output. The following subsections briefly introduce some of the methods that can be and have been used for uncertainty analysis. Readers should refer to the references cited for more details of the individual methods.

15.4.1 First-order variance estimation method

This method estimates uncertainty in terms of the variance of system output which is evaluated using the variances of individual contributing factors. Consider that a project quantity Y is related to a number of input parameters X_j's as

$$Y = g(\mathbf{X}^T) = g(X_1, X_2, \ldots, X_n) \tag{15.1}$$

where $\mathbf{X}$ is an n-dimensional column vector of input parameters in which all X_j's are subject to uncertainty, the superscript "T" represents the transpose, and $g(\,)$ denotes a functional relationship. In this method, the function $g(\,)$ is approximated by the first-order Taylor series expansion with respect to the mean values of stochastic input parameters X_j's as

$$Y \approx g(\boldsymbol{\mu}) + \mathbf{s}^T \cdot (\mathbf{X} - \boldsymbol{\mu}) \tag{15.2}$$

in which μ is the vector of mean values of $\mathbf{X}$ and $\mathbf{s}$ is an n-dimensional vector containing the sensitivity coefficients which can be evaluated as

$$s_j = \left[\frac{\partial g(\mathbf{X})}{\partial X_j}\right]_{X=\mu}, \quad j = 1, 2, \ldots, n \tag{15.3}$$

The sensitivity coefficients s_j represent the rate of change of model output with respect to unit change of input parameters X_j when all parameters are at their mean values. When all stochastic input parameters are independent, the mean and variance of the model output Y can be approximated as

$$\mu_Y \approx g(\boldsymbol{\mu}) \tag{15.4}$$

$$\sigma_Y^2 = \sum_{j=1}^{n} s_j^2\, \sigma_j^2 \tag{15.5}$$

in which σ is the standard deviation. From equation 15.5, the ratio, $s_j(\sigma_j/\mu_j)^2/(\sigma_Y/\mu_Y)^2$, indicates the proportion of overall uncertainty in the model output contributed by uncertainty associated with input parameter X_j. Further details of this method can be found in Yen et al. (1986).

The method can be expanded to include the second-order terms in equation 15.2 and to incorporate the correlation between stochastic input parameters. In general, the first two moments are used in uncertainty analysis for practical engineering design. To estimate higher-order moments of Y, the method can be implemented straightforwardly only when the stochastic input parameters are uncorrelated. The analysis does not require knowledge of the pdf of input parameters. Information regarding the pdf of the input parameters is generally difficult to obtain. In many engineering analyses and designs, model parameters have to be quantified subjectively based on engineering judgment and professional experience.

15.4.2 Rosenblueth's and similar Point Estimate (PE) methods

Rosenblueth's (1975) method is also a computationally easy technique in uncertainty analysis. Essentially, the method is based on a Taylor series expansion about the means of input variables X_j's. It can be used to estimate statistical moments of any order of a model output involving several stochastic input variables which are either correlated or uncorrelated (Harr 1987). Considering only the first two moments, the method estimates the kth order moment about the origin for the model output $\mathbf{Y}$ by approximating the total probability mass of a random variable X_j to be concentrated at two points which are located one standard deviation away from the mean of the variable.

Consider n stochastic input parameters which are correlated. The kth moment of $\mathbf{Y}$ about the origin can be approximated as

$$E(\mathbf{Y}^k) \approx \sum p(\delta_1, \delta_2, \ldots, \delta_n)\, \mathbf{Y}^k(\delta_1, \delta_2, \ldots, \delta_n) \tag{15.6}$$

in which the parameter δ_j is a sign indicator (+ or −) representing the input parameter X_j having the value of $\mu_j + \sigma_j$ or $\mu_j - \sigma_j$, respectively; $p(\delta_1, \delta_2, \ldots, \delta_n)$ is determined as

$$p(\delta_1, \delta_2, \ldots, \delta_n) = \frac{1 + \sum\sum_{j \neq j'} \delta_j \delta_{j'}\, r_{jj'}}{2^n} \tag{15.7}$$

where $r_{jj'}$ is the correlation coefficient between input parameters X_j and $X_{j'}$. The number of terms in the summation of equation 15.6 is 2^n which corresponds to the total number of possible combinations of + and − for all n stochastic input parameters. For example, consider Y being a function of two correlated variables X_1 and X_2. The kth moment of Y about the origin can be approximated as

$$E(Y^k) \approx p_{++} Y_{++}^k + p_{--} Y_{--}^k + p_{+-} Y_{+-}^k + p_{-+} Y_{-+}^k \tag{15.8}$$

where

$$\begin{aligned} p_{++} &= p_{--} = (1 + r_{12})/4 \\ p_{+-} &= p_{-+} = (1 - r_{12})/4 \end{aligned} \tag{15.9}$$

and

$$\begin{aligned} Y_{++}^k &= Y^k(\mu_1 + \sigma_1, \mu_2 + \sigma_2) \\ Y_{+-}^k &= Y^k(\mu_1 + \sigma_1, \mu_2 - \sigma_2) \\ Y_{-+}^k &= Y^k(\mu_1 - \sigma_1, \mu_2 + \sigma_2) \\ Y_{--}^k &= Y^k(\mu_1 - \sigma_1, \mu_2 - \sigma_2) \end{aligned} \tag{15.10}$$

Once the moments about the origin of model output Y are estimated, the central moments can be computed by

$$\mu_k = \sum_{i=0}^{k} {}_kC_i(-i)^i \mu^i \mu'_{k-i} \qquad (15.11)$$

where ${}_kC_i = k!/[(k-i)!i!]$, μ_k and μ'_k are the kth order central moment and moment about the origin, respectively. Later Rosenblueth (1981) improved his two-point method to incorporate the asymmetry of the random variables.

Rosenblueth's PE method requires 2^n evaluations when n random variables are involved. Harr (1989) proposed an alternative probabilistic PE method which reduces the required model evaluations from 2^n to $2n$. The method is a second-moment method that is capable of taking into account the first two moments (that is, the mean and variance) of the involved random variables and their correlations. Skew coefficients of the variables are ignored by the method. Hence, the method is appropriate for treating random variables that are symmetric. For problems that involve only single random variable, Harr's PE method is identical to Rosenblueth's method with zero skew coefficient. The theoretical basis of Harr's PE method is built on orthogonal transformations of the correlation matrix.

Recently, Li (1992) proposed a computationally practical PE method that allows incorporation of the first four moments of individual random variable and the correlations among the variables. Li's PE method is a three-point representation to which Rosenblueth's two-point representation is a special case. The algorithm for Li's three-point representation requires $(n^2 + 3n + 2)/2$ model evaluations. The amount of computation for problems involving multivariate normal variables is about the same as Harr's method.

15.4.3 Integral transformation techniques

Integral transform techniques, such as the well-known Fourier transform, Laplace transform, and exponential transform, can also be used for uncertainty analysis. Another useful but less known transform technique to the water engineering community is the Mellin transform (Epstein 1948; Park 1987). In all these transforms, the pdf of the random variable, $f(x)$, is known. A brief description of these integral transforms is given in the following.

(a) Fourier Transform – The Fourier transform of a function $f(x)$ is defined as

$$\mathfrak{F}_f(\omega) = \int_{-\infty}^{\infty} f(x)e^{i\omega x}dx \qquad (15.12)$$

where $i = \sqrt{-1}$. If the function $f(x)$ is the pdf of a random variable X, the resulting Fourier transform $\mathfrak{F}(\omega)$ is called the characteristic function. Using the characteristic function, the r-th moment about the origin of the random variable X can be obtained as

$$E[X^r] = (-1)^r \left[\frac{d^r \mathfrak{F}_f(\omega)}{d\omega^r}\right]_{\omega=0} \qquad (15.13)$$

The Fourier transform is particularly useful when random variables are related linearly. In such cases, the convolutional property of the Fourier transform can be applied to derive the characteristic function of the resulting random variable. More specifically, consider that $Y = X_1 + X_2 + \ldots + X_k$ and all X's are independent random variables with known pdf: $f_i(x)$. The characteristic function of Y can be obtained as

$$\mathfrak{F}_Y(\omega) = \mathfrak{F}_{f_1}(\omega)\,\mathfrak{F}_{f_2}(\omega)\ldots\mathfrak{F}_{f_k}(\omega) \qquad (15.14)$$

which is the multiplication of characteristic functions of each individual random variable X_i. The resulting characteristic function for Y can be used in equation 15.13 to obtain statistical moments of any order for Y. Further, inverse transform of $\mathfrak{F}_Y(\omega)$ can be made to derive the pdf of Y.

(b) Laplace and Exponential Transforms – The Laplace and exponential transforms of a function $f(x)$ are defined, respectively, as

$$\mathfrak{L}_f(s) = \int_0^{\infty} e^{-sx} f(x)dx \qquad (15.15)$$

$$\mathfrak{E}_f(s) = \int_{-\infty}^{\infty} e^{sx} f(x)dx \qquad (15.16)$$

The transformations given by equations 15.15 and 15.16 of a pdf are called moment generating functions. Similar to the characteristic function, statistical moments of a random variable X can be derived as

$$E[X^r] = (-1)^r \left[\frac{d^r \mathfrak{E}_f(s)}{ds^r}\right]_{s=0} \qquad (15.17)$$

However, the moment generating function of a random variable may not exist for all functions.

(c) Mellin Transform – The Mellin transform is particularly attractive for models involving nonnegative, independent input parameters where the functional relation between model output and input parameters is multiplicative. In water resources analyses, one often selects such a model that the two requirements are satisfied. The Mellin transform, $M_f(s)$, of a function $f(x)$, is defined as (Giffin 1975; Springer 1979)

$$M_f(s) = M[f(x)] = \int_0^{\infty} x^{s-1} f(x)dx, \quad x > 0 \qquad (15.18)$$

When $f(x)$ is a pdf, one can immediately recognize the relationship between the Mellin transform of a PDF and the moments about the origin as

$$\mu'_{s-1} = E(X^{s-1}) = M_f(s) \qquad (15.19)$$

for $s = 1, 2, \ldots$ As can be seen, the Mellin transform provides an alternative way to find the moments of any order of a nonnegative random variable. The Mellin transform possesses a powerful convolutional property and some useful operational characteristics (Giffin 1975; Springer 1979). An extensive tabulation of Mellin transforms of mathematical functions is given by Bateman (1954). Knowing the distribution of each stochastic input parameter involved and the multiplicative form of the model, one can easily obtain the exact moments of model output about the origin without extensive simulation. In principle, the pdf of the model output can be derived by inverting the Mellin transform of the model output. However, such inverse transform involves integration in complex variable space which is a rather formidable task. Therefore, for all practical purposes, one is satisfied with knowing the moments of the model output.

The restrictions of nonnegative and uncorrelated random variables are not absolutely required. The removal of such restrictions will increase mathematical manipulation tremendously (Springer 1979). When applying the Mellin transform, there is one potential weakness, namely, under certain combinations of distributions and functional forms, the resulting transform may not be analytic for all s values. This could occur especially when quotients or variables with negative exponents are involved.

The Fourier transform of a pdf leads to the characteristic function while the Laplace and exponential transforms yield the moment generating function. The Laplace transform is the same as the exponential transform except that the Laplace transform is only applicable to nonnegative random variables. Tables of characteristic functions and moment generating functions can be found in standard mathematical handbooks (e.g., Abramowitz and Stegun 1972) or special publications (Bateman 1954). For instance, for uniform distribution the characteristic function is $(e^{isb} - e^{isa})/(b - a)t$ and the moment generating function is $(e^{sb} - e^{sa})/(b - a)t$, and for Gumbel (EVI) distribution $e^{i\alpha s}\Gamma(1 - \beta s)$ and $e^{\alpha s}\Gamma(1 - i\beta s)$ with $s < 1/\beta$. Once the characteristic function or moment generating function is derived, statistical moments of a random variable can be computed. It is known that the characteristic function of a pdf always exists while the moment generating function may not. Another disadvantage of the moment generating function is that functions for exponential transform are not extensively tabulated. One nice thing about integral transforms is that, if such transforms of a pdf exist, the relationship between the pdf and its integral transform is unique (Kendall, Stuart, and Ord 1987).

In dealing with a multivariate problem in which a random variable is a function of several random variables, the convolutional property of these integral transforms becomes analytically powerful, especially when stochastic input parameters in the model are independent. Fourier and exponential transforms are powerful in treating the sum and difference of random variables, while the Mellin transform is attractive to the quotient and product of random variables (Giffin 1975; Springer 1979).

15.4.4 Monte Carlo simulation

The Monte Carlo simulation technique, in principle, generates random values of stochastic input parameters according to their respective probabilistic characteristics. A large number of random parameter sets are generated to compute the corresponding model output Y. Then, analyses are performed upon the simulated model output to determine the statistical characteristics of model output such as the mean, the variance, the pdf, and the confidence interval.

The basic idea of Monte Carlo simulation is simple. The core of the computation lies on the random number generation based on a specified probability law. However, the main concern of using Monte Carlo simulation is its computational intensiveness. There are variations of Monte Carlo simulation that allow one to reduce computation time and to reduce bias (Rubinstein 1981). Another implementation concern of Monte Carlo simulation is the generation of correlated random numbers. There are several computationally efficient techniques to generate random values of correlated normal variables (Nguyen and Chowdhury 1985; Quimby 1986). Correlated lognormal random variables pose no difficulty. However, difficulty arises when one deals with a mixture of correlated non-normal and normal random variables.

Another form of Monte Carlo simulation which recently has been studied extensively in statistical literature is called the bootstrap resampling technique. The technique was first proposed by Efron (1979a, 1979b, 1982) to deal with variance estimation of sample statistics based on observations. The technique is applicable when sample observations are available. The technique intends to be a more general and versatile procedure for sampling distribution problems without having to rely heavily on the normality conditions on which classical statistical inferences are based. In fact, it is not uncommon to observe non-normal data in water resources problems. Similar to Monte Carlo simulation, the bootstrap technique is also computationally intensive – a price to pay to break away from dependence on normality theory; such a concern is diminishing as the capability of computers increases.

Since the bootstrap resampling technique was introduced, it has rapidly caught the attention of statisticians and those who apply statistics in their research work. The bootstrap technique and its variations have been applied to various statistical problems such as bias estimation, regression analysis, time-series analysis, and others. An excellent overall review and summary of bootstrap techniques, their variations, and other resampling procedures such as the jackknife method are given by Efron (1982). An application of the bootstrap procedure to assess the confidence interval of optimal risk-based design parameters is given by Tung (1993).

15.5 REMARKS ON UNCERTAINTY ANALYSIS TECHNIQUES

Each of the four uncertainty analysis techniques described in this chapter has different levels of sophistication, computational complexity, and data requirements. The first-order variance estimation method needs only the mean and standard deviation of the random parameters involved in water resources system. Correlation between random model parameters can be easily incorporated to estimate the variance of the model output. However, such correlations cannot be used to estimate the mean value by the method without including the second-order term in equation 15.2. For estimating the higher-order moments of the model output, the method requires knowledge about higher-order moments of individual model parameters. The use of first-order variance estimation method does not require the complete information regarding to the pdf's of the stochastic parameters. Computationally, the method requires calculation of sensitivity coefficients which are the first-order partial derivatives of model output with respect to all stochastic model parameters. If a water resources model is complex or non-analytical, computations of sensitivity coefficients could be cumbersome. Higher-order moments of a model output involving stochastic input parameters can be estimated by the method with straightforward, but not necessarily simple, extension when input parameters are uncorrelated. Estimations of higher-order moments by the method involving correlated stochastic parameters require knowing the product moments that can be obtained with additional computations when data for model parameters are available.

Like the first-order variance estimation method, point estimate method for uncertainty analysis requires knowledge of the mean, variance, and correlation of stochastic input parameters. It also can consider the asymmetry of model input parameters. Computationally, Rosenblueth's two-point estimate method could be simpler than the first-order variance estimation method. The main factor dictating the computational effort of PE methods is the number of stochastic input parameters in a given model. As discussed previously, Harr's (1989) two-point method and Li's (1992) three-point method require less computation but have other restrictions.

In uncertainty analysis, the methods described in this chapter provide approximations to the statistical moments of a model output in terms of those of the random variables in the model. Both the first-order variance estimation and Rosenblueth's PE method provide approximations to the statistical moments of a model output in terms of those of the model inputs. The FOVE method has been known to yield rather accurate estimations of the first two moments if nonlinearity of the model or the uncertainty of random variables are not too large. As the nonlinearity or parameter uncertainty increases, the accuracy of the FOVE method deteriorates rapidly as the order gets higher. Investigation of the performance of Rosenblueth's PE method under a multivariate normal condition was made by Nguyen and Chowdhury (1985) in a geotechnical application. It was found to be more accurate than the FOVE method. Chang, Tung, and Yang (1995) conducted a systematic evaluation of the relative performance of three probabilistic point estimation methods, namely, Rosenblueth, Harr, and modified Harr, by applying them to different models of varying degrees of nonlinearity. The model parameters are assumed to be correlated or independent normal random variables. In general, all three PE algorithms are capable of yielding rather accurate estimation for the first two moments, especially when the model is close to being linear. However, Rosenblueth's method produces better estimation for the skew coefficient of model output than Harr's procedure when a model is nonlinear involving independent normal variables. For a model having correlated normal random variables, the modified Harr's method outperforms the other two competitors in estimating the first three moments of model output. In the evaluation, it is generally observed that the accuracy of moments estimation decreases as the order of the moment increases and as the degree of model nonlinearity increases. The performance of Rosenblueth's method under nonnormal conditions is not available in the literature and should be investigated.

Use of integral transform methods for uncertainty analysis requires knowledge about the pdfs of stochastic input parameters in the model. When stochastic parameters are uncorrelated and strictly multiplicative or additive, the convolutional property of an appropriate integral transform can be applied to obtain the exact statistical moments of the model output. In a general case, where stochastic parameters are correlated and model functional form is more complicated, integral transforms would become analytically and computationally difficult. This appears to be the main factor that severely restricts the practical usefulness of integral transforms for uncertainty analysis in water resources problems. Another potential shortcoming of using integral transform for uncertainty analysis is that integral transform does not always exist analytically under some conditions.

Simulations, similar to methods of integral transform, require knowledge of the pdfs of all stochastic model parameters involved. Depending on the information desired in uncertainty analysis, the number of simulated samples required for model output varies. In general, the number of simulations required for accurate estimation of the mean and standard deviation of a model output is significantly less than that required to estimate the PDF or confidence limit of the output. The bootstrap resampling technique, a variation of Monte Carlo simulation, can be performed nonparametrically without having to know the pdfs of stochastic model parameters. However, it would require sample observations of model parameters to be able to assess the uncertainty of the model output. Simulation of uncorrelated random parameters is straightforward. To date, generation of multivariate correlated random variables is largely limited to all normal or all log-normal cases.

In addition to the techniques just mentioned, several other probabilistic methods can be used to assist in uncertainty estimation. For instance, the Bayesian technique is useful in updating changing uncertainty knowing prior uncertainty. Fuzzy set theory is helpful in quantifying abstract parameters for uncertainty analysis. Adoption of these methods for uncertainty analysis of water resources systems should also be explored.

15.6 ANALYSIS OF RELATIVE CONTRIBUTIONS OF UNCERTAINTIES

Uncertainty analysis of the parameters pertinent to a water resources system provides a means to assess the relative uncertainty contribution of these parameters to the entire system. Accordingly, measures to reduce or minimize the effects of component uncertainties of certain contributing parameters to improve the total system uncertainty can be contemplated.

The relative uncertainty contribution of parameter X_j to the total system uncertainty can be expressed as

$$U_{Xj} = \eta_{Xj}\Omega_{Xj}^2/\Omega_t^2 \tag{15.20}$$

in which Ω_t and Ω_{Xj} are the coefficient of variation of the system and X_j, respectively, and

$$\eta_{Xj} = s_j^2\left(\frac{\overline{X}_j^2}{[g(\mu)]^2}\right) \tag{15.21}$$

where $g(\,)$ and s_j have been defined in equations 15.1 and 15.3. In the case of the first-order variance estimation method or integral transform technique, if the function $g(X)$ is differentiable, the sensitivity coefficient s_j can easily be determined. Otherwise, alternate techniques such as response surface function may be used to estimate s_j.

As an example of assessing the relative uncertainties of the contributing factors of a system, the result of design of a culvert given by Yen (1993) using the first-order technique is reproduced in Tables 15.1 and 15.2 as an illustration. For simplicity of demonstration, the flood is estimated by using the rational method and the culvert capacity is computed by using the Manning formula. In these tables, Ω_t^2 is the sum of squared coefficient of variation of the load (flood), Ω_L^2 and that of the resistance (culvert capacity), Ω_c^2. For example, column (5) of Table 15.1 shows the percentage contribution of each random factor in the rational formula to the overall uncertainty in the computed flood discharge. As can be seen, the runoff coefficient and the rain intensity are the most significant parameters in their contributions to the overall uncertainty in estimated flood in the example, whereas uncertainty in the basin area is the least important. Furthermore, from column (6) of Table 15.1 and column (7) of Table 15.2, one observes that uncertainty associated with the estimation of flood dominated that linked to the culvert capacity. With this type of information, one is able to identify the important parameters so that attention can be focused to effectively reduce the overall system uncertainty.

15.7 CONCLUDING REMARKS

Formal quantitative uncertainty analysis, which is not yet a standard practice in water resources planning, design, and operation, is a powerful tool for effective and rational management of water resources. Which of the techniques described in this

Table 15.1. *Uncertainty of loading (flood) of example culvert design*

Factor (1)	Parameter (2)	Mean (3)	Coef. of variation Ω (4)	Ω^2/Ω_L^2 % (5)	Ω^2/Ω_t^2 % (6)
Runoff coefficient	C	0.90	0.250	45.1	39.9
Rain intensity	i_T	171 mm/h	0.252	45.9	40.5
Basin area	A	2.24 ha	0.05	1.8	1.6
Model error	λ_L	1.0	0.10	7.2	6.4
Flood	Q_L	0.94 m^3/s	0.372	100.0	88.4

Table 15.2. *Uncertainty of resistance (culvert capacity) of example culvert design*

Factor (1)	Parameter (2)	Mean (3)	Coef. of variation Ω (4)	η (5)	$\eta\Omega^2/\Omega_C^2$ % (6)	$\eta\Omega^2/\Omega_t^2$ % (7)
Head	$H_u - z_d - \frac{D}{2}$	1.68 m	0.170	0.250	39.7	4.62
Manning's	n	0.014	0.117	0.453	34.1	3.95
Length	L	61 m	0.001	0.113	0.0	0.00
Diameter	D	0.61 m	0.004	6.000	0.5	0.06
Entrance loss coefficient	K_u	0.40	0.306	0.002	1.0	0.12
Exit loss coefficient	K_d	1.00	0.020	0.014	0.0	0.00
Model error	λ_C	1.0	0.067	1.0	24.7	2.86
Culvert capacity	Q_C	0.81 m^3/s	0.135		100.0	11.6

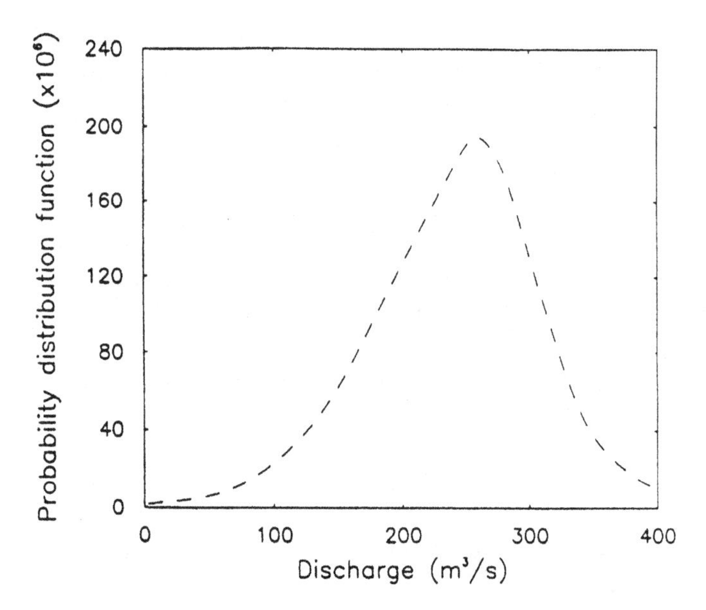

Figure 15.1. Probability distribution function for predicted peak discharge of Vermilion River watershed at Pontiac for May 5, 1965 rainstorm (after Yen 1989).

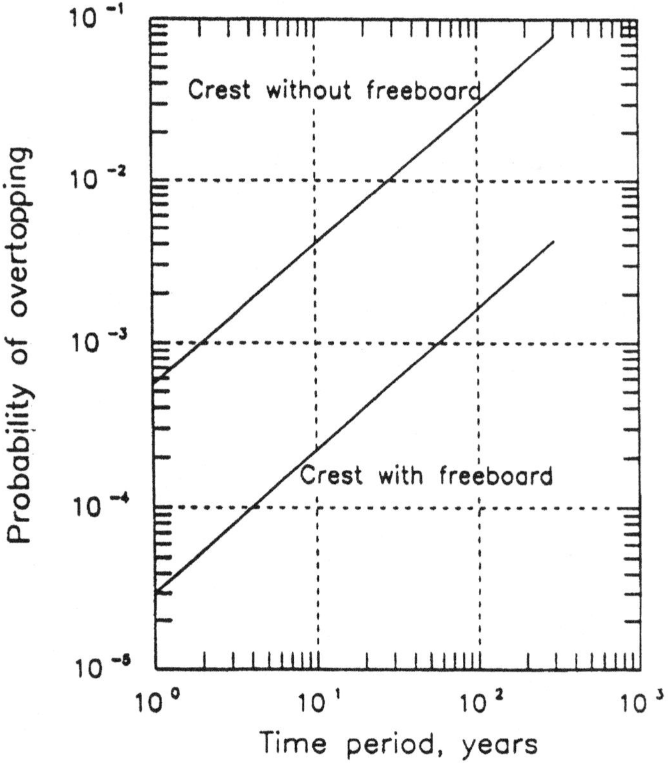

Figure 15.2. Example illustrating freeboard as reduction of overtopping risk.

chapter is most suitable for what kind of water resources problems depends on the nature of the problem, the objectives sought, the data available, the simulation model used, and the computational facilities available. In addition to the example of accessing the relative contributions of uncertainties of different contributing parameters described in the preceding subsection, the techniques can be applied to numerous other water resources problems. For instance, by considering uncertainties, forecast of flood from a given rainstorm can be made with a probability density function as depicted in Figure 15.1 (Yen 1989; Melching, Yen, and Wenzel 1990) instead of a single predicted peak value as conventionally done. This probability information is precisely the type of pdf mentioned in Krzysztofowicz (2001) and desired by the U.S. National

Weather Service. A similar uncertainty analysis can help in determining the worth and effectiveness of raingauges and radars in a network in flood forecasting discussed by Moore (2001).

Tang, Mays, and Yen (1975) demonstrated how uncertainties can be accounted for in sewer design and suggested the use of failure risk instead of return period as a more rational indicator of design protection level. Cheng, Yen, and Tang (1993) used first-order uncertainty analysis to investigate overtopping risk of a dam. These studies would fit well with the risk management ideas mentioned by Plate (2001) and Shamir (2001). Yen (1979) showed that the traditional safety factor can now be determined scientifically considering uncertainties of contributing factors. Likewise, the arbitrarily chosen values of freeboard used in many hydraulic designs can actually be assessed with uncertainty analysis, as demonstrated in Figure 15.2 (Cheng et al. 1993). Mays (1993) reported a study on analysis of component uncertainties of the nodes, links, and pumps of a water distribution network and how these component uncertainties were combined to yield the network system uncertainty.

It appears that the tools of quantitative uncertainty analysis are now available for water resources planning, design, and operation. Further research and improvement of the uncertainty analysis techniques is undeniably desired and needed. Nevertheless, the existing tools are readily adoptable and applicable to numerous kinds of water resources problems. Perhaps now is the time to explore this young promising field for its practical applications.

REFERENCES

Abramowitz, M. and Stegun, I. A., ed. (1972) *Handbook of Mathematical Functions with Formulas, Graphs, and Mathematical Tables*, 9th ed. Dover Publications, New York.

Bateman, H. (1954) *Tables of Integral Transforms*, Vol. I. McGraw-Hill, New York.

Burges, S. J. and Lettenmaier, D. P. (1975). Probabilistic methods in stream quality management. *Water Resources Bulletin* 11(1): 115–30.

Chang, C. H., Tung, Y.-K., and Yang, J. C. (1995) Evaluating performance of point estimates methods. *Journal of Mathematical Modeling* 19(2): 95–105.

Cheng, S. T., Yen, B. C., and Tang, W. H. (1993) Stochastic risk modeling of dam overtopping. In *Reliability and Uncertainty Analyses in Hydraulic Design*, edited by B. C. Yen and Y. K. Tung, ASCE, New York, 123–32.

Efron, B. (1979a) Bootstrap methods: Another look at the jackknife. *The Annals of Statistics* 3: 1189–1242.

(1979b) Computers and theory of statistics: Thinking the unthinkable. *SIAM Review* 21: 460–80.

(1982) The jackknife, the bootstrap, and other resampling plans. *CBMS 38*, SIAM-NSF.

Epstein, B. (1948) Some application of the Mellin transform in statistics. *Annals of Mathematical Statistics* 19: 370–79.

Giffin, W. C. (1975) *Transform Techniques for Probability Modeling*. Academic Press, Orlando.

Harr, M. E. (1987) *Reliability-Based Design in Civil Engineering*. McGraw-Hill, New York.

(1989) Probabilistic estimates for multivariate analyses. *Applied Mathematical Modeling* 13(5): 313–18.

Kendall, M., Stuart, A., and Ord, J. K. (1987) *Kendall's Advanced Theory of Statistics, Vol. 1: Distribution Theory*, 5th ed. Oxford University Press, Oxford.

Krzysztofowicz, R. (2001) Probabilistic hydrometeorological forecasting. In *Risk, Reliability, Uncertainty, and Robustness of Water Resources Systems*, edited by J. J. Bogardi and Z. W. Kundzewicz, Cambridge University Press, New York, 41–6.

Li, K. S. (1992) Point estimate method for calculating statistical moments. *Journal of Engineering Mechanics*, ASCE, 118(7): 1506–11.

Mays, L. W. (1993) Methodologies for reliability analysis of water distribution systems. In *Reliability and Uncertainty Analyses in Hydraulic Design*, edited by B. C. Yen and Y. K. Tung, ASCE, New York, 233–68.

Melching, C. S., Yen, B. C., and Wenzel, H. G., Jr. (1990) A reliability estimation in modeling watershed runoff uncertainties. *Water Resources Research* 26(10): 2275–86.

Moore, R. J. (2001) Aspects of uncertainty, reliability, and risk in flood forecasting systems incorporating weather radar. In *Risk, Reliability, Uncertainty, and Robustness of Water Resources Systems*, edited by J. J. Bogardi and Z. W. Kundzewicz, Cambridge University Press, New York, 30–40.

Nguyen, V. U. and Chowdhury, R. N. (1985) Simulation for risk analysis with correlated variables. *Geotechnique* 35(1): 47–58.

Park, C. S. (1987) The Mellin transform in probabilistic cash flow modeling. *The Engineering Economist* 32(2): 115–34.

Plate, E. J. (2001). Risk management for hydraulic systems under hydrological loads. In *Risk, Reliability, Uncertainty, and Robustness of Water Resources Systems*, edited by J. J. Bogardi and Z. W. Kundzewicz, Cambridge University Press, New York, 209–20.

Quimby, W. F. (1986). Selected topics in spatial statistical analysis: Nonstationary vector kriging, large scale conditional simulation of three dimensional Gaussian random fields, and hypothesis testing in a correlated random fields. Ph.D. thesis, Dept. of Statistics, University of Wyoming, Laramie.

Rosenblueth, E. (1975) Point estimates for probability moments. *Proceedings*, National Academy of Science 72(10): 3812–14.

(1981) Two-point estimates in probabilities. *Applied Mathematical Modeling* 5: 329–35.

Rubinstein, R. Y. (1981) *Simulation and the Monte Carlo Method*. John Wiley and Sons, New York.

Shamir, U. (2001) Risk and reliability in water resources management: Theory and practice. In *Risk, Reliability, Uncertainty, and Robustness of Water Resources Systems*, edited by J. J. Bogardi and Z. W. Kundzewicz, Cambridge University Press, New York, 162–68.

Springer, M. D. (1979) *The Algebra of Random Variables*. John Wiley and Sons, New York.

Tang, W. H., Mays, L. W., and Yen, B. C. (1975) Optimal risk based design of storm sewer networks. *Journal of Environmental Engineering Division*, ASCE 101(EE3): 381–98.

Tung, Y. K. (1993) Confidence intervals of optimal risk-based hydraulic design parameters. In *Reliability and Uncertainty Analyses in Hydraulic Design*, edited by B. C. Yen and Y. K. Tung, ASCE, New York, 81–96.

Yen, B. C. (1979) Safety factors in hydrologic and hydraulics engineering design. In *Reliability in Water Resources Management*, edited by E. A. McBean, K. W. Hipel, and T. E. Unny, Water Resources Publications, Highlands Ranch, CO, 389–407.

(1989) Flood forecasting and its reliability. In *Taming the Yellow River: Silt and Floods*, edited by L. M. Brush, M. G. Wolman, and B.-W. Huang, Kluwer Academic Publishers, Dordrecht, 163–96.

(1993) Risk consideration in storm drainage. In *Urban Storm Drainage*, edited by C. Cao, B. C. Yen, and M. Benedini, Water Resources Publications, Highlands Ranch, CO, 237–52.

Yen, B. C. and Ang, A. H.-S. (1971) Risk analysis in design of hydraulic projects. In *Stochastic Hydraulics* (Proceedings of First Internat.

Symp.), edited by C. L. Chiu, University of Pittsburgh, Pittsburgh, PA, 694–709.

Yen, B. C., Cheng, S. T., and Melching, C. S. (1986). First-order reliability analysis. In *Stochastic and Risk Analysis in Hydraulic Engineering*, edited by B. C. Yen, Water Resources Publications, Highlands Ranch, CO, 1–36.

Yen, B. C. and Tung, Y. K. (1993) Some recent progress in reliability analysis for hydraulic design. In *Reliability and Uncertainty Analyses in Hydraulic Design*, edited by B. C. Yen and Y.-K. Tung, ASCE, New York, 35–80.

Yevjevich, V. (1972) *Probability and Statistics in Hydrology.* Water Resources Publications, Highlands Ranch, CO.

16 Managing water quality under uncertainty: Application of a new stochastic branch and bound method

B. J. LENCE* AND A. RUSZCZYŃSKI**

ABSTRACT

The problem of water quality management under uncertain emission levels, reaction rates, and pollutant transport is considered. Three performance measures – reliability, resiliency, and vulnerability – are taken into account. A general methodology for finding a cost-effective water quality management program is developed. The approach employs a new stochastic branch and bound method that combines random estimates of the performance for subsets of decisions with iterative refinement of the most promising subsets.

16.1 INTRODUCTION

Devising successful and cost-effective water quality management strategies can be difficult because the inputs to, and the behavior of, the system being managed are never entirely predictable. Decision makers do not know what conditions will exist in the future nor how these conditions will affect the impact of their decisions on the environment. Vincens, Rodriguez-Iturbe, and Schaake (1975) classify uncertainty in modeling hydrologic systems into three categories: uncertainty in the model structure (Type I uncertainty); uncertainty in the model parameters (Type II uncertainty); and uncertainty resulting from natural variability (Type III uncertainty). For water quality systems, uncertainty in the pollutant transport model, the model reaction rates, and the natural variability of emission rates and receiving water conditions, such as streamflow, temperature, and background pollutant loadings from unregulated pollution sources, contribute to difficulties in predicting the future behavior of the system (Beck 1987). This chapter develops an approach for identifying water quality management solutions under Type II and Type III uncertainty. It is based on an application of the stochastic branch and bound method of Norkin, Ermoliev, and Ruszczyński (1994) to water quality management, which is modified to account for the performance indicators of reliability, resiliency, and vulnerability.

Common methods for accommodating input uncertainty in environmental quality management problems include chance-constrained optimization, combined simulation and optimization and multiple realization based approaches. Each of these approaches may be used to develop the trade-off between total cost of optimal waste management and system reliability. However, these common techniques may not be used to estimate the reliability under all types of input uncertainty, for example, under cases when the pollutant emission rates may also vary stochastically, and they become increasingly difficult to apply as the number and type of random inputs increases.

In general, the frequency, duration, and magnitude of violations of a given environmental quality standard are indices of pollution control performance that represent the reliability, resiliency, and vulnerability, respectively, of the management decision. The reliability criterion describes how likely the environmental standards may be achieved. The resiliency and the vulnerability criteria give indications of the degree to which the system is expected to recover from a failure sojourn and the severity of the consequences of environmental quality violations, respectively. Most studies that account for uncertainty in environmental management modeling include reliability, but not these other important indices. However, each of these measures may offer important insights and information to decision makers when formulating successful environmental quality programs. The importance of these performance indicators is illustrated for water resources management systems by Glantz (1982); Hashimoto, Stedinger, and Loucks (1982); and Fiering (1982a–d).

In the following section, the elements of a water quality management model are developed for cases with stochastic input information. In such models, emission rates, as well as factors that affect pollution transport and impact, may be random. Next, the water quality management model is formulated as a probabilistic problem that maximizes reliability and

* University of British Columbia, Department of Civil Engineering, 2324 Main Mall, Vancouver, British Columbia V6T 1Z4, Canada.

** Department of Management Science and Information Systems, Rutgers University, 94 Rockefeller Rd, Piscataway, NJ 08854, USA.

resiliency and minimizes vulnerability under a total cost constraint. The decision variables are the discrete design waste treatment levels of the dischargers in the system. Given a specified set of decision variable values, the objective function for this model may be estimated using Monte Carlo simulation. In the fourth section, the stochastic branch and bound method of Norkin et al. (1994) is described and the approach for estimating the bounds for this technique is presented. The method is based on a branch and bound algorithm in which the branches, or partitions, are subsets of discrete decision variables for waste treatment levels and the bounds are estimates of the upper and lower limits of the reliability, resiliency, and vulnerability, for a given branch. In the fifth section, the stochastic branch and bound method is demonstrated for water quality management using a case study based on the Willamette River in Oregon. Finally, a summary of the work is presented, including insights drawn from the case study and suggestions for future work.

16.2 ELEMENTS OF A WATER QUALITY MANAGEMENT MODEL

A water quality management program is defined as a selection of technologies $x = (x_1, x_2, \ldots, x_m)$ such that $x_i \in X_i$, $i = 1, 2, \ldots, m$. It is characterized by its cost

$$c(x) = \sum_{i=1}^{m} c_i(x_i) \tag{16.1}$$

and ambient quality levels at monitoring points $j = 1, \ldots, n$:

$$S_j^l(x, \omega) = A_j^l(e_1^l(x_1, \omega), \ldots, e_m^l(x_m, \omega); \omega) \tag{16.2}$$

Where $i = 1, \ldots, m$ represents the emission sources; $l = 1, \ldots, L$ represents the pollutants being controlled; $j = 1, \ldots, n$ represents the monitoring points; $c_i(x_i)$ represents the cost, including the capital and operation and maintenance costs of performing the given technology; $e_i^l(x_i, \omega)$ represents the random emission level of pollutant l; ω denotes an elementary event in some probability space (Ω, $\boldsymbol{F}$, $\boldsymbol{P}$), S_j^l represents the ambient water quality for pollutant l at the monitoring point j; and A_j^l represents the transfer functions that describe the effect of reactions involving pollutant l that take place between the pollution sources and the monitoring points along the stream.

The random emission level accommodates the fact that biological, chemical, and physical treatment technologies face stochastic inflows and operational variability even under the most stable conditions. The transfer functions relate the pollution abatement decisions to the instream water quality levels. These transfer functions are random, depend on ω and the pollutants being emitted, and are developed based on pollutant transport simulation models, pollutant characteristics, streamflow, stream velocity, stream temperature, reaction rates, and background water quality levels. They may be linear or nonlinear with respect to the emission levels, depending on the pollutant simulation model used. For simulation models that are linear with respect to the pollutant emission levels (e.g., the Streeter-Phelps BOD-DO model), the transfer functions may be a matrix of constants that represent the impact obtained by simulating water quality improvement along the river per unit change in the emission levels, for a given set of stream conditions.

Adding the time dimension provides a more exact description of the relations between emissions and ambient quality levels. In such a model, emissions are stochastic processes $e_i^l(x_i, \omega, t)$ where $t \in \{0, 1, 2, \ldots\}$ denotes discrete time intervals. Then, clearly, the ambient quality levels at monitoring points are stochastic processes, too. Values of their realizations at each time interval t can be written as

$$S_j^l(x, \omega, t) = A_j^l(e_1^l(x_1, \omega), \ldots, e_m^l(x_m, \omega); \omega, t) \tag{16.3}$$

where A_j^l is a causal operator, that is, such an operator whose values depend on the past emission levels $e_i^l(x_i, \omega, \tau)$ for $\tau \in \{0, 1, \ldots, t\}$, but not on the future ones.

16.3 A PROBABILISTIC WATER QUALITY MANAGEMENT MODEL

A probabilistic form of the water quality management problem is given here which has an objective function based on a combination of performance indicators, that is, reliability, resiliency, and vulnerability, and maintains a limit on total cost, or budget. Assume that there are some quality standards $\bar{S}_j^l$ for pollutants l at monitoring points j. Let us define the state of the system as the vector $S = (S_j^l)_{\substack{j=1,\ldots,n \\ l=1,\ldots,L}}$ and the set of satisfactory states:

$$G = \{S \in \boldsymbol{R}^{nl}; 0 \le S_j^l \le \bar{S}_j^l, j = 1, \ldots, n, l = 1, \ldots, L\} \tag{16.4}$$

One would like to have water quality levels S_j^l below the standards $\bar{S}_j^l$ that is,

$$S(x, \omega, t) \in G, t = 1, \ldots T, \tag{16.5}$$

but requiring that this is satisfied for all possible events $\omega \in \Omega$ may be extremely conservative and could lead to a very expensive worst-case design. To arrive at meaningful and practically useful formulations typically water quality management programs are designed to exploit the probabilistic nature of the problem. Sets of management decisions are selected based on measures of system performance that indicate the extent of environmental damage under critical hydrological and background water quality conditions. Hashimoto et al. (1982)

discuss reliability, resiliency, and vulnerability applied to water resources systems. They derive mathematical expressions for these criteria and utilize the expressions to evaluate the possible performance of water supply conditions for a water supply reservoir. For water quality systems, these measures indicate the acceptable frequency, duration, and magnitude of water quality violation. That is, they may reflect what we know about the acceptable effects of frequent water quality violation, of different lengths, and at different degrees of contamination, on species in a region.

16.3.1 Reliability

Given the quality standards $\bar{S}$ define the reliability of the system as the probability of the event that the state remains in the set of satisfactory states in the planning horizon:

$$R_1(x) = \boldsymbol{P}\{S(x,\omega,t) \in G, t = 1, \ldots T\} \qquad (16.6)$$

This allows us to formulate the reliability maximization problem:

$$\max_{x \in X} R_1(x) \qquad (16.7)$$

subject to

$$\sum_{i=1}^{m} c_i(x_i) \le \bar{c} \qquad (16.8)$$

where $\bar{c}$ is a prescribed budget level. By varying $\bar{c}$, one can develop the cost-reliability trade-offs for water quality management.

16.3.2 Reliability under violation length limit

The notion of reliability can be relaxed by allowing violations of a short duration. For example, if violations of only one period in length are allowed, a performance measure, which is the probability that a failure sojourn will not last more than one period, $R_2(x)$, may be introduced.

$$R_2(x) = \boldsymbol{P}\{S(x,\omega,t) \in G \text{ or } S(x,\omega,t+1) \in G, t = 1, \ldots, T-1\} \qquad (16.9)$$

16.3.3 Resiliency

The characteristic of resiliency of the system measures the ability of the system to recover from failure states and can be defined as the conditional probability that $S(x, \omega, t+1) \in G$, if $S(x, \omega, t) \notin G$. To be more precise, let

$$\tau(x,\omega) = \{1 \le t \le T;\, S(x,\omega,t) \notin G\} \qquad (16.10)$$

and define resiliency as

$$R_3(x) = \boldsymbol{P}\{S(x,\omega,t+1) \in G \text{ for all } t \in \tau(x,\omega) | \tau(x,\omega) \ne 0\} \qquad (16.11)$$

Thus, the resiliency may be described as the system's average recovery rate and equivalently defined as given in Hashimoto et al. (1982), as:

$$R_3(x) = [R_2(x)]/[1 - R_1(x)] \qquad (16.12)$$

16.3.4 Vulnerability

Classical water quality management has relied on setting strict standards and designing management programs to meet these standards with some level of reliability. This approach assumes that below some allowable standard, the water quality of the system is acceptable, and that above that standard, the system is infinitely damaged. While simplifying the problem, this may not represent what happens in reality. In some river systems, a hierarchy of water quality standards may be more acceptable for describing the allowable degree of water quality degradation. Furthermore, the allowable frequency and duration of water quality violation may be different for different levels of contamination, for example, the allowable frequency and duration of water quality violation may decrease with increases in water quality standard levels, as is the case with the U.S. Environmental Protection Agency (USEPA) chronic and acute ambient standards for ammonia nitrogen levels (USEPA 1992). Therefore, another possibility of defining performance measures is to introduce a hierarchy of quality standards

$$\bar{S} \le \bar{S}^{(1)} \le \bar{S}^{(2)} \le \ldots \bar{S}^{(H)} \qquad (16.13)$$

for which the corresponding satisfactory states are:

$$G^{(h)} = \{S \in \boldsymbol{R}^{nl};\, 0 \le S \le \bar{S}^{(H)}\} \qquad (16.14)$$

Let us define the events:

$$\begin{aligned}\Theta^{(1)}(x) &= \{\omega: S(x,\omega,t) \in G^{(1)}, t = 1, \ldots, T\}\\ \Theta^{(2)}(x) &= \{\omega: S(x,\omega,t) \in G^{(2)}, t = 1, \ldots, T\} \backslash \Theta^{(1)}(x)\\ \Theta^{(H)}(x) &= \{\omega: S(x,\omega,t) \in G^{(H)}, t = 1, \ldots, T\} \backslash \Theta^{(H-1)}(x)\end{aligned} \qquad (16.15)$$

The performance measure (negatively related to the vulnerability) of the system can be defined as

$$V(x) = \sum_{h=1}^{H} W_h \boldsymbol{P}\{\Theta^{(h)}(x)\} \qquad (16.16)$$

If the coefficients W_h, $h = 1, \ldots, H$, satisfy the inequalities: $W_1 \ge W_2 \ge \ldots \ge W_H$, this expression for vulnerability is the opposite of the classical expression for vulnerability given in Hashimoto et al. (1982) and may be maximized as a performance indicator for certainty of system outcome.

In general, all these performance measures may be included into the objective function for the optimization problem in equations 16.7–16.8 by formulating a composite objective:

$$\max_{x \in X}[F(x) = \gamma_1 R_1(x) + \gamma_2 R_2(x) + \gamma_3 R_3(x) + \gamma_4 V(x)] \quad (16.17)$$

where $\gamma_1, \ldots, \gamma_4$ are some positive weights. However, it is likely that the presence of one objective may eliminate the need for another. For example, if resiliency, $R_3(x)$ is selected as an objective, the inclusion of $R_2(x)$ may not be necessary.

The main difficulty associated with the stochastic problem in equations 16.17 and 16.8 is that it involves functions defined as probabilities of some events. The values of these functions cannot be calculated analytically for realistic models. For example, calculating reliability would require evaluating a multidimensional integral over the set implicitly defined by equation 16.5, potentially involving nonlinear models of emissions and transfer. Except for some special cases, such as models with one source and one receptor and with linear transfer functions, the only tool available for identifying reliability is simulation. In the simulation approach, for selected technologies $x_1, \ldots, x_m$, one can execute the transfer function models with some randomly drawn uncertain parameters $\tilde{\omega}$ and evaluate the function

$$\chi_1(x, \tilde{\omega}) = \begin{cases} 1 & \text{if } S(x, \tilde{\omega}, t) \in G, t = 1, \ldots, T \\ 0 & \text{otherwise} \end{cases} \quad (16.18)$$

The reliability is the expected value of this function

$$R_1(x) = \boldsymbol{E}\{\chi_1(x, \tilde{\omega})\} \quad (16.19)$$

And the reliability may be estimated by the Monte Carlo method

$$R_1(x) \approx (1/N) \sum_{s=1}^{N} \chi_1(x, \omega^s) \quad (16.20)$$

where $\omega^1, \ldots, \omega^N$ are independent observations (realizations) of ω. However, the number of simulations N necessary to evaluate the reliability at only one management program x with a sufficient accuracy can be very large. The objective of the water quality management problem is to find the best set of waste treatment decisions among all possible options, which requires that the objective be evaluated for many candidate solutions, and makes a straightforward simulation of the combinatorial problem computationally burdensome. In this work, an approach is developed that is capable of determining the best water quality management program without examining all possible programs and without calculating the objective function value (e.g., the reliability) for each of them exactly. The approach adapts the stochastic branch and bound method of Norkin et al. (1994) to the water quality management problem.

16.4 THE STOCHASTIC BRANCH AND BOUND METHOD

The stochastic branch and bound method of Norkin et al. (1994) is based on the classical integer programming branch and bound algorithm in which the partitions, or branches, are based on subsets of discrete decision variables for waste treatment levels. At each partition, or branch, the upper and lower bounds of the system objective functions are estimated using Monte Carlo simulation.

The set of all possible waste abatement strategies $X = X_1 \times X_2 \times \ldots \times X_m$, are split into disjoint subsets

$$X^p = X_1^p \times X_2^p \times \ldots \times X_m^p, \quad p \in P \quad (16.21)$$

such that $U_{p \in P} X^p = X$. And at each subset X^p, estimates of the upper and lower bounds on the objective function in equation 16.17, $\xi(X^p)$ and $\eta(X^p)$, respectively, are determined under budgetary constraints. Here the number of simulations, N, controls the accuracy of the method. The stochastic upper bounds are used to select the record set: that is the X^p for which the upper bound, $\xi(X^p)$, is the largest. The record set, as the most promising set of possible programs, is partitioned into smaller subsets, and new stochastic bounds are evaluated, etc., until a singleton is achieved. At each stage, an approximate solution $\tilde{x}$ is selected as an element of the set with the largest lower bound, $\eta(X^p)$. Since the bounds are random, the record set is random; consequently, all objects generated by the method are random.

The experience gained in Norkin et al. (1994) suggests that stopping after achieving the first singleton is a reasonable strategy; it leads to a good solution, and guarantees finding the best solution if the method is run in a regenerative fashion. Different approaches can be used to determine the most efficient way to partition the initial and subsequent sets of waste abatement strategies. Various partitioning techniques are examined in Hägglö (1996), but the technique applied here, in a preliminary analysis of the application of the stochastic branch and bound method for water quality management of BOD waste effluents, is the heuristic ranking method proposed by Hägglö (1996). This method determines the ranked importance of the emission sources for improving the probability that the water quality goals are met. The rank of an emission source is determined by examining the active constraints from the linear programs used to generate the upper bounds. The rank of the emission source is the rank of its ratio between the transfer function values in the active constraints and the cost for technology improvement, compared to all other emission sources.

16.4.1 Reliability bounds

For a set X^p choose a point $x^p \in X^p$ such that $c(x^p) \leq \bar{c}$. Then, define

$$\eta_q^N(X^p) = (1/N)\sum_{s=1}^{N} \chi_q(x^p, \omega^s),\ q = 1,2 \qquad (16.22)$$

where $\omega^1, \ldots, \omega^N$ are independent observations of ω. These random variables are stochastic lower bounds for the values of the functions $R_q(x)$ in X^p, $q = 1, 2$.

Generating stochastic upper bounds is more involved. The key observation is the inequality:

$$\hat{R}_1(X^p) = \max_{\substack{x \in X^p \\ c(x) \leq \bar{c}}} \boldsymbol{P}\{S(x, \omega, t) \in G, t = 1, \ldots, T\}$$
$$\leq \boldsymbol{P}\{\exists x \in X^p; c(x) \leq \bar{c}, S(x, \omega, t) \leq \bar{S}, t = 1, \ldots, T\} \qquad (16.23)$$

Let us generate a random estimate $\zeta^N(X^p)$ of the right hand side of the above inequality. Consider the problem with a fixed event ω:

$$\min_{x \in X^p} c(x) \qquad (16.24)$$

subject to

$$S(x, \omega, t) \leq \bar{S}, t = 1, \ldots, T \qquad (16.25)$$

and denote by $\hat{c}_1(X^p,\omega)$ its optimal value. The following equality holds:

$$\boldsymbol{P}\{\exists x \in X^p; c(x) \leq \bar{c}, S(x, \omega, t) \leq \bar{S}, t = 1, \ldots, T\}$$
$$= \boldsymbol{P}\{\hat{c}_1(x, \omega) \leq \bar{c}\} \qquad (16.26)$$

Define the function

$$\psi_1(x^p, \omega) = \begin{cases} 1 & \text{if } \hat{c}_1(X^p, \omega) \leq \bar{c} \\ 0 & \text{otherwise} \end{cases} \qquad (16.27)$$

Combining equations 16.23 and 16.26 yields

$$\hat{R}_1(X^p) \leq \boldsymbol{E}\{\psi_1(x^p, \omega)\} \qquad (16.28)$$

Therefore, for independent observations, $\omega^1, \ldots, \omega^N$ *of* ω, the random variables

$$\xi_1^N(X^p) = (1/N)\sum_{s=1}^{N} \psi_1(x^p, \omega^s) \qquad (16.29)$$

satisfy the relations:

$$\boldsymbol{E}\{\xi_1^N(X^p)\} \geq \hat{R}_1(X^p) \qquad (16.30)$$

and, by the law of large numbers,

$$\lim_{N \to \infty} \xi_1^N(X^p) = \boldsymbol{E}\{\xi_1^N(X^p)\}\ a.s. \qquad (16.31)$$

The bounds in equation 16.29 are relatively easy to calculate. Indeed, for a fixed $\omega = \omega^s$, the problem in equations 16.24–16.25 is a deterministic cost minimization problem that can be solved by mathematical programming methods.

A stochastic upper bound for the reliability with a violation length limit as in equation 16.9 can be calculated similarly. Equation 16.25 is replaced by

$$S(x, \omega, t) \leq \bar{S} \text{ or } S(x, \omega, t+1) \leq \bar{S}, t = 1, \ldots, T-1 \qquad (16.32)$$

The optimal value $\hat{c}_2(X^p, \omega)$ for equations 16.24 and 16.32 is used to define the indicator function ψ_2 (similarly to equation 16.27), and one defines

$$\xi_2^N(X^p) = (1/N)\sum_{s=1}^{N} \psi_2(x^p, \omega^s) \qquad (16.33)$$

Again,

$$\boldsymbol{E}\{\xi_2^N(X^p)\} \geq \hat{R}_2(X^p) \qquad (16.34)$$

and, by the law of large numbers,

$$\lim_{N \to \infty} \xi_2^N(X^p) = \boldsymbol{E}\{\xi_2^N(X^p)\}\ a.s. \qquad (16.35)$$

so equation 16.33 is a stochastic upper bound for R_2.

16.4.2 Resiliency bounds

Lower and upper bounds on the reliability measures $\hat{R}_q(X^p)$, $q = 1,2$, together with the expression in equation 16.12, can be used to define stochastic resiliency bounds. Using the bounds in equation 16.22 calculated at the same point $x^p \in X^p$ we can construct the stochastic lower bound on resiliency

$$\eta_3^N(X^p) = [\eta_2^N(X^p)]/[1 - \eta_1^N(X^p)] \qquad (16.36)$$

In a similar way, one defines a stochastic upper bound,

$$\xi_3^N(X^p) = [\xi_2^N(X^p)]/[1 - \xi_1^N(X^p)] \qquad (16.37)$$

16.4.3 Vulnerability bounds

The stochastic reliability bounds can be generalized in a straightforward way to obtain vulnerability bounds. To obtain a lower bound on vulnerability, define for $h = 1, \ldots, H$ the indicator functions

$$\chi^h(x^p, \omega) = \begin{cases} 1 & \text{if } \bar{S}_k^{l(h-1)} \leq S_j^l(X^p, \omega) \leq \bar{S}_j^{l(h)} \\ 0 & \text{otherwise} \end{cases} \qquad (16.38)$$

Then, for a selected $x^p \in X^p$ and independent observations, $\omega^1, \ldots, \omega^N$ of ω, the random variables

$$\lambda_V^N(X^p) = (1/N)\sum_{s=1}^{N}\sum_{h=1}^{H} W_h \chi^{(h)}(x^p, \omega^s) \qquad (16.39)$$

are stochastic lower bounds on the vulnerability function. This follows directly from the law of large numbers.

To construct a stochastic upper bound, consider the deterministic problems (with a fixed ω)

$$\min_{x \in X^P} c(x) \tag{16.40}$$

subject to

$$S(x, \omega, t) \leq \bar{S}^{(h)}, t = 1, \ldots, T \tag{16.41}$$

for the family of quality standards $h = 1, \ldots, H$. Let $\hat{c}^{(h)}(X^P, \omega)$ denote its optimal value. Define for $h = 1, \ldots, H$, the functions

$$\psi^h(x^P, \omega) = \begin{cases} 1 & \text{if } \hat{c}^{(h)}(X^P, \omega) \leq \bar{c} \\ 0 & \text{otherwise} \end{cases} \tag{16.42}$$

By the ordering of the quality standards, $\psi^{(h)} \leq \psi^{(h+1)}$, $h = 1, \ldots, H - 1$. Define

$$\Delta^{(h)}(x^P, \omega) = \psi^{(h)}(x^P, \omega) - \psi^{(h-1)}(x^P, \omega) \tag{16.43}$$

where one sets $\chi^{(o)}$ $(x^P, \omega) = 0$. Then, in an identical way as for the reliability bounds, one obtains for all $h = 1, \ldots, H$, the relations

$$\max_{\substack{x \in X^P \\ c(x) \leq \bar{c}}} \sum_{h=1}^{H} W_h \boldsymbol{P}\{\boldsymbol{Q}^{(h)}(x)\} \leq \sum_{h=1}^{H} W_h \boldsymbol{E}\{\Delta^{(h)}(x^P, \omega)\} \tag{16.44}$$

Indeed, by selecting x after the event ω is known, one can only improve the quality standard (move to the inner subset), which by the monotonocity of the weights implies the above inequality.

Therefore, for independent observations $\omega^1, \ldots, \omega^N$ of ω, the random variables

$$\xi_V^N(X^P) = (1/N) \sum_{s=1}^{N} \sum_{h=1}^{H} W_h \Delta^{(h)}(x^P, \omega^s) \tag{16.45}$$

are stochastic upper bounds on the vulnerability function.

16.4.4 Using multiple scenarios

If the probability of the event of interest is very close to one, Monte Carlo estimates of the form in equation 16.29 (for the case of reliability) will frequently be equal to one for small N. A large number of observations will be necessary to obtain different estimates for different subsets. One way to overcome this difficulty is the use of many observations not only in the averaging formula in equation 16.29 (or similar), but within the key inequality of the form in equation 16.23. Similar use of multiple scenarios can be made for estimating $R_2(x)$, resiliency, and vulnerability. This approach was employed in the preliminary analysis of the application of the stochastic branch and bound method for management of BOD emissions on the Willamette River, which is described in the following section.

16.5 APPLICATION OF THE STOCHASTIC BRANCH AND BOUND METHOD FOR MANAGING BOD DISCHARGES IN THE WILLAMETTE RIVER

The stochastic branch and bound method is applied here for managing point sources of BOD wastes and their impacts on instream DO for an example basin of the Willamette River in Oregon, United States. The 298 km Middle Fork of the river is analyzed and receives waste emissions from eight major tributaries and ten BOD waste dischargers. Cost data (in 1978 US$), waste load characteristics of the dischargers, discharger locations, streamflow, and temperature data, and water quality simulation model inputs, such as decay rates, and velocity and reaeration rate versus flow relationships were based on Takyi and Lence (1994). All emission sources have waste treatment options available that remove BOD at removal levels of between 35 percent and 95 percent, and these may be selected in discrete increments of 5 percent.

The water quality model used to develop the transfer functions is based on the Camp-Dobbins modification of the Streeter-Phelps equation for the coupled reactions of BOD decay and reaeration and its effect on DO. Benthic oxygen demand and the background DO deficit are assumed to be zero. The seven-day average low flow and the highest mean monthly temperature for the months of June through September are used for this analysis. The transfer functions used in the water quality management model are based on the water quality simulation model and describe the unit decrease in DO (usually in mg/l) at monitoring points in the stream as a result of unit increases of BOD effluent (in mg/l or kg/day) at the emission sources. They are linear functions of the waste treatment levels of the emission sources. The river segment is divided into eighteen reaches and thirty-five monitoring points are used.

The goal of the simple least-cost water quality management model for BOD-DO is to minimize the total cost of waste treatment while meeting lower bounds on the level of allowable DO in the stream. In this case, the equality standards described in the third section are lower bounds on water quality. This requires a change in the sign of the inequality in equation 16.4, but the application of the stochastic branch and bound method remains the same.

The stochastic inputs to the water quality simulation model are the seven-day averaged low flows at five gauging stations in

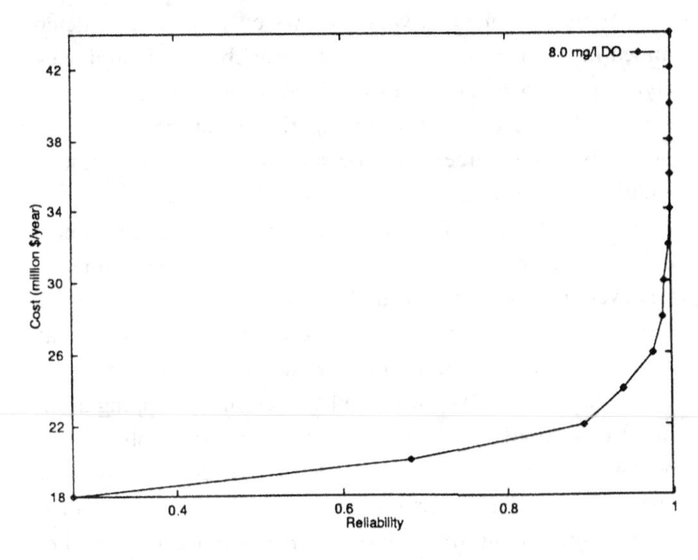

Figure 16.1. Cost versus reliability of meeting the 8.0 mg/l standard.

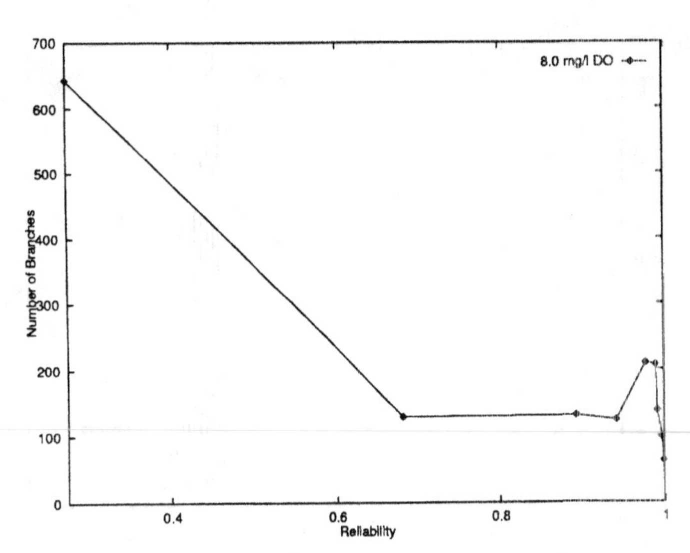

Figure 16.2. Number of iterations of the stochastic branch and bound method for different cost-reliability levels.

the river, the highest monthly mean temperatures in the river, based on the Harrisburg gauging station data, and the stream velocities and reaction rates for each reach of the river under the varying streamflow conditions. For each simulation used in the calculation of the bounds, a random seven-day averaged low flow and stream temperature are generated based on the two-parameter log-normal distributions for the seven-day averaged low flow and the mean monthly temperature, respectively. The stream velocity and re-aeration rates are computed based on the generated flows and functional relationships between velocity and flow and re-aeration rate (at 20°C) and flow, respectively, and normally distributed zero mean noise terms, as described by Takyi and Lence (1994).

In this preliminary demonstration, the stochastic branch and bound method is applied for maximizing reliability for maintaining a DO water quality standard of 8.0 mg/l. The allowable total cost (i.e., capital and operating costs) for the entire river basin is limited to no more than US$20 million/year. The adequate number of simulations required for each bound calculation depends on the complexity of the water quality management problem and on the quality of the uncertain input information. For this example, the number of simulations used is 500, which was determined to be adequate by gradually increasing the number of simulations until the statistical properties of the input and output information converged for an experimental trial.

Figure 16.1 shows the total cost of waste treatment, above the cost of primary treatment (i.e., 35 percent BOD removed), versus reliability of meeting the 8.0 mg/l standard. It may be used by decision makers in selecting the best choice of management solution, given their preferences for the objectives of efficiency and certainty of system outcome.

The stochastic branch and bound method results in an efficient use of computational resources. The number of iterations (splittings) needed to obtain the solution (understood here as the first singleton), for one set of simulations and varying budget level, is illustrated in Figure 16.2. We see that for reliability levels very close to one, less computational effort is required to reach the first singleton, because it is more difficult to differentiate the quality of different subsets on the basis of random simulations. For this reason the quality of the singleton obtained is not good in this case. The use of the multiple scenarios approach improves the quality of the singleton (Häggöf 1996), but still more research is needed to find a proper approach for the case of a very high reliability. The total CPU time needed to solve a problem was in the range of two hours on a SUN Sparc Server 1000 with two CPUs and 128MB memory.

Finally, Figure 16.3 illustrates the operation of the method run in a regenerative fashion. Two hundred different runs of the method were made with different seeds for the random number generator used, and the algorithm was stopped at the first moment at which the record set was a singleton. The quality of the singletons thus obtained was then evaluated by a prolonged simulation. Figure 16.3 illustrates the distribution of the solutions thus obtained. The best singleton (with the reliability level of 0.79) was obtained only once, but it is interesting to note that all the solutions selected in this way have a rather good quality.

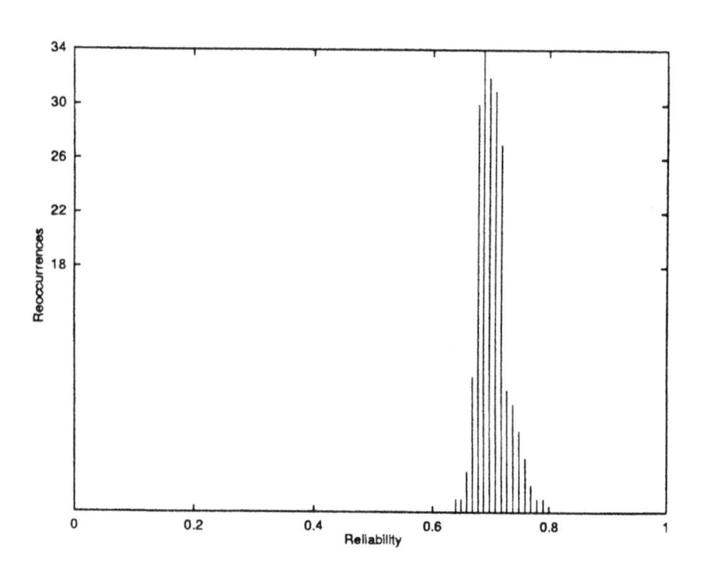

Figure 16.3. Quality distribution of the first singleton.

16.6 CONCLUSIONS

The stochastic branch and bound method of Norkin et al. (1994) is an attractive approach for solving the multi-objective problem of maintaining water quality in a river system while minimizing costs and maximizing the certainty of system outcome. This method is extended in this chapter for including the performance indicators of reliability, resiliency, and vulnerability in the classical water quality management problem. The method is demonstrated for maximizing reliability in an example river basin, and is shown to be efficient and accurate, at least in this preliminary application. This suggests that the method may be effective for addressing other water quality and water resources problems that require management solutions that are robust to uncertainties in the input information.

There are a number of theoretical, implementational, and application-specific issues associated with the stochastic branch and bound method that are as yet unaddressed. First, research on lower bounds and on partitioning strategies needs to be advanced. Since it is unlikely that general approaches exist for identification of bounds and partitioning strategies for all problems, application-specific approaches need to be developed. The heuristic procedures for determining the lower bound solution and the variables on which to branch should, ideally, exploit the natural ordering of technologies in terms of their cost-benefit properties. The notion of benefit, though, needs to be analyzed in a more precise way in this context. Moreover, it should be stressed that the existing theory of the stochastic branch and bound method has been developed for the case of deterministic branching, which allows only static (i.e., determined in advance) ordering of the waste abatement strategies. Dynamic ordering strategies (i.e., where the choice is dependent on the outcomes of some experiments at the given node of the branching tree) are stochastic in nature, and require additional theoretical work.

Second, stopping strategies need to be investigated in more detail. The theory presented in Norkin et al. (1994) guarantees that every recurrent record singleton, that is, a singleton set that turns out to be the record set infinitely many times, is optimal. Approaches are needed to identify such sets sufficiently early with a reasonable level of reliability. Certainly, stopping at the first record singleton is premature, but this approach should also be investigated in more detail. One might consider such an approach a random selection of a potentially interesting alternative. By running the method in a regenerative fashion (i.e., restarting it with a different random number of seeds) one can identify a larger number of record set candidates and then select the best one by performing extensive simulations for each of them.

Third, in the case of very high reliability the basic upper bound estimates may frequently lead to upper bounds equal to one, which make it difficult to differentiate the quality of different subsets on the basis of random simulations. This is a highly undesirable outcome, since it does not allow for ranking the sets. This is a situation when the use of multiple scenario estimates, as employed in the example presented here, may prove useful. The basic idea is to look for decisions that are good for many scenarios simultaneously, so the chance of being successful is lower. Research currently being conducted in this direction focuses on how to determine the number of observations used to generate upper bounds, whether they should be dependent on the estimated reliability, and whether they should be allowed to change it in the course of computation (Hägglöf 1996).

Acknowledgments

The research in this study was carried out when the authors were at the International Institute for Applied Systems Analysis (IIASA), Laxenburg, Austria, and the kind support of the Institute, involving the Optimization Under Uncertainty Project and the Institute Scholars Program, is acknowledged. The first author was supported in part by a research leave grant from the University of Manitoba and a grant from the Natural Sciences and Engineering Research Council of Canada (Award No. OGP00A16343) provided funds for data collection. Views or opinions expressed herein do not necessarily represent those of IIASA, its Nation Member Organizations, or other organizations supporting the work.

REFERENCES

Beck, M. B. (1987) Water quality modeling: a review of the analysis of uncertainty. *Water Resources Research* 23(8): 1393–1442.
Fiering, M. B. (1982a) A screening model to quantify resilience. *Water Resources Research* 18(1): 27–32.
(1982b) Alternative indices of resilience. *Water Resources Research* 18(1): 33–39.
(1982c) Estimates of resilience indices by simulation. *Water Resources Research* 18(1): 41–50.
(1982d) Estimating resilience by canonical analysis. *Water Resources Research* 18(1): 51–57.
Glantz, M. H. (1982) Consequences and responsibilities in drought forecasting: the case of Yakima, 1977. *Water Resources Research* 18(1): 3–13.
Hashimoto, T., Stedinger, J. R., and Loucks, D. P. (1982) Reliability, resiliency, and vulnerability criteria for water resource system performance evaluation. *Water Resources Research* 18(1): 14–20.
Hägglöf, K. (1996) The implementation of the Stochastic Branch and Bound Method for applications in river basin water quality management, Working Paper WP-96-089, International Institute for Applied Systems Analysis, Laxenburg.
Norkin, V. I., Ermoliev, Y. M., and Ruszczynski, A. (1994) On optimal allocation of indivisibles under uncertainty, Working Paper WP-94-021, International Institute for Applied Systems Analysis, Laxenburg.
Takyi, A. K. and Lence, B. J. (1994) Incorporating input information uncertainty in a water quality management model using combined simulation and optimization. In Proceedings of the International UNESCO Symposium. *Water Resources Planning in a Changing World*, Karlsruhe, 189–98.
United States Environmental Protection Agency (1992) National Criteria for Unionized Ammonia, Washington, D.C.
Vincens, G. J., Rodriguez-Iturbe, I., and Schaake, J. C. (1975) A Bayesian framework for the use of regional information in hydrology. *Water Resources Research* 11(3): 405–14.
Wagner, B. J. and Gorelick, S. M. (1987) Optimal groundwater quality management under parameter uncertainty. *Water Resources Research* 23(7): 1162–74.

APPENDIX: NOTATION

$A_j^l(e_1^l(x_1, \omega), \ldots, e_m^l(x_m, \omega); \omega)$ = transfer function that describes the effect of reactions involving pollutant l that takes place between the pollution sources and monitoring point j; related to treatment technologies x_i, where $i = l, \ldots, m$

$A_j^l(e_1^l(x_1, \omega), \ldots, e_m^l(x_m, \omega); \omega, t)$ = transfer function for pollutant l and monitoring point j at any time t

$c_i(x_i)$ = cost of performing technology x_i

$\bar{c}$ = prescribed budget level

$\hat{c}_1(X^p, \omega)$ = optimal cost for the problem described in equations 16.24–16.25, the least-cost problem for all time periods, given a fixed X^p, and parameter input event ω

$\hat{c}_2(X^p, \omega)$ = optimal cost for the problem described in equations 16.24 and 16.32, the conditional least problem, given a fixed X^p, and parameter input event ω

$\hat{c}^{(h)}(X^p, \omega)$ = optimal cost for the problem described in equations 16.40–16.41, the least-cost problem for all time periods and all water quality standards, given a fixed X^p and parameter input event ω

$e_i^l(x_i, \omega)$ = random emission level of pollutant l, due to treatment technology x_i

G = set of satisfactory water quality states

h = hierarchy index for water quality standards

H = total number of intervals for the hierarchy of water quality standards

i = emission source

j = monitoring point

l = pollutant type

L = total number of pollutant types

m = total number of emission sources

n = total number of monitoring points

N = number of Monte Carlo realizations of $\tilde{\omega}$

p = disjoint subset of discrete waste treatment options, or waste abatement strategies for the entire river basin, that may be examined in the stochastic branch and bound method

P = total number of disjoint subsets of discrete waste treatment options, or waste abatement strategies for the entire river basin, that may be examined in the stochastic branch and bound method

$R_1(x)$ = reliability, or the probability of the event that the state of the water quality system remains in a set of satisfactory states in the planning horizon, under waste treatment technologies x

$R_2(x)$ = probability that a system failure sojourn will not last more than one period, under waste treatment technologies x

$R_3(x)$ = resiliency, or the conditional probability that the water quality system state will be satisfactory in time $t + 1$, given that it is unsatisfactory in time t, under waste treatment technologies x

$\hat{R}_q(X^p)$ = estimate of the maximum value of $R_q(X^p)$, where $q = 1, 2, 3$

$S_j^l(x, \omega)$ = ambient water quality for pollutant l at monitoring point j due to treatment technologies $x = 1, \ldots, X$

$S_j^l(x, \omega, t)$ = ambient water quality for pollutant l at monitoring point j at any time t

$\bar{S}_j^l$ = water quality standard for pollutant l at monitoring point j

$\bar{S}^h$ = h^{th} water quality standard

t = time period

T = total number of time periods

$V(x)$ = vulnerability, or the weighted probability that a satisfactory water quantity state is achieved for all standards in a hierarchy of standards, under waste treatment technologies x

W_h = weight that describes the importance of water quality standard $\bar{S}^h$

x_i = a waste treatment technology at emission source i

X_i = available waste treatment technologies at emission source i

X^p = one subset of discrete waste treatment options for the entire river basin

$\chi_1(x, \tilde{\omega})$ = number of times that the water quality system is in a satisfactory state, under waste treatment technologies x and input event $\tilde{\omega}$

$\chi_2(x, \tilde{\omega})$ = number of times that the water quality is in a satisfactory state in time period t or in time period $t + 1$, under waste treatment technologies x and input event $\tilde{\omega}$

$\chi^{(h)}(x^p, \omega)$ = number of times satisfactory water quality occurs over h hierarchy intervals of standards given a fixed X^p and parameter input event ω

$\Delta^{(h)}(x^p, \omega)$ = the difference between $\chi^{(h)}(x^p, \omega)$ and $\chi^{(h-1)}(x^p, \omega)$

γ_1 = weight in overall water quality management objective function for $R_1(x)$

γ_2 = weight in overall water quality management objective function for $R_2(x)$

γ_3 = weight in overall water quality management objective function for $R_3(x)$

γ_4 = weight in overall water quality management objective function for $V(x)$

$\eta^N(X^p)$ = lower bound for equation 16.17

$\eta_q^N(X^p)$ = stochastic lower bound on $R_q(X^p)$, where $q = 1, 2, 3$

$\eta_V^N(X^p)$ = stochastic lower bound on $V(X^p)$

$\Theta^h(x)$ = occurrence of a satisfactory water quality state for the hth standard, under waste treatment technologies, x, given $\Theta^{h-1}(x)$

$\tau(x, \omega)$ = an unsatisfactory water quality state

ω = elementary event in some probability space $(\Omega, \boldsymbol{F}, \boldsymbol{P})$

$\tilde{\omega}$ = randomly drawn water quality simulation parameters

ω^s = independent observations of ω, or Monte Carlo realizations of ω, where $s = 1, \ldots, N$

$\xi^N(X^p)$ = upper bound for equation 16.17

$\xi_q^N(X^p)$ = stochastic upper bound on $R_q(X^p)$, where $q = 1, 2, 3$

$\xi_V^N(X^p)$ = stochastic upper bound on $V(X^p)$

$\psi_1(x^p, \omega)$ = number of times $\hat{c}_1(X^p, \omega)$ is less than the prescribed budget, $\bar{c}$ for N realizations of ω and a fixed X^p

$\psi_2(x^p, \omega)$ = number of times $\hat{c}^{(h)}(X^p, \omega)$ is less than the prescribed budget, $\bar{c}$ for N realizations of ω and a fixed X^p

$\psi^{(h)}(x^p, \omega)$ = number of times $\hat{c}^{(h)}(X^p, \omega)$ is less than the prescribed budget, $\bar{c}$ for N realizations of ω and a fixed X^p

17 Uncertainty in risk analysis of water resources systems under climate change

BIJAYA P. SHRESTHA*

ABSTRACT

A three-phase system framework is identified to assess uncertainty in the risk analysis of water resources systems under climate change. The uncertainty arises among others from the hydrometeorological inputs, such as precipitation and temperature, that are used for predicting extreme events (floods and droughts) under climate change. These inputs play an important role in the three system phases, namely planning, design, and operation. In this study, the hydrometeorological inputs under climate change are obtained by using downscaling models that use information from a general circulation model (GCM) output. The uncertainty is assessed by a new technique using a fuzzy approach. The methodology is illustrated by an example of uncertainty assessment in the risk analysis of a water resources system under climate change in a semi-arid region.

17.1 THREE-PHASE SYSTEM FRAMEWORK

A water resources system, such as a reservoir built for flood control or for water supply during severe droughts, may be described in three phases, namely planning, design, and operation. Planning phase of a water resources system must take into account multiple users, multiple purposes, and generally multiple criteria and/or objectives. Furthermore, in the planning phase, the emphasis also needs to be given to the system sustainability. A sustainable system should not only meet the present demands but should also consider future generations' demands. The notion of sustainability which is often described by non-numerical, qualitative, and philosophical means, may be found in more detail, for example, in Haimes (1992), Gleick et al. (1995), Plate (1993), and WCED (1987). Planning of a water resources system generally deals with building a suitable mathematical model considering multiple objectives and all the requirements to be met; alternatives (often called Pareto optimal solutions) are generated and finally a set of "satisfying choices" is found using a suitable multiple criteria tool for the multiple users of the system. Among the set of "satisfying choices" the most preferred choice (say ranked one) is often selected by decision makers for designing the system. Design phase is basically the engineering part where more detailed study of every parameter of the system is carried out (sometimes even with building a model for prototype) and then the system is physically built. In the final or operational phase, decisions considering the operation of the system are actually made, for example, rules to release water from a reservoir system for different purposes (or different users). These purposes may be hydropower, municipal, industrial, and irrigation demands, flood control and navigation, and environmental: water quality for fish and wildlife preservation, recreational needs, and downstream flow regulation. The decision-making procedure in the operational phase may itself be seen at three levels, namely operational, implementation, and policy level. In the United States, at the operational level many water users have permits that describe the conditions under which they can utilize water, at the implementation level the deployment of such a permit system (as well as the contents of particular permits) is usually determined by a water agency, and at the policy level the rules under which such a water agency operates usually are determined by a state legislature (Duckstein, Shrestha, and Waterstone 1996).

17.2 RISK AND UNCERTAINTY

Risk of a system may be defined simply as the possibility of an adverse and unwanted event. Risk may be due solely to physical phenomenon such as health hazards or to the interaction between man-made systems and natural events, for example, a flood loss due to an overtopped levee. Risk, or more appropriately "engineering risk," for water resources systems

* CH2M Hill, Sacramento CA, USA.

in general has also been described in terms of a figure of merit which is a function of performance indices, say for example, reliability, incident period, and repairability as in Shrestha, Duckstein, and Stakhiv (1996). Performance indices measure a system performance while figure of merit constitutes a super criterion and reflects overall system performance. In an "incident mode" failure when load exceeds resistance, the above-mentioned performance indices – reliability, incident period, and repairability – can be defined as the measure of the system reliability, the average recurrence time when the system goes to failure (or unwanted) mode, and the average recovery time from the occurrence of failure mode, respectively. Then one can say "high engineering risk" is a combination of low reliability and incident period, and large (poor) repairability.

$$RI(Z, a(1), a(2), a(3)) = C(PI^1(Z) < a(1), PI^2(Z) < a(2), PI^3(Z) > a(3)) \quad (17.1)$$

where RI and PI are the engineering risk and performance index of the system Z, respectively, C is a combination function, for example, which may be a joint probability distribution; $a(j)$s $j = 1, \ldots, 3$ are threshold values which may be deterministic (crisp) or expressed as fuzzy numbers.

Uncertainty, on the other hand, means simply the lack of certainty mainly due to inadequate or imprecise information. Uncertainty in a water resources system may be classified into three different types:

1. parameter uncertainty which is associated with the parameters of a selected mathematical model,
2. model uncertainty which is due to selection of an appropriate model,
3. model imprecision or model prediction uncertainty. The imprecision uncertainty can be avoided using a better model that predicts a more accurate result. However, more accurate models in most cases would have greater data requirements, which sometimes cannot be fulfilled due to shortage in data on hand. For example, in the present study the semi-arid regions of the southwestern United States (Arizona and New Mexico) have been studied for which the data set of precipitation lacks the number of rainy days. Thus, imprecision uncertainty cannot be completely avoided in such a scenario.

This study focuses assessment on the uncertainty in the risk analysis of water resources systems where the uncertainty arises from the hydrometeorological inputs, such as precipitation and temperature, that are used for predicting extreme events (floods and droughts) under climate change. The following section describes hydrometeorological inputs and climate change.

17.3 HYDROMETEOROLOGICAL INPUT AND CLIMATE CHANGE

The study by the Intergovernmental Panel on Climate Change (IPCC 1994) has concluded that the recent variations in seasonal weather patterns may not merely be natural fluctuations. A climate change may have resulted from a rapid increase of trace gases concentrations in the atmosphere due to anthropogenic activities, such as industrial pollution, car emissions, deforestation. Trace gases, especially carbon dioxide (CO_2), methane (CH_4), chlorofluorocarbons (CFCs), and nitrous oxides (NO_x), have the property of permitting the fairly free passage of short wavelength solar radiation from the sun to the earth's surface, but absorbing the reflected radiation (at low temperatures and long wavelength) from the earth, thus keeping earth much warmer than it would be in their absence. Analogous to the effect of glass in a greenhouse, this mechanism has been known as the "greenhouse effect" and the gases, as greenhouse gases (GHGs).

In this study, a $2 \times CO_2$ scenario of climate change resulting from the so-called greenhouse effect is considered. IPCC (1994) defines a scenario as a coherent, internally consistent, and plausible description of a possible future state of the world. Synthetic, analogue, and general circulation model (GCM) scenarios are three major ways of looking at the possible future states. Synthetic scenarios describe techniques where particular climatic elements are changed by a realistic but arbitrary amount (often according to a qualitative interpretation of climate model predictions for a region). Analogue scenarios are constructed by identifying recorded climatic regimes which may serve as descriptors of the future climate in a given region. These records can be obtained either from the past (temporal analogue) or from another region at present (spatial analogues). GCM's scenarios are obtained from three-dimensional numerical models of the global climate system (including atmosphere, oceans, biosphere, and cryosphere) and are the only credible tools currently available for simulating the physical processes that determine the global climate (IPCC 1994). GCMs produce estimates of climatic variables for a regular network of grid points across the globe. Results from about twenty GCMs have been reported to date. However, these estimates are uncertain because of some significant weaknesses of these models, which include (after IPCC 1994):

1. Poor model representation of cloud processes.
2. A coarse spatial resolution, at best employing grid cells of about 200 km horizontal dimension in model runs for which outputs are widely available for impact analysis.
3. Generalized topography, disregarding some significant orographic features.

4. Problems in the parameterization of subgrid scale atmospheric processes such as convection and soil hydrology.
5. A simplified representation of land-atmosphere and ocean-atmosphere interactions.

While GCM outputs represent the features of general circulation quite well (e.g., Simmons and Bengtsson 1988), they fail to provide a description of regional or local climatic conditions primarily because of low spatial resolution. Therefore, a methodology is necessary to distribute large-scale climate changes over smaller areas (regional/river basin scale), namely, downscaling which may thus be seen as a procedure of climate inversion (Kim et al. 1984). Giorgi and Mearns (1991) distinguish three approaches to downscaling:

1. Empirical techniques that use the so-called similarity hypothesis (e.g., Hansen et al. 1984). The temporal and spatial analogue scenarios described earlier fall into this category. For example, paleoclimatic information (temporal analogue) is used in Glantz (1988). The main problem with temporal analogues is a poor knowledge of the physical mechanisms and boundary conditions giving rise to the warmer climate. Aspects of these were almost certainly different in the past from those involved in GHG-induced warming. The problem with spatial analogues is due to the fact that different latitudes have different radiation processes.
2. Meso-scale numerical modeling techniques using GCM outputs as initial and boundary conditions (Dickinson et al. 1989; Beniston and Sanchez 1992). Model parameterizations and substantial computations limit the use of these techniques.
3. Semi-empirical techniques (Brinkmann 1993; Bogardi et al. 1993; Matyasovszky et al. 1994a, 1994b; Shrestha 1995) that use local observational data to link large-scale circulation patterns (CPs) to local climatic variables. Here, the large-scale forcing of local variables is described throughout the stochastic modeling of CPs and local forcing is incorporated into the linkage model. There are two major assumptions. The first assumption is that GCM-produced CP types may be considered the same as the types obtained from historical data. The other one is that the CP-hydrologic variables linkage will remain the same as the historical one in climate change situations. The second assumption may not be fully satisfied because of the land surface-biosphere-atmosphere feedbacks. These interactions, however, are poorly understood and it is difficult if not impossible to predict how such a linkage may change. Furthermore, present models of the linkage seem to be invariant under broadly varying space-time conditions (the same type of CP may produce rain in several seasons over a large region).

Semi-empirical techniques of downscaling climatic variables seem to be promising because of their advantages over the other two approaches (e.g., feasible computations, flexibility and good results for Europe: Bárdossy and Plate 1992). Such a semi-empirical technique has been used in this study. Specifically, the global effect on local precipitation and temperatures is described by atmospheric CP types. The Max Planck Institute (MPI) GCM (model T21, Cubash et al. 1992) outputs for the $1 \times CO_2$ and $2 \times CO_2$ scenarios are used to represent present and climate change scenarios.

17.4 MODELING AND SIMULATION

Precipitation and temperature processes are modeled by using multivariate autoregressive (MVAR) processes. These models are conditioned on both CP type and average local pressure heights of 500 hPa (h) which provide linkage so as to be able to generate precipitation and temperature under climate change scenario in the semi-arid region. The time series of CP type is modeled by a Markov chain. The conditional precipitation probability on a CP type and h is modeled by a linear regression with dependent indicator variable, rainfall amount conditioned on a CP by a gamma distribution. Temperature conditioned on a CP type and h is modeled by a two-sided normal distribution. Transformation schemes are developed so that MVAR model is built with normal variables while actual variables are gamma (for precipitation amount) and two-sided normal distribution (temperature). For the sake of brevity, details of modeling and simulation are omitted. Instead, schematic charts are shown in Figures 17.1, 17.2, and 17.3 so as to describe a complete view about the modeling and simulation of precipitation and temperature under a $2 \times CO_2$ climate change scenario. Interested readers can find more mathematical details in Shrestha (1995).

17.5 ANALYSIS OF RESULTS

The precipitation model assumes that only rainfall probability is related to h. Therefore, rainfall amount distribution remains the same as the one of the observed scenario for both historical simulated output and $2 \times CO_2$ scenarios, and the validation should show that the probabilities in the observed and historical simulated cases do not differ significantly.

The simulation results show that the rainfall probabilities in both Arizona and New Mexico (along the upper Rio Grande) increase significantly in all four seasons. Figure 17.4 shows the comparative results of the probability of rainfall in Arizona

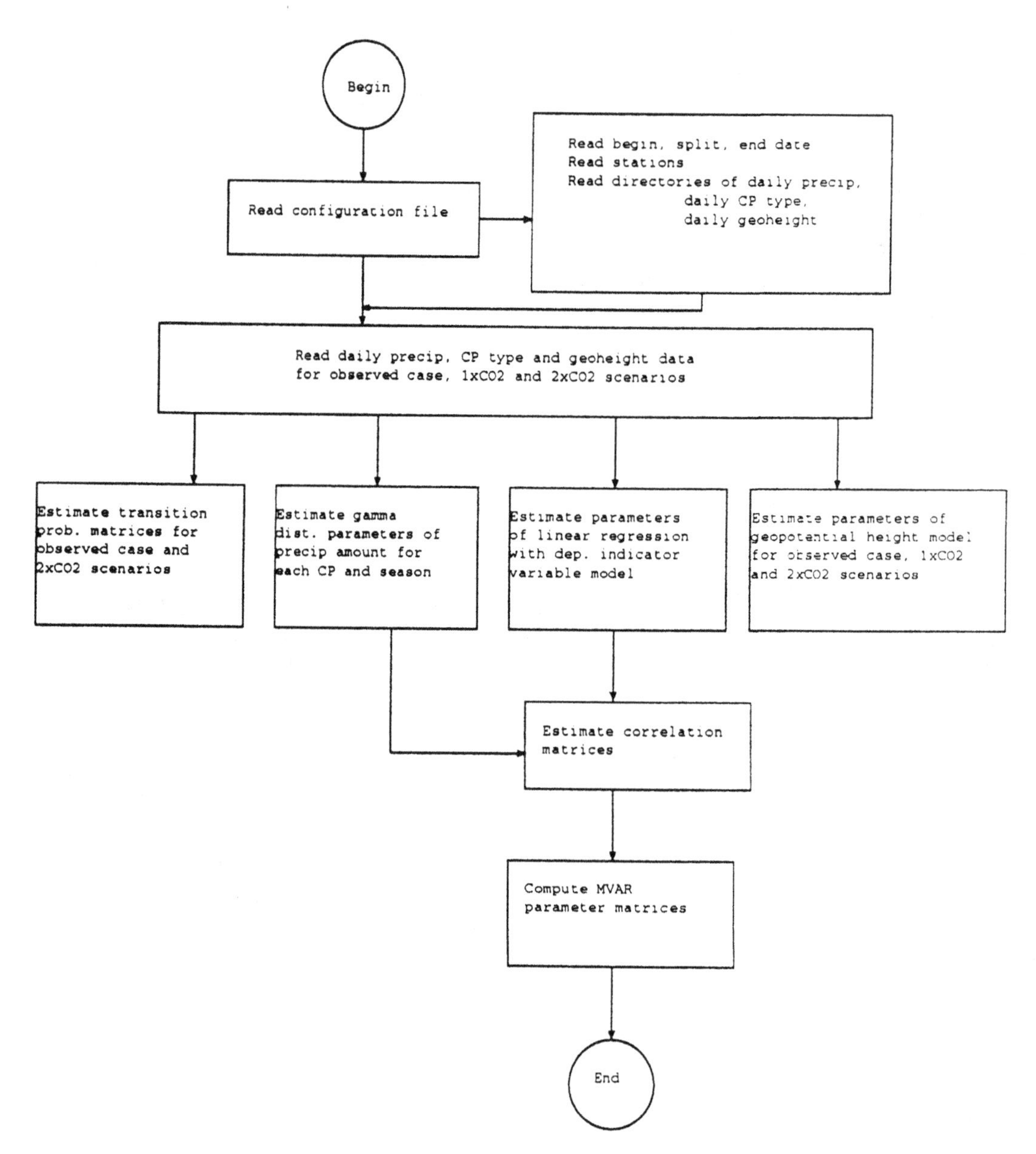

Figure 17.1. Flow chart for estimating parameters of the precipitation model.

stations for the winter season. The increase in summer probability is more evident than that in other seasons, but there seems to be a significant increase in rainfall probability in all seasons. The amount of increase is found to be dependent on the particular station, which confirms the results found in eastern Nebraska (Matyasovszky et al. 1994a). This necessitates downscaling techniques to be used since a large area-wide result does not reflect fluctuations in local findings.

The temperature for the future scenario increases in all seasons, all stations, and all CPs. The amount of temperature increase is greater than 1°C during the spring and fall seasons. A less significant increase is found in summer. The amounts of temperature increase are found to be dependent both on the particular station and the season. This again corroborates the results found in eastern Nebraska (Matyasovszky et al. 1994b) and shows the necessity of downscaling. Figure 17.5 displays output results for mean values of maximum temperature in Arizona for winter.

More details of the results of simulation can be found in Shrestha (1995).

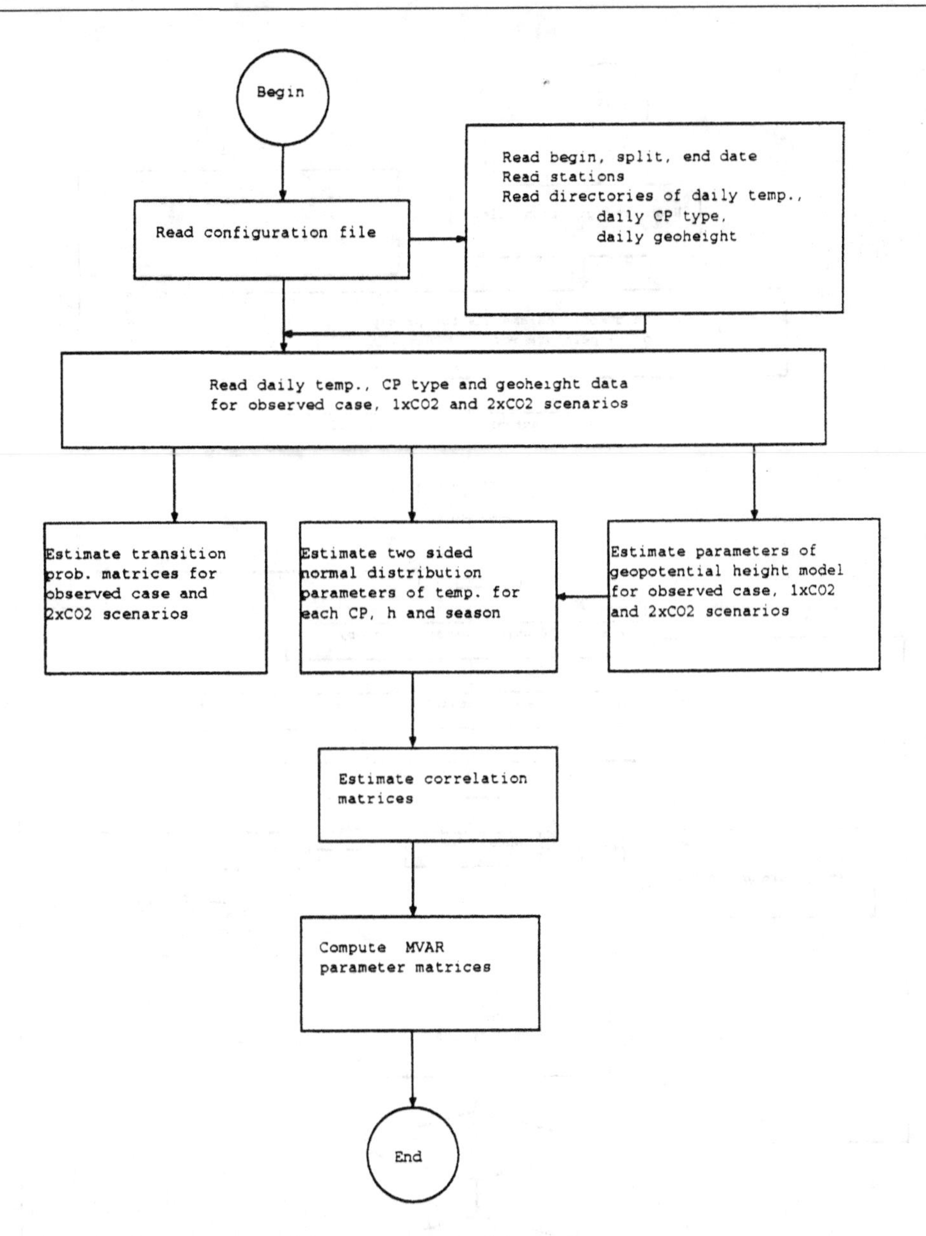

Figure 17.2. Flow chart for estimating parameters of the temperature model.

17.6 UNCERTAINTY OF RISK ANALYSIS

In this section, an example of uncertainty in the risk analysis of flood protection under climate change will be presented. This example, which is a case study of a semi-arid region of the southwestern United States, follows closely the one described in Duckstein and Bogardi (1991). Consider the case of a binary decision to be made along a reach of the Santa Cruz River in southern Arizona, whether or not to protect against the maximum peak floods occurring during a twenty-year horizon. Within the framework of this example, uncertainty and/or imprecision may be found in the following elements of the risk analysis procedure (after Duckstein and Bogardi 1991):

1. randomness or natural uncertainty of flood peaks
2. informational uncertainty on flow model and loss functions
3. levee resistance to floods due, for example, to nonhomogeneity of soils
4. the failure phenomenon itself, which may be "small" or "large"

The uncertainty in the risk analysis of the levee system due to climate change fits more of types (1) and (2).

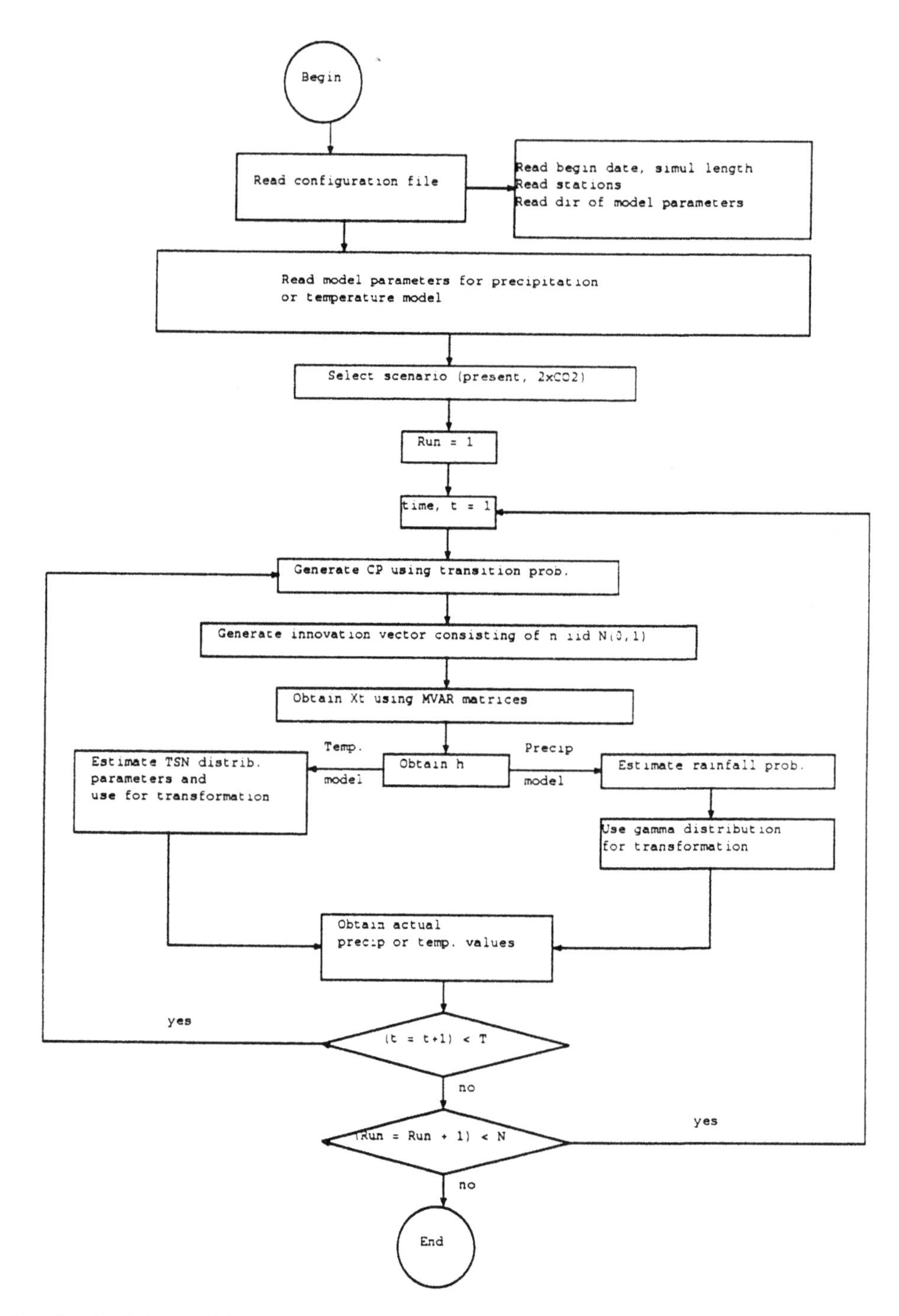

Figure 17.3. Flow chart for simulation model.

As described in that paper, it is possible to describe uncertainty of types (2), (3), and (4) above by either probabilistic, Bayes, or fuzzy set analysis, or else by a mix of these approaches.

The decision-theoretic formulation of the problem may be described with the following elements: two states (X) of nature:

(i) $X_1 : Q \geq Q_0$ and $X_2 : Q < Q_0$

where Q is the peak flood over the time horizon and Q_0 is the protection level

(ii) a Bernoulli model of occurrence of state X_i, $i = 1, 2$:

$$P(X_1) = P(Q \geq Q_0 | p) = p$$
$$P(X_2) = P(Q < Q_0 | p) = 1 - p$$

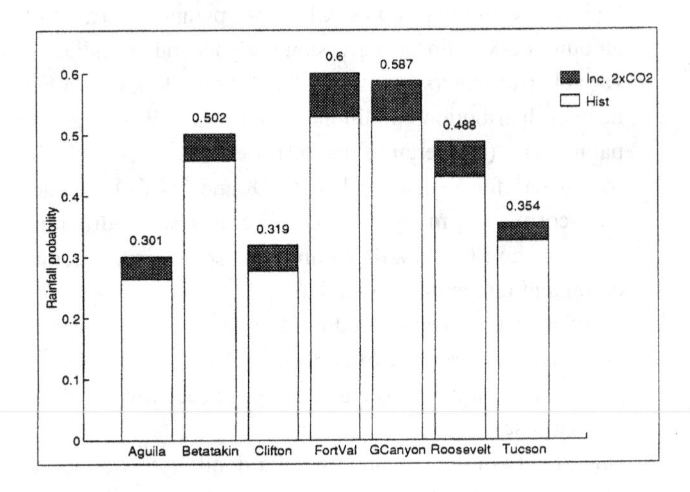

Figure 17.4. Rainfall probability in Arizona (winter).

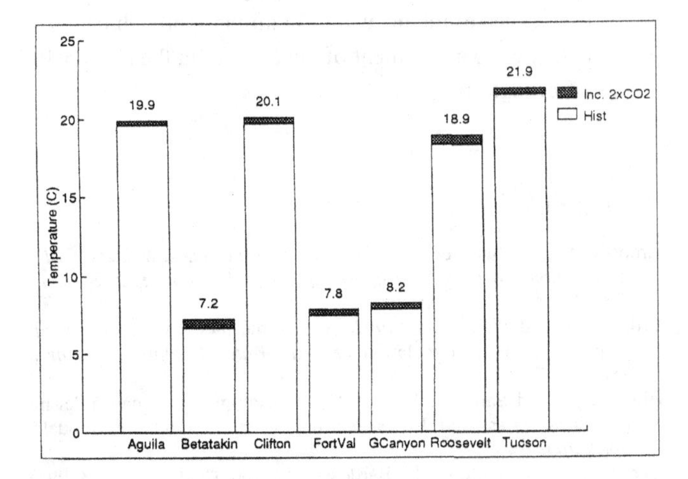

Figure 17.5. Means of maximum temperature in Arizona (winter).

Table 17.1. *Economic loss matrix*

i	X_i	a_1	a_2	$p(X_i)$
1	$X_1 : Q \geq Q_0$	C_1	C_2	p
2	$X_2 : Q < Q_0$	C_1	0	$1 - p$

(iii) two actions: a_1: protect and a_2: do not protect, and
(iv) protection costs C_1 and annualized flood losses C_2.

The decision problem can be displayed in the loss matrix given in Table 17.1.

Next, let the following uncertainties be taken into account and represented in the form of fuzzy numbers as in the example given in Duckstein and Bogardi (1991).

1. the probability p of exceedance, reflecting sample and model/technique uncertainties as well as fuzziness in Q_0.
2. economic losses C_1 and C_2, because of the difficulty in estimating annualized construction, operation, and maintenance costs C_1, and flood losses C_2.

Let the following triangular fuzzy numbers (TFN) (Bárdossy and Duckstein 1995) represent, respectively, the fuzzy elements of the decision problem:

$\hat{p} = (7, 8, 10) \times 10^{-4}$ corresponding to a protection level of about 400 cm in the Santa Cruz River basin, in southern Arizona

$\hat{C}_1 = (3, 4, 5)$ corresponding to a cost of "about 4" units

$\hat{C}_1 = (3.5, 4.5, 5.5) \times 10^3$

The expected losses $(\hat{L})$ corresponding to actions a_1 and a_2 can now be calculated as

$$\hat{L}(a_1) = \hat{C}_1(\cdot)\hat{p}(+)\hat{C}_1(\cdot)(\hat{1}(-)\hat{p}) \tag{17.2}$$

$$\hat{L}(a_2) = \hat{C}_2(\cdot)\hat{p} \tag{17.3}$$

where $\hat{1} = (1, 1, 1)$, and (+), (−), (·) indicate the fuzzy arithmetic operations of addition, subtraction, and multiplication, respectively (Bárdossy and Duckstein 1995). Substituting the values

$$\hat{L}(a_1) = (3, 4, 5)(\cdot)(7, 8, 10) \times 10^{-4}(+)(3, 4, 5)(\cdot)(1, 1, 1)$$
$$(-)(3, 4, 5)(\cdot)(7, 8, 10) \times 10^{4} = (3, 4, 5)$$

$$\hat{L}(a_2) = (3.5, 4.5, 5.5)(\cdot)(7, 8, 10) \times 0.1 = (2.45, 3.6, 5.5)$$

To determine the optimum action a_1, the two fuzzy numbers $\hat{L}(a_1)$ and $\hat{L}(a_2)$ must be compared, which is done here simply by using the fuzzy mean $m(A)$ of fuzzy number A defined on the domain of real numbers $D(x)$:

$$m(A(x)) = \frac{\int_{D(x)} x\mu_A(x)}{\int_{D(x)} \mu_A(x)} \tag{17.4}$$

where μ_A is the membership function (see Bárdossy and Duckstein 1995). For the TFN $A(a, b, c)$, fuzzy mean is simply given by

$$m(A(x)) = \frac{a+b+c}{3}$$

Accordingly, as the fuzzy means of $\hat{L}(a_1) = (3, 4, 5) = 4 > \hat{L}(a_2) = (2.45, 3.6, 5.5) = 3.85$, the resulting action $a = a_2$ (do not protect) is preferred to the action $a = a_1$ (protect).

Next, consider the same example under a climate change scenario. Our results and analysis showed that precipitation probability in southern Arizona where Santa Cruz lies increases under a $2 \times CO_2$ climate change scenario. Consequently, the annual peak flows are more likely to be higher. Consider that the cost for the "do not protect" scenario remains the same and increased probability p is now $\hat{p} = (7.7, 8.7, 10.7) \times 10^{-4}$. Then the expected loss becomes

$$\hat{L}(a_1) = (3,4,5)(\cdot)(7.7,8.7,10.7)\times10^{-4}(+)(3,4,5)(\cdot)(1,1,1)$$
$$(-)(3,4,5)(\cdot)(7.7,8.7,10.7)\times10^{-4} = (3,4,5)$$
$$\hat{L}(a_2) = (3.5,4.5,5.5)(\cdot)(7.7,8.7,10.7)\times0.1 = (2.70,3.92,5.89)$$

Then following the results of comparison of the fuzzy means of $\hat{L}(a_1) = (3, 4, 5) = 4 < \hat{L}(a_2) = (2.70, 3.92, 5.89) = 4.17$, the action $a = a_1$ (protect) is preferred to action $a = a_2$ (do not protect).

Thus, taking climate change effect into account leads to a reversal of the decision.

17.7 DISCUSSION AND CONCLUSIONS

Uncertainty in the risk analysis of a water resources system: a flood protection levee, under a $2 \times CO_2$ climate change scenario in the semi-arid region of southern Arizona, United States, has been assessed. The results showed that, in the absence of climate change effect, the action "do not protect" (or do not build a flood levee) has been found to be a preferred decision. When climate change effect is taken into account, the reversal of the decision, "protect" (or build a flood levee) has been found to be the preferred action. The uncertainty in the risk analysis of flood levee here has been assessed using a fuzzy approach, which is often an appropriate tool to model uncertainty due to imprecision.

Though the flood levee is used to illustrate the methodology, one can use a similar approach to water resources systems serving drought alleviation. Simulation results of temperature under a $2 \times CO_2$ climate change scenario showed that temperature increases under a climate change scenario. Such a temperature increase would be quite substantial: the various demands for water would increase, at the same time as the timing of water supply would shift from late to early spring because of earlier snowmelt, which plays a very important role in the western United States. In such a case more frequent and more severe shortages could occur. The shortages or drought may become more severe due to increasing demands resulting from increasing population and industrial growth.

The major concluding points may be made as follows:

1. Two major hydrometeorological inputs, precipitation and temperature, have been modeled under a climate change scenario represented by $2 \times CO_2$. The methodology uses a space-time stochastic modeling of daily precipitation and temperature conditioned on CP type and h over the local region. The linkage to future change is provided through the conditioning part (both on CP type and h).
2. The precipitation and temperature models require the parameter estimation of: transitional probability matrices for a Markov chain (CP type transition model), gamma distribution conditioned on a CP type parameter (rainfall amount model), linear regression with dependent indicator variable parameters (rainfall probability model), two-sided normal distribution conditioned both on a CP type and h parameters (temperature model), seasonal trend fitting parameters for h (historical, $1 \times CO_2$ and $2 \times CO_2$ scenarios), correlation matrices among the data set at different locations and the MVAR parameters (space-time precipitation and temperature models).
3. Simulation has been carried out to generate daily precipitation and temperature of historical simulated (validation purpose) and a $2 \times CO_2$ climate change scenario (predicting purpose).
4. Uncertainty in the risk analysis of a flood protection levee has been assessed using a fuzzy approach. Taking into account the climate change effect may entirely change (that is, even reverse) the decision. A similar approach can also be extended to assessment of uncertainty in the risk analysis of drought purpose systems.

REFERENCES

Bárdossy, A. and Duckstein, L. (1995) *Fuzzy Rule-Based Modeling with Applications to Geophysical, Biological and Engineering Systems.* CRC Press, Boca Raton, Florida.

Bárdossy, A. and Plate, E. J. (1992) Space-time model of daily rainfall using atmospheric circulation patterns. *Water Resources Research* 28(5): 1247–59.

Beniston, M. and Sanchez, J. P. (1992) An example of climate relevant processes unresolved by present-day general circulation models. *Speedup* 6: 19–25.

Bogardi, I., Matyasovszky, I., Bárdossy, A., and Duckstein, L. (1993) Application of a space-time stochastic model for daily precipitation using atmospheric circulation patterns. *J. Geoph. Res.* 98(D9): 1653–67.

Brinkmann, W. A. R. (1993) Development of an airmass-based regional climate change scenario. *Theor. Appl. Climatol.* 47: 129–36.

Cubash, U., Hasselmann, K., Hock, H., Maier-Reimer, E., Mikolajewicz, U., Santer, B. D., and Sausen, R. (1992) Time-dependent greenhouse warming computations with a coupled ocean-atmosphere model. *Climate Dynamics* 8: 55–69.

Dickinson, R. E., Errico, R. M., Giorgi, F., and Bates G. T. (1989) A regional climate model for the western United States. *Climatic Change* 15: 382–422.

Duckstein, L. and Bogardi, I. (1991) Reliability with fuzzy elements in water quantity and quality problems. In *Water Resources Engineering Risk Assessment*, edited by J. Ganoulis. NATO ASI Series, *G-29*, Springer-Verlag, Heidelberg, 231–51.

Duckstein, L., Shrestha, B. P., and Waterstone, M. (1996) Policy and engineering decision-making under global change: case of upper Rio Grande river basin. In *Risk-Based Decision-Making in Water Resources* VII, edited by Y. Y. Haimes, D. Moser, and E. Z. Stakhiv.

Giorgi, F. and Mearns, L. O. (1991) Approaches to the simulation of regional climate change: a review. *Rev. Geophys.* 29: 191–216.

Glantz, M. H. (ed.) (1988) *Societal Responses to Regional Climate Change: Forecasting by Analogy*. Westview Press, Boulder.

Gleick, P. H., Loh, P., Gomez, S. V., and Morrison, J. J. (1995) Water and sustainability. In *California Water 2020, a Sustainable Vision*, Pacific Institute for Studies in Development, Environment, and Security, 23–28.

Haimes, Y. Y. (1992) Sustainable development: a holistic approach to natural resource management. *IEEE Transactions on Systems, Man, and Cybernetics* 22(3): 413–17.

Hansen, J., Lacis, A., Rind, D., Russel, G., Stone, P., Fung, I., Ruedy, R., and Lerner, J. (1984) Climate sensitivity: analysis of feedback mechanisms. In Climate processes and climate sensitivity. *Geophys. Monograph* 29: 130–63.

IPCC (1994) *Climate Change Impacts and Adaptations: IPCC Technical Guidelines for Assessing Climate Change Impacts and Adaptations*, edited by T. R Carter, M. L. Parry, H. Harasawa, and S. Nishioka. University College London and Center for Global Environmental Research, 59.

Kim, J.-W., Chang, J.-T., Baker, N. L., Wilds, D. S., and Gates, W. L. (1984) The statistical problem of climate inversion: determination of the relationship between local and large-scale climate. *Mon. Wea. Rev.* 112: 2069–77.

Matyasovszky, I., Bogardi, I., Bárdossy, A. and Duckstein, L. (1994a) Comparison of two general circulation models to downscale temperature and precipitation under climate change. *Water Resour. Res.* 30: 3437–48.

(1994b) Local temperature estimation under climate change. *Theor. Appl. Climatol.* 50: 1–13.

Plate, E. J. (1993) Sustainable development of water resources: a challenge to science and engineering. *Water International* 18: 84–94.

Shrestha, B. P. (1995) Downscaling precipitation and temperature under climate change over semi-arid regions of southwestern USA. Unpublished Dissertation, Department of Systems and Industrial Engineering, University of Arizona, Tucson.

Shrestha, B. P., Duckstein, L., and Stakhiv, E. Z. (1996) Fuzzy rule-based modeling of reservoir operation. *J. of Water Resources Planning and Management, ASCE* 122(4).

Simmons, A. J. and Bengtsson, L. (1988) Atmospheric general circulation models: their design and use for climate studies. In *Physically Based Modeling and Simulation of Climate and Climatic Change*, vol. II, *NATO ASI Series E*, 254, edited by M. Schlesinger. Kluwer Academic, Norwell, Mass., 627–52.

WCED – World Commission on Environment and Development (1987) *Our Common Future*, Oxford University Press, Oxford.

18 Risk and reliability in water resources management: theory and practice

URI SHAMIR*

ABSTRACT

The existence of a gap between theory and practice of risk and reliability in water resources management has been widely recognized. An overview of stages of risk management is offered. It consists of: identification of failure modes; evaluation of likelihood of each failure mode; listing the consequences of each failure; risk valuation and mitigation. The results of these stages are synthesized, to form the basis for selection of an optimal plan or policy. Finally, the role of forecasting in risk management practice is reviewed.

18.1 INTRODUCTION

There is a gap between the theory and practice of incorporating considerations of risk and reliability into management of water resources systems. The reasons for this are:

1. Criteria for defining risk and quantifying reliability have not been standardized and accepted; and
2. Methodologies for incorporating reliability measures and criteria into procedures and formal models for management of water supply systems are still not well developed, and the complexity of those methods and models that do exist make them difficult to use in engineering practice.

Reliability criteria should be defined from the point of view of the consumer, and reflect the costs of less-than-perfect reliability. Here "cost" means the measurable economic losses due to failures, such as the loss of crops that are not irrigated on time or the loss of production in a factory, and any other loss incurred by failure to supply an adequate quantity of good-quality water at the time it is required. Losses are very difficult to measure and quantify, especially those associated with the quality of life of urban consumers.

* Water Research Institute, Technion, Israel Institute of Technology, Haifa, Israel.

Therefore, somewhat intuitive and arbitrary measures of reliability are used. One that is standard for urban systems is: the distribution systems must be looped, establishing more than one supply path to each consumer so that if one fails another will provide at least some of the service. More sophisticated but still somewhat arbitrary criteria also exist, for example: the maximum number of hours per year that a consumer is disconnected from the supply cannot exceed a prescribed value.

In recent years, research has been directed to the development of practical methods for incorporating reliability directly into models for planning, design, and operation of supply systems. This chapter includes an overview of risk management, and an attempt to provide some operational criteria and methodologies for incorporating reliability criteria into management of water resources systems.

The phases of risk management are (Plate 2001):

- List the potential modes of failure of the system.
- Calculate or estimate the likelihood of each failure mode.
- Describe the physical consequences of each failure mode.
- Assess the damages and losses due to each failure mode, and the value, on a selected scale, assigned to each damage or loss.
- Generate the options for mitigating each of the failure modes, wholly or partially.
- Compute the costs of each option.
- Synthesize all of the above into a framework for management of the water resources system, and select the optimal policy for planning, design, and operation.

We shall now consider each of these phases. To help in making the discussion more concrete, we refer to a project in which a reservoir is to be constructed on a river, for water supply. The reservoir is designed to regulate the flows, and is operated to provide water to a distribution system. The water is delivered through a treatment plant to the distribution system.

18.1.1 Potential failure modes of the system

Examples:

- The sum of the inflow into the reservoir plus the amount stored in it is low, resulting in $[S < D]$ = [Supply < Demand].
- Demands are higher than forecasted, resulting in $[D > S]$.
- Pollution of the water in the reservoir prevents its use, even with the existing treatment plant.
- The treatment plant fails, and the water cannot be used until the plant is back in operation.
- An element of the distribution system fails, eliminating or reducing service to some customers.
- The dam fails, causing damage or loss of life downstream.

Each failure mode will require a different model for reliability analysis (Shamir and Howard 1981). The models range from aggregate to detailed, as follows.

Lumped supply – lumped demand. The entire system is depicted as having a single source and a single demand, each a random variable, with a known joint probability distribution of supply and demand. The joint *pdf* is integrated over the region where $[S < D]$, to yield $P[S < D]$. When the *pdf* has a simple form, the computations can be carried out analytically (Shamir and Howard 1981), by the so-called "statistical-analytical approach." A modification of this approach, which considers also the uncertainty of the probability distributions, especially their tails, has been developed by Ben-Haim (1996).

The *pdf* of the demand should be made conditional upon the *pdf* of the supply, in one of two ways. First, the two variables may both depend on the same external condition, such as hot dry weather. In this case, there is a negative correlation between demand and supply: demands are high when supply is low. The other dependence has to do with demand management. It is wrong to assume that demands are not affected by the level of supply. Public appeals to reduce consumption at times of droughts, followed by compulsory measures to reduce consumption when the situation becomes worse, are very relevant to the analysis. In this case the correlation between demand and supply is positive.

One might argue that consideration of demand management should be relegated to the phase in which management options are considered, but in fact it is necessary to consider this aspect already at the time the probability $P[S < D]$ is computed.

The analytical approach is still feasible when the number of sources and demands is more than one, say two or three. However, once this number increases, the computations become very cumbersome, since they require integration of joint probability functions of several random variables over complicated domains.

Articulated supply and/or demand. The system is described as a set of several elements (up to a few tens), in series-parallel connection (Shamir and Howard 1981; Yang et al. 1996a). Fault tree methods are useful in this case. The computations are based on the probability of failure of each element, and allow explicit evaluation of the residual capacity of the system when each failure occurs.

Reduced network model of the distribution system. A skeletonized network is used to describe the system, and analytical methods are used to compute topological measures on the network, such as connectivity and reachability (Wagner, Shamir, and Marks 1988a), which are considered to be measures of the system's reliability. The hydraulic performance of the system is not computed accurately for each failure mode (because it is too costly to do), but may be estimated in some fashion.

A different approach has been proposed by Ostfeld and Shamir (1996), applicable also to multi-quality supply systems. It is based on optimization of the system, under constraints that specify the performance of the system in full operation as well as at times of failure. To accomplish this, two or more "backup subsystems" are defined, such that when a component of the system fails, one of the backup subsystems is still capable of supplying the consumers. The performance of each backup subsystem is specified in constraints, which specify the quantities to be supplied, the pressures, and the water quality at the consumer nodes. This level of service need not be the same as for the full system, with no failure, although it can be the same if so desired. The cost of system operation under normal conditions, without a failure, plus the cost (possibly including penalties for reduced service) at times of failure weighted by the expected time of their operation (which is a small fraction of the total time) are incorporated into the objective function.

Full network model. A fully detailed model of the network is used in a simulation mode (Wagner, Shamir, and Marks 1988b; Yang et al. 1996b). In each run, one or more components are removed (failed), and the residual capability of the remaining system is computed with a network solver. The results of simulations are analyzed, to yield the values of the selected reliability criteria for the system.

Wagner et al. (1988b) incorporated into the model a function approximating the quantity extracted from the system by a consumer, as a function of the pressure at the connecting node. This amounts to a reduced service at the time a failure occurs. Yang et al. (1996b) simulate the performance of the system over time, each element subject to failures according to its MTBF (mean time between failures), and to repair according to its MTTR (mean time to repair).

18.1.2 Likelihood of failures

Examples:

- Using hydrological records, possibly extended through simulation with a longer meteorological record, estimate the probability distribution of cumulative inflows for the relevant time horizons, and of the resulting shortage (see the section below on forecasting).
- Using pump maintenance and pipe break records, estimate the probability of failure of system components, its duration and time to repair, then compute for each failure mode the resulting shortage in supply.
- Using information on past demands, and a model for the effect of various demand management strategies, estimate the probabilities of different demand levels, dependent on the supply level, and estimate the joint probabilities of supply and demand.
- Using flow records, possibly extended by paleohydrological data, estimate extreme (high or low) values of flow and river stage.

The above procedures are relevant for cases in which failures are caused by deviations around an expected value. A different approach is used when one is to design for an event of a prescribed return period or frequency of occurrence. The procedure followed in this case is:

- Calculate the magnitude of the event of the given return period.

When the return period does not exceed the length of the historical record, the magnitude of the design event can be computed with some confidence. An example is the five-year flow required for minor drainage facilities. The more difficult cases are when the return period is very long, and the estimate of the computed event is highly uncertain. This is the case with the PMF, or even the 100-year flood. Klemeš (2001) expresses severe criticism of the procedures for estimation of the magnitude of very rare hydrological events, principally the design flood. He does so with considerable sarcasm, in reference to the behavior of old-time bureaucracies.

Design events should actually be selected according to the same procedure as any other analysis of risk. But because the probabilities of such events are highly uncertain (Klemeš 2001), and the consequences very severe, the preferred procedure is to set the return period, arbitrarily, and put all the emphasis on computing the magnitude of the corresponding event. Logical questions may be posed, such as:

- What are the actual expected consequences and damages that would result from the event whose return period has been prescribed.
- What is the sensitivity of the damage to variations in the return period.
- What is the sensitivity of the cost of the system to variations in the return period.

Figure 18.1 shows schematically the cost of a system versus its level of reliability, which is an expression of the magnitude of the rare event it is designed to sustain. The shape of the curve is typical. At low values of reliability there is a steep part, where sizeable increments of the system are required in order to achieve the basic level of reliability. Then there is a flatter portion of the curve, where a modest increase in cost achieves a large increment of reliability. At higher levels of reliability, the curve bends upward, indicating that the same increment of reliability costs more and more. When perfect reliability is approached (obviously, no perfect reliability is possible, although this might be stated as a goal) the cost of the system soars. It is reasonable to expect that the optimal level of reliability lies in the area where the curve bends

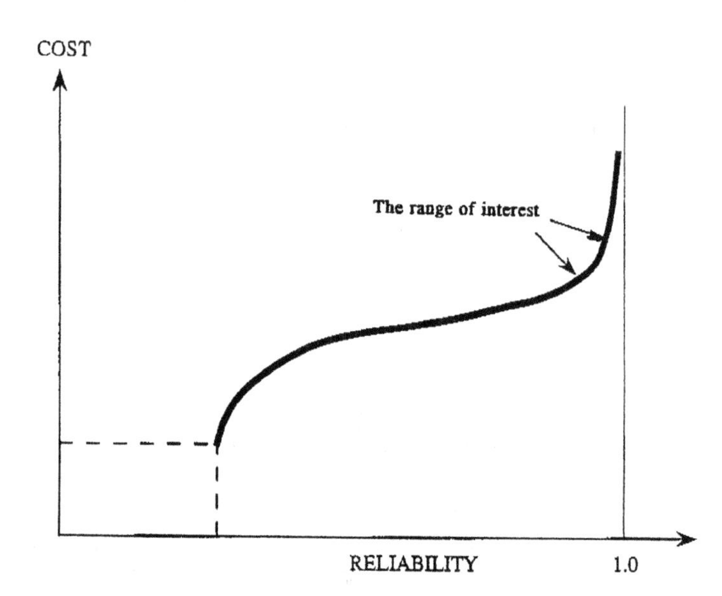

Figure 18.1. Selection of the design reliability according to the cost of reliability.

sharply upward. If the damages due to failure are not, or cannot be, computed, the cost-versus-reliability curve may be a good device for aiding decision making. The marginal cost of added reliability can be used to guide decision makers to selecting the solution.

18.1.3 Physical consequences of failure modes, and the associated damages and losses

The consequences are first measured in physical terms, such as the amount of water not supplied, the number of hours customers go without water, the area inundated, the number of people evacuated, the number of lives lost. The next step, which is considerably more difficult, is conversion of these consequences into "damage values."

18.1.4 Valuation of risk

The risk is expressed as a product of the damage times its probability of occurrence. Assessing the damage is one of the most difficult phases of risk management. In this respect, it is instructive to differentiate between different types of failures. Let us return to the example of the design of a water supply reservoir.

Damages due to shortages in supply of water to production systems, be they agriculture or industry, can be estimated as the loss of production value. Although there may be intangible components, such as loss of market position and loss of reputation, one can still turn to the market place and to rulings of the courts (with all their failings in reflecting true values!) to determine the value of shortage, or at least some estimate thereof.

When the shortage is to the domestic sector, the damage is less obvious. It is common to assume that domestic consumption is rigid and must be met with very high reliability. In reality, the reliability required may be too high, since there is more elasticity in urban water consumption than often assumed.

18.1.5 Options for mitigating each failure mode, and their cost

Examples:

- Sizing the reservoir for higher supply reliability.
- Adding capacity to the treatment plant.
- Adding redundancy in the distribution system.
- Designing the dam to withstand a larger flow.
- Using demand management options, to reduce consumption when there is a shortfall in supplies.

18.1.6 Synthesize all of the above into a framework for management of the water resources system, and select the optimal policy

To fix ideas, consider the following case. A reservoir of size D is to be constructed on a river, for water supply. The annual flow in the river, q, which will be the inflow to the reservoir, has been measured over a period of years. A probability density function has been fitted to the data. The central part of $f_Q(q)$ may be known with an acceptable degree of confidence. It is based on measured data, possibly extended through hydrologic simulation, with a longer series of meteorological data. The tails of the distribution have a much greater degree of uncertainty. The tail to the left may be known with greater confidence, but the one to the extreme right is always highly uncertain.

The *pdf* of the inflow and the benefit/loss function are shown in Figure 18.2.

The net benefit from water supply is denoted $B[q|D]$. It is a function of the inflow, q, and depends on the size of the reservoir, D. The unconditional benefit function, $B[D]$, lies above the curve $B[q|D]$, tangent to it at the point $q = q_D$, where the benefit is $B[D] = B[q_D|D]$.

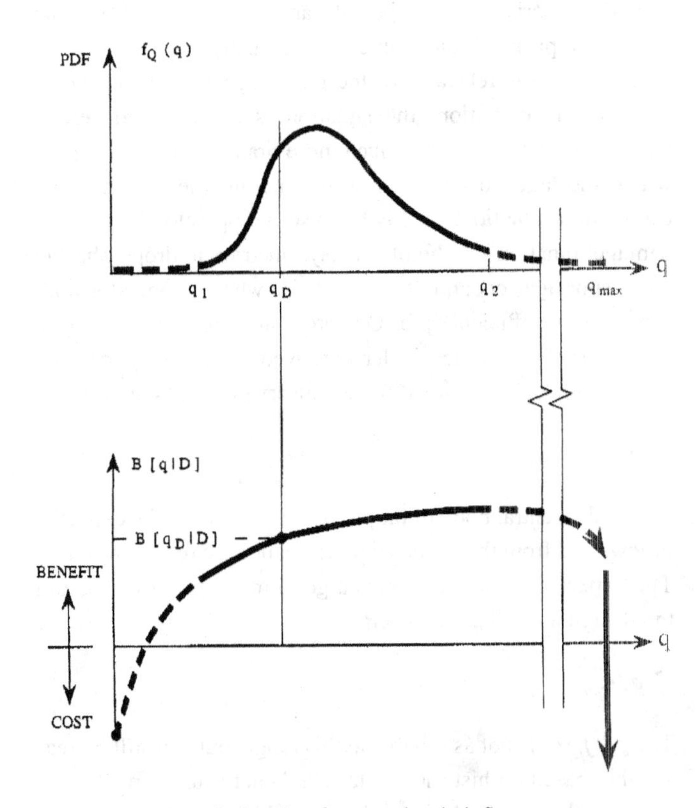

Figure 18.2. Benefit function for a stochastic inflow.

In examining the functions $f_Q(q)$ and $B[q|D]$ we can observe several ranges, as follows:

$0 < q < q_1$

This is the left tail of $f_Q(q)$, the region of low flows, droughts, and shortages. Reliability of the *pdf* in this range is low but may not be too bad, since the record usually does contain some low values, and, at least, there is a low limit (zero) to the possible values. For water supply, this is the most critical range of the possible outcomes.

The actual value of $B[q|D]$ in this range (which moves rapidly from benefits to losses, as q approaches zero) is difficult to forecast, because it depends on the recourse taken by consumers at the time a shortage occurs. The actual damage to consumers may be a lot lower than frequently assumed. For example, urban water demands are usually treated as fixed and inflexible, signaling a very high loss due to shortage. In reality, however, urban consumers adjust to shortages, when they occur as a result of natural causes (not so much when they are due to human error or mismanagement!), without much real damage.

The cost that accrues to the water supply utility due to shortage may be much greater than the actual real losses to individual consumers and to society as a whole. They are the result of claims made through the judicial system and pressures exerted through the political system. Hence, the value of $B[q|D]$ that actually materializes may be substantially different from that assumed a priori, through an economic analysis.

The above is relevant for the planning and design phases. In real-time operation, the situation is somewhat different. Operators tend to assign a much more dramatic drop in $B[q|D]$ when shortages occur than would be justified by a strict economic evaluation. This is because the operators' "personal damage function" is highly nonsymmetric: it drops sharply when shortages occur, but does not rise when an operator does a good (i.e., efficient) job. Operators are fired for shortages; they are rarely rewarded for having saved some money or water. This is why operators tend to be conservative in their actions.

$q_1 < q < q_2$

This is the central body of the *pdf*. Deviations from the expected inflows and from the expected performance are relatively small. The expected value criterion is a good measure for evaluating the design as well as the operation.

$q_2 < q$

The *pdf* $f_Q(q)$ is not as reliable in this range, but can still be reasonably based on historical data. The benefit function, $B[q|D]$, remains close to its design value, possibly with some increase (or decrease) from the design value $B[q(D)|D]$. As q increases further above the limit q_2, the reliability of the *pdf* decreases, and $B[q|D]$ begins to drop. The strength of the expected value criterion decreases.

$q_2 << q$

As q grows further beyond q_2, reliability of the statistical calculations drops, as does the reliability of the benefit function $B[q|D]$. This range is frequently ignored in the computation of $EV\{B[q|D]\}$.

$q \dashrightarrow q_{max}$

q_{max} is the extreme value for which the system must be designed; it is the "load" that the system has to withstand without failing in a catastrophic mode. The damage due to the occurrence of q_{max} is not included in the computation of the benefit (cost) of the project. The implicit value assigned to $B[q_{max}|D]$ is essentially (negative) infinite, and it does not figure in the economic valuation of the decision.

On the other hand, the cost of the project or facility is highly dependent on the value selected for q_{max}. For our example, this is the dam design for the PMF (or some fraction thereof). The use of the expected value of benefit/cost is not appropriate, since it would include the product of a very large (uncertain) consequence by a very low (difficult to estimate) probability. The product in this case has little significance.

Nevertheless, there is great value in attempting to assign values to the consequence of the extreme event and to its probability. Critics warn that attempting to assign probabilities to such rare events is folly, in the statistical and practical sense. They must, however, suggest another way for making decisions.

One way to estimate the magnitude of the extreme event, without necessarily assigning it a probability, is to use a physically based approach. This is what is done with the PMF. A PMP is estimated, on the basis of the most extreme conditions which can be reached in the atmosphere, supported by transportation of selected observed extreme storms to the watershed in question. When snowmelt is significant, a temperature profile is selected on the basis of historical data, to melt the snowpack. The runoff from this PMP and snowmelt is then computed, with the initial conditions on the watershed assumed to be those that will produce the largest flows.

Once the PMF has been computed, it is instructive to see where it falls relative to the extrapolated *pdf* of extreme flows. A degree of confidence can be achieved if the two approaches are convergent. Continuous hydrological simulation is a better way to approach the estimation of extreme flows than a single event calculation.

18.2 THE ROLE OF FORECASTING

Forecasting is used to modify the a priori probability distribution of future time series: of hydrological data and of demands. Figure 18.3 indicates its two phases, identified as "forecast" and "possible futures," in real-time forecasting.

The first phase extends from time NOW over some period, during which future flows can be viewed in a sense as a projection from the current state, using data from the immediate past: river flows upstream, precipitation in the recent past, snow pack, temperatures, demands. Data used in this phase can extend from some time in the past all the way to time NOW. However, often there is a gap between measurement of the data and its availability for the forecasting model.

Hydrological forecasting is carried out with a watershed simulation model. The model tracks the state of the watershed, and keeps a running record of such variables as soil moisture, groundwater levels, snow pack, and reservoir levels. The model is run every time new data become available. It also uses records and forecasts of the relevant meteorological variables, such as temperatures and precipitation, to simulate the response of the watershed and compute the inflows into the reservoirs. As shown in Figure 18.3, there is a confidence band of the forecasted inflow, whose width increases with time.

The period over which this forecast is useful depends on the size and response of the watershed, and the quality of the meteorological forecast. Typically, the larger the watershed, the longer the dependable "forecast period." It can range from hours (something in the order of seventy-two hours) for a small- to medium-size watershed, to several days or even a few weeks on a large river basin.

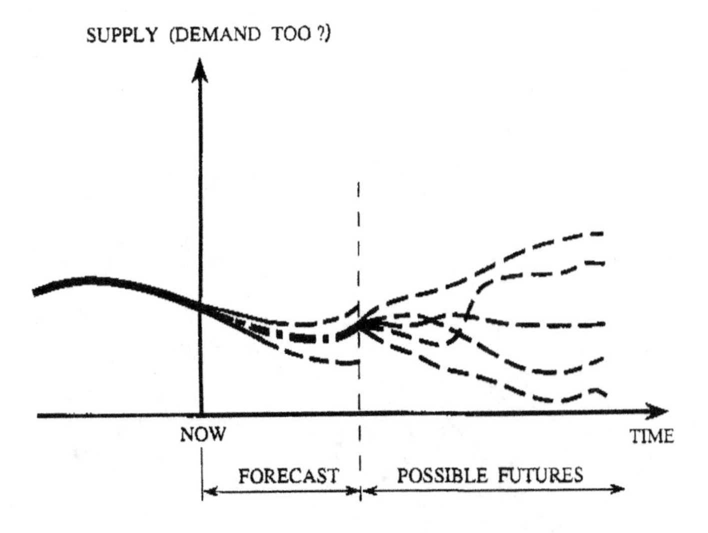

Figure 18.3. The role of forecasts.

The reliability of this forecast deteriorates as time progresses, and we move into the second phase. Here, all "possible futures" are equally likely. Each starts from the final state of the first phase forecast (which is itself a random variable, as seen in Figure 18.3) and represents the inflows that would have been experienced if one particular year of historical meteorological data were to occur from that date on.

These trajectories can be used to estimate the probability distribution of cumulative inflows to any selected point in future time, thus providing the data needed for analyzing the performance of the water supply system. Given this forecast, the operator and decision maker have an improved basis for evaluating alternative management policies.

Thus, real-time forecasting has an important effect on operational reliability of the water resources system. But the capacity for real-time forecasting to improve system reliability should also be figured as an element in the planning and design phases. To demonstrate, let us look at the design of dam safety for the PMF.

The PMF is calculated as an accumulation of all possible worst-case conditions: the largest storm possible is dropped on a saturated watershed, at the same time a large snowmelt occurs. This results in the largest flow deemed possible, for a return period of hundreds to thousands of years (depending on whether lives are in danger). The PMF is also used to dictate the freeboard for the relevant season of year.

If the operator has access to reliable measured data and a good weather forecast, backed by a calibrated and tested watershed model, he may be able delay vacating space in the reservoir in anticipation of a storm. Soil moisture may not be as high, the snow pack not as deep, the forecasted storm not as severe, the forecasted temperature not as high – as used in the PMF calculations. Thus, a good forecast provides extra reliability, or, alternatively, an opportunity for less wasteful operation with the same level of safety.

REFERENCES

Bao, Y. and Mays, L. W. (1990) Model for water distribution system reliability. *Journal of Hydraulic Engineering, ASCE* 16(9): 119–37.

Ben-Haim, Y. (1996) *Robust Reliability in the Mechanical Sciences.* Springer-Verlag, Berlin.

Biem, G. K. and Hobbs, B. (1988) Analytical simulation of water supply capacity reliability: 2. Markov-chain approach and verification of the models. *Water Resources Research* 24(9): 1445–49.

Casti, J. L. (1996) Risk, natural disasters, and complex system theory. Lecture presented at Third Kovacs Colloquium on Risk Reliability, Uncertainty and Robustness of Water Resources Systems. UNESCO, Paris, September.

Damelin, E., Shamir, U., and Arad, N. (1972) Engineering and economic evaluation of the reliability of water supply. *Water Resources Research* 8(4): 861–77.

Duan, N., Mays, L. W., and Lansey, K. E. (1990) Optimal reliability-based design of pumping and distribution systems. *Journal of Hydraulic Engineering, ASCE* 116(2): 249–68.

Fujiwara, O. and Da Silva, A. U. (1990) Algorithm for reliability-based optimal design of water networks. *Journal of Environmental Engineering, ASCE* 116(3): 575–86.

Fujiwara, O. and Tung, H. D. (1991) Reliability improvement for water distribution networks through increasing pipe size. *Water Resources Research* 27(7): 1395–402.

Hobbs, B. and Biem, G. K. (1988) Analytical simulation of water supply capacity reliability: 1. Modified frequency duration analysis. *Water Resources Research* 24(9): 1431–44.

Kessler, A., Ormsbee, L., and Shamir, U. (1989) A methodology for least-cost design of invulnerable water distribution networks. *Civil Engineering Systems* 6: 20–28.

Klemeš, V. (2001) Risk analysis: The unbearable cleverness of bluffing. In *Risk, Reliability, Uncertainty, and Robustness of Water Resources Systems*, edited by J. J. Bogardi and Z. W. Kundzewicz, Cambridge University Press, New York, 22–29.

Minard, R. A. (1996) Comparative risk and the states. In *Resources*, 122, Resources for the Future.

Ostfeld, A. and Shamir, U. (1993) Incorporating reliability in optimal design of water distribution networks – review and new concepts. *Reliability Engineering and System Safety* 42: 5–11.

(1996) Design of optimal reliable multi-quality water supply systems. *Journal of Water Resources Planning and Management, ASCE* 122(5): 322–33.

Plate, E. J. (2001) Risk management for hydraulic systems under hydrologic loads. In *Risk, Reliability, Uncertainty, and Robustness of Water Resources Systems*, edited by J. J. Bogardi and Z. W. Kundzewicz, Cambridge University Press, New York, 209–20.

Shamir, U. (1987) Reliability of water supply systems. In *Engineering Reliability and Risk in Water Resources*, edited by L. Duckstein and E. Plate, NATO ASI Series, E. M. Nijhoff, Dorderecht, 233–48.

Shamir, U. and Howard, C. D. D. (1981) Water supply reliability theory. *Journal of the American Water Works Association* 73(3): 379–84.

(1985) Reliability and risk assessment for water supply systems. In *Computer Applications in Water Resources*, edited by H. Torno, Proceedings of the Specialty Conference of the Division of Water Resources Planning and Management, ASCE, Buffalo, NY, June, 1218–28.

Wagner, J. M., Shamir, U., and Marks, D. H. (1988a) Water distribution reliability: analytical methods. *Journal of the Water Resources Planning and Management Division, ASCE* 114(3): 253–75.

(1988b) Water distribution reliability: simulation methods. *Journal of the Water Resources Planning and Management Division, ASCE* 114(3): 276–94.

Wagner, J. W., Shamir, U., and Nemati, H. R. (1992) Groundwater quality management under uncertainty: stochastic programming with recourse and the value of information. *Water Resources Research* 28(5): 1233–46.

Yang, S.-L., Hsu, N.-S., Louie, P. W. F., and Yeh, W. W.-G. (1996a) Water distribution network reliability: connectivity analysis. *Journal of Infrastructure Systems, ASCE* 2(2): 54–64.

(1996b) Water distribution network reliability: stochastic simulation. *Journal of Infrastructure Systems, ASCE* 2(2): 65–72.

19 Quantifying system sustainability using multiple risk criteria

DANIEL P. LOUCKS*

ABSTRACT
Measuring the relative sustainability of water resources systems is much needed though very difficult. Being able to quantify sustainability makes it possible to evaluate and compare development alternatives, plans, and policies in order to choose the preferred solutions. It also makes it possible to include sustainability as one of multiple objectives in system design and operation. Commonly used multiple risk criteria – measures of reliability, resilience, and vulnerability – are combined into an aggregate index quantifying relative system sustainability. The procedure is illustrated in an example of regional development alternatives.

19.1 INTRODUCTION

Ever since the concept of sustainability, as expressed in the Brundtland Commission's report *Our Common Future* (WCED 1987), was introduced, professionals from many disciplines have been trying to define and measure it. This has turned out to be more difficult than expected. Nevertheless, this chapter attempts to do so, that is, to define sustainability in a manner that can help us better address some of the many issues and challenges that accompany the Commission's concept of sustainability. At the same time this definition should allow us to measure or quantify, at least relatively, the extent to which sustainability is being, or may be, achieved. We need such measures if we are to evaluate our development alternatives and monitor our water resources systems, and indeed our economy, our environment, and our social systems to see if they are becoming increasingly sustainable.

This chapter focuses on the relative sustainability of renewable water resources systems. This limited focus on water permits us to redefine sustainability in a way that makes it easier to quantify sustainability and include it as one of multiple objectives to be achieved, or at least considered, when making decisions regarding the design and operation of water resources systems. Commonly used measures of reliability, resilience, and vulnerability, based on subjective judgments concerning what is acceptable or unacceptable with respect to multiple system performance indicators, can be used as measures of changes in relative sustainability over time associated with particular systems and their management.

* Civil and Environmental Engineering, Cornell University, Ithaca, NY 14853, USA.

19.2 SUSTAINABILITY: SOME ISSUES AND CHALLENGES

The word sustainability has assumed a variety of meanings. While it can imply different things to different people, it always includes a consideration of the future. The Brundtland Commission (WCED 1987) was concerned about how our actions today will affect "... the ability of future generations to meet their needs." Their notion of sustainability commands us in this generation not to take actions aimed at meeting our current needs that would limit or constrain the ability of future generations to meet their needs. Because there are disagreements on just what this definition or statement suggests we do (or indeed how we today can know what the needs of future generations will be), there are also disagreements on just how sustainability can or should be achieved.

Do we enhance the welfare of future generations by preserving or enhancing the current state of our natural environmental resources and ecological systems? If so, over what space scales? What do we do about our nonrenewable resources, for example, the water that exists in many of our deep groundwater aquifers, which are not being replenished by nature? The concept of the preservation of nonrenewable resources now and in the future would imply that those resources should never be consumed. If permanent preservation seems as unreasonable to you as it does to me, then how much of a nonrenewable resource should be consumed, and when?

If sustainability applies only to human living conditions and standards, as some argue, then perhaps some of today's stock of natural resources should be consumed. The amount consumed today could be used to increase our standard of living, improve our technology, enhance our knowledge, create a greater degree of social stability and harmony, and contribute to our culture. All of this might provide future generations with an improved technology and knowledge base that would enable them to further increase their standard of living using even less natural, environmental, and ecological resources. Of course, it is impossible to know whether this substitution of natural resources for other capital, intellectual, and social resources will happen – or that even if it does happen, whether it would necessarily lead to higher levels of sustainable development, eventually. We, together with the public, can only guess and act accordingly.

Thus, the debate over the definition of sustainability is among those who differ over just what it is that should be sustained and just how to do it. Without question, determining who in this debate has the better vision of what should be sustained and how we can reach a path of sustainable development will continue to challenge us all. But this challenge need not delay our attempts to achieve a more sustainable water resources infrastructure and its operation.

An explicit consideration of the needs or desires of future generations may require us to give up some of what we could consume and enjoy in this generation. How should trade-offs among our own immediate "needs" and those of future generations be made? A standard economic approach to making these intergenerational trade-offs involves the discounting of future benefits, costs, and losses. This type of analysis requires converting all future benefits, costs, and losses to equivalent present-day values to account for inflation and the time-value of money. Simply put, people are willing to pay more for something today than for the promise that the same will be given to them at some specified date in the future. Money available today can be invested and earn interest over time. As a consequence, investments increase in value over time. Their future values, when discounted to the present (using a specific discount [interest] rate) will be what they would appear to be worth today.

Since future economic benefits, costs, and losses and future non-economic statistical or risk-based measures of performance, when discounted to the present time, decrease in amount or significance, any impact in the future will be weighted less than the same impact occurring at the present time. Consider then the result of such a procedure if the supply of a supporting environmental resource is limiting, such as the quality of the soil of an irrigation area or the assimilative capacity of a water body. Should our decisions today allow this resource to be diminished in the future just because that future loss, when discounted to today's values, is nil? Most would argue that the concept of discounting is valid. But, they would generally agree that discounting should be applied with safeguards where the integrity of our life-supporting resources such as fertile soils, potable water, clean air, biodiversity, and other environmental and ecological systems are concerned.

Then how can any safeguards or constraints on the traditional benefit-cost analysis be applied, and by whom, to ensure that we who live and consume today will adequately consider the needs and desires of those who follow us in the future? How can we possibly know what those future needs and desires are? To what extent do people understand what impacts their actions today might even be having fifty or a hundred years from now?

Resource economists have been telling us over the past century why it is so difficult for us as individuals to manage and use our environmental resources now in a way that will benefit all of us today, let alone that it might be of benefit to those in the distant future. Any study of recent history also shows how difficult it has been for governments to modify either a free-market system or a centrally planned and controlled economy (by means of taxes and subsidies or by laws and regulations) in attempts to ensure any sustainable use of common property environmental resources. But these difficulties should not be excuses for ignoring sustainability issues. Rather, we who are involved in natural resources management need to work to ensure that the public and those who make decisions are aware of the temporal as well as spatial sustainability impacts and trade-offs associated with those decisions.

When considering trade-offs of natural, capital, and social resources that affect the welfare of humans and other living organisms over time, one must also address the question of spatial scale and resource mobility. As previously asked, should each square kilometer of land be sustainable? Should each watershed or country or province or state be sustainable? Might some large regions (e.g., the Aral Sea and its basin) be sacrificed in order to enhance the economic survival of a larger region or country? Opportunities for resource transfers and trade-offs and for the achievement of sustainability are generally greater the larger the space scale. And yet, concern only with the sustainability of larger regions may overlook the unique attributes of particular local economies, environments, ecosystems, and possible limits on ecosystem adaptation, resource substitution, and human health.

Given these and no doubt many other questions and issues, it is evident why there has been so much difficulty in achieving a consensus on just what is meant or implied by sustainability or sustainable development. Most will probably agree that sustainability involves an explicit focus on at least maintaining if not increasing the quality of life of all individuals over time. Sustainability also addresses the challenge of developing regional economies that can ensure a desired and equitable standard of living for all the inhabitants and their descendants. We are not sure how this will be done, or even if it can be done. It will, without question, require some truly interdisciplinary research over a considerable period of time to address and answer many of these questions and issues.

For water resources managers, this concept of sustainability challenges us to develop and use better methods for explicitly considering the possible needs and expectations of future generations along with our own. We must develop and use better methods of identifying development paths that keep more options open for future populations to meet their, and their descendants', needs and expectations. Finally we must create better ways of identifying and quantifying the amounts and distribution of benefits and costs (however many ways they might be measured) when considering trade-offs in resource use and consumption among current and future generations as well as over different populations within a given generation.

The issue of intergenerational equity is central to the concept of sustainability. Intergenerational equity requires that each generation manage its resources in ways to ensure that future generations can meet their demands for goods and services, at economic and environmental costs consistent with maintaining or even increasing per capita welfare through time. However, how do we know that future generations will value environmental resources as we do? Conditions for sustainability will vary for specific regions, and these differences will increase as the area of the region decreases. Mobility of populations and resources also affects sustainability. It is important, then, to consider the spatial or regional dimensions of sustainability, and the institutional conditions and arrangements that determine the relations among regions.

Irrespective of the definitions used, a number of questions remain:

- Should resource use and population be controlled?
- Are resource limits critical?
- Can technology be improved fast enough to eliminate a possible crisis caused by resource degradation if not depletion?
- What is the appropriate spatial scale for examining these questions?
- Resource substitution and spatial mobility of resources and people need to be considered. Do we permit the devastation of certain regions in favor of others?
- The same question as the last can be asked with respect to time scales.
- Over what periods should we expect or permit decreases in overall welfare due to, for example, fluctuations in renewable resource supplies?
- What is irreversible?

Clearly there is no general agreement on exactly how to answer these questions. But they are important when we are trying to define sustainability precisely (Pezzey 1992). However defined, nevertheless, sustainability involves the notion of trade-offs over time. How to identify these trade-offs and their implications for us today is the essence of the debate taking place regarding sustainable development.

19.3 DEFINING SUSTAINABILITY OF WATER RESOURCES SYSTEMS

Sustainable development can be defined in terms of resource use, including water resources use (UN 1991). Under this definition an alternative development policy might be considered sustainable if positive net benefits derived from natural resources (including water supplies) can be maintained in the future. However, there are difficulties in measuring net benefit for some resource uses. For example, how does one evaluate the benefits of wetlands, fisheries, water quality for recreation, ecosystem rehabilitation and preservation?

Goodland, Daly, and Serafy (1991) view sustainable development as a relationship between changing human economic systems and larger, but normally slower-changing, ecological systems. In this relationship human life can continue indefinitely and flourish, and the supporting ecosystem and environmental quality can be maintained and improved. Sustainable development, then, is one in which there is an improvement in the quality of life without necessarily causing an increase in the quantity of resources consumed. But the idea of sustainable growth, that is, the ability to get quantitatively bigger, continually, is an impossibility. On the other hand sustainable development, that is, to improve qualitatively and continually, may be possible.

Economists and ecologists differ somewhat in their definitions of sustainable development. Much of this debate revolves around the concept of social capital. Should we be focusing on the preservation or enhancement of natural resources or should we be looking at the entire mix of resources (the environment, human knowledge, man-made capital, etc.) that comprise what is called social capital. It also revolves around the general problem of understanding which elements of social capital our future generations are likely to value the most.

Sustainable development has also been defined in terms of financial viability. To be financially sustainable, all costs associated with a water resources development policy should be recovered. The service provided by a water development project must be able to pay for the project. In fact, the revenues should exceed the costs, and thereby provide for the improvement and maintenance of the project. An indication of the financial sustainability of a project paid for by a development bank could be the ability of that project to continue to deliver service or welfare after the initial funding has been spent.

Falkenmark (1988) focuses on the role water plays in sustainable development. She identifies various conditions for sustainability. Soil permeability and water retention capacity have to be secured to allow rainfall to infiltrate and be used in the production of biomass on a large enough scale for self-sufficiency. Drinkable water has to be available. There has to be enough water to permit general hygiene. Fish and other aquatic biomass have to be preserved and remain edible.

It is evident that there is no clear or commonly accepted definition of sustainability, even when focusing on the single topic of water. Can we as professionals involved in water resources planning and management come to some agreement on how we can define sustainability of water resources systems, and then how we might measure the extent to which our systems are sustainable? Probably not, since even this much more narrow focus on water resources systems is still incredibly broad. But we can certainly become involved in a debate over how it might be best done. This chapter offers a proposed method, based on the topic of this Colloquium – measures of risk and uncertainty.

Before discussing such risk-based measures of sustainability, I think it is useful to try to define a little more precisely what sustainability means with respect to water resources systems. It is a definition that encompasses all the concepts just discussed and that still focuses on the central point of the Brundtland Commission (WCED 1987) definition – a concern for the future. First of all, the word "needs" in the Brundtland Commission's definition is bothersome. Even if we could define our own water resource "needs," let alone those of future generations, we may not be able to meet them, at least at reasonable or acceptable economic and social costs.

Hence, the following proposed definition:

> Sustainable water resources systems are those designed and managed to fully contribute to the objectives of society, now and in the future, while maintaining their environmental and hydrological integrity.

Do we wish our decisions and actions in this generation to be viewed favorably by future generations? Will future generations find fault with what we decide to do in this generation that may affect what they can do and enjoy in their generation? If we placed all our preferences on future generations, we might define as sustainable those actions that minimize the regret of future generations. Clearly we have our current interests and desires too, and indeed there may be trade-offs between what we wish to do for ourselves now versus what we think future generations might wish us to do now for them. These issues and trade-offs must be debated and decisions made in the political arena. There is no scientific theory to help us identify which trade-offs, if any, are optimum.

Thus, sustainability is not a scientific concept, but rather a social goal. It implies an ethic. Public value judgments must be made about which demands and wants should be satisfied today and what changes should be made to ensure a legacy for the future. Different individuals have different points of view, and it is the combined wisdom of everyone's opinions that will shape what society may consider sustainable.

19.4 MEASURING SUSTAINABILITY OF WATER RESOURCES SYSTEMS

This section focuses on the measurement or quantification of sustainability. Although the focus is on quantification, mathematical or technical languages alone are not sufficient to fully measure sustainability. Sustainability involves other aspects that deserve intensive discussion, and it requires a willingness to go beyond the scope of what may be quantifiable or measurable. But unless we can measure or describe in precise terms what we are trying to achieve, it becomes rather difficult (if not impossible) to determine how effective we are in doing what we wish to do, or even in comparing alternative plans and policies with respect to their relative sustainability.

19.4.1 Efficiency, survivability, and sustainability

To begin a discussion concerning how one might measure and include sustainability in planning models, it is perhaps useful first to distinguish among several planning objectives that focus on future conditions. These objectives are called efficiency, survivability, and sustainability (after Pezzey 1992).

To compare these three objectives, assume there is some way we can convert whatever decisions we make into a common metric (unit of measure) called welfare. Each possible decision that could be made today, denoted by a different value of the index k, will result in a time series of net welfare values, $W(k,y)$, for each period y from now on into the future. Assume there is a minimum level of welfare needed for survival, W_{min}.

A decision k is efficient if it maximizes the present value of current and all future net welfare values. Using a discount rate of "r" per period, an *efficiency* objective involves a search for the alternative k that will

$$Maximize \sum_y W(k, y)/(1+r)^y \tag{19.1}$$

Clearly, as the discount rate r increases, the welfare values $W(k, y)$ obtained now will increasingly contribute more to the present value of total welfare than will the same welfare values obtained sometime in the distant future. In other words, as the discount or interest rate r increases, what happens in the future becomes less and less important to those living today. This objective, while best satisfying present or current demands, may not always assure a survivable or sustainable future.

Efficiency involves the notion of discounting. There is a time value of assets. Those who need and could benefit from the use of a given resource today are likely to be willing to pay more for it today than for the promise of having it some time in the future. Yet high discount or interest rates tend to discourage the long-term management of natural resources and the protection of long-term environmental assets. Low discount rates, however, may favor investment in projects that are less likely to survive economically, and that are less likely to invest in environmental protection and the technology needed for efficient resource use and recycling. Thus, the relationship between interest rates, resource conservation, and sustainable development is ambiguous (see, for example, Norgaard and Howarth 1991).

An alternative decision k can be considered *survivable* if in each period y (on into the future) the net welfare, $W(k, y)$, is no less than the minimum required for survival, W_{min}. Hence, if

$$W(k, y) \geq W_{min} \tag{19.2}$$

then alternative k is survivable for all periods y.

A survivable alternative, and there may be many, is not necessarily efficient (in the normal sense).

Next, consider an alternative development path that is sustainable. Here we will consider an alternative as sustainable if it assures that the average (over some time period) welfare of future generations is no less than the corresponding average welfare available to previous generations. Welfare could involve or include opportunities for resource development and use. A *sustainable* alternative k assures that there are no long-term decreases in the level of welfare of future generations. In other words, if

$$W(k, y+1) \geq W(k, y) \tag{19.3}$$

the alternative k is sustainable for all periods y.

Equivalently, a sustainable alternative k is one that assures a nonnegative change in welfare,

$$dW(k, y)/dy \geq 0 \tag{19.4}$$

in each period y. There may be many development paths that meet these sustainability conditions.

The duration of each period y must be such that natural variations in a resource, like water, are averaged out over the period. Results of recent climate change studies suggest these periods may have to be longer than what past historical precipitation and streamflow records would suggest due to the increased likelihood of more frequent and longer periods of extremes.

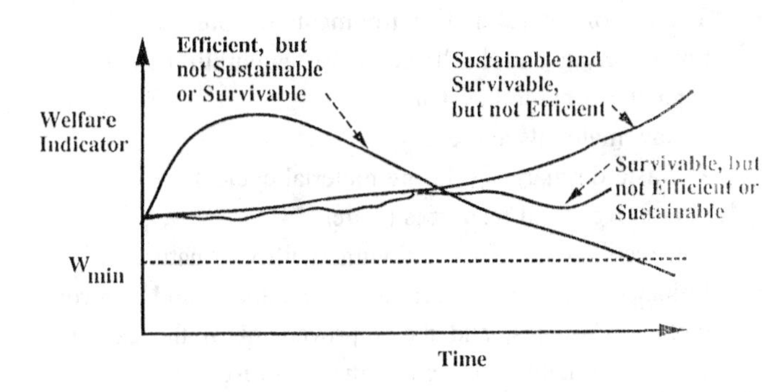

Figure 19.1. Plots of social welfare (or other quality-of-life indicator) resulting from three alternative development scenarios over time.

Any constraints on resource use for sustainability reasons must also be judged based on their impacts on economic efficiency and externality conditions (Toman and Crosson 1991). Before setting them, resource managers must understand how sustainability constraints could affect development paths and policies, especially in regions where substitution among multiple resources is possible but affected by uncertainty and endogenous technical change.

Figure 19.1 illustrates various development paths of net welfare that represent examples of efficient, survivable, and/or sustainable development. These development paths are only examples. While not illustrated, it might be possible in some situations to identify a development path that is efficient, survivable, and sustainable at the same time. In most cases, however, some trade-offs are required among these three development objectives.

19.4.2 Weighted criteria indices

The Delft Hydraulics Laboratory in The Netherlands has proposed a procedure that can be used to measure or quantify the extent to which projects may contribute to sustainable development (Baan 1994). This procedure consists of a checklist, the responses to which are very subjective. Five main criteria have been identified. Each of the five criteria is subsequently divided into subcriteria.

The five main criteria and their respective subcriteria are:

- Socioeconomic aspects and impacts on growth, resilience, and stability.
  - Effects on income distribution
  - Effects on cultural heritage
  - Feasibility in socioeconomic structure

- The use of natural and environmental resources including raw materials and discharge of wastes within the carrying capacity of natural systems.
  - Raw materials and energy
  - Waste discharges (closing material cycles)
  - Use of natural resources (water)
  - Effects on resilience and vulnerability of nature
- Enhancement and conservation of natural and environmental resources, and the improvement of the carrying capacity of natural and environmental resources.
  - Water conservation
  - Accretion of land or coast
  - Improvement and conservation of soil fertility
  - Nature development and conservation of natural values
- Public health, safety, and well-being
  - Effects on public health
  - Effects on safety (risks)
  - Effects on annoyance/hindrance (smell, dust, noise, crowding)
  - Effects on living and working conditions
- Flexibility and sustainability of infrastructure works, management opportunities for multifunctional use, and opportunities to adapt to changing circumstances.
  - Opportunities for a phased development
  - Opportunities for multifunctional use and management and to respond to changing conditions
  - Sustainable quality of structures (corrosion, wear)
  - Opportunities for rehabilitation of the original situation (autonomous regeneration, active reconstruction, and restoration)

Each subcriterion is given equal weighting. The sum of numerical values given to each subcriterion is a sustainability index. The higher the index value, the greater the contribution of a project to sustainable development. Users of this index must then decide, based on the index value, whether to accept, reject, or urge the potential client to modify the project.

19.4.3 Weighted statistical indices

Alternatively, sustainability indices can be defined as separate or weighted combinations of reliability, resilience, and vulnerability measures of various economic, environmental, ecological, and social criteria. To do this it is first necessary to identify the appropriate economic, environmental, ecological, and social criteria to be included in the overall measure of relative sustainability. These criteria must be able to be expressed quantitatively or at least linguistically (such as "poor," "good" and "excellent") and be determined from time series of water resources system variables (such as flow, velocity, water-surface elevation, hydropower production or consumption).

Criteria that can be expressed in monetary units can be considered economic criteria. This might, for example, include the present value of the economic costs and benefits derived from hydropower, irrigation, industry, and navigation. Economic criteria usually include distributional as well as efficiency components. Who pays and who benefits is as (if not more) important as how much the payments or benefits are or will be.

Environmental criteria may include pollutant and other biological and chemical constituent concentrations in the water as well as various hydraulic and geomorphologic descriptors at designated sites of the water resources system. Ecological criteria could include the extent and depth of water in specified wetlands, the diversity of plant and animal species in specified floodplains, and the integrity or continuity of natural ecosystems that can support habitats suitable for various aquatic (including fish) species. Social criteria may include the frequency and severity of floods and droughts that cause hardship or dislocation costs not easily expressed in monetary units and the security of water supplies for domestic use. They might also include descriptors of recreation opportunities provided by the rivers, lakes, and reservoirs, their operation or regulation, and the relative quality or attractiveness of the scenery provided for those living next to or using the water resources system.

Once the water resources system is simulated using hydrologic inputs representative of what one believes could occur in the future, the time-series values of these system performance criteria can be derived. These time-series values themselves can be examined in any comparison of alternative water resources system designs and/or operating policies. Alternatively, they can be summarized using the statistical measures of reliability, resilience, and vulnerability. The relative sustainability of the system with respect to each of these criteria is higher the greater the reliability and resilience, and the smaller the vulnerability. There are often trade-offs between these three statistical measures of performance.

To illustrate this procedure, consider any selected criterion called C. Its time series of values from a simulation study are denoted as C_t, where the simulated time periods t extend to some future time T. To define reliability one must identify the ranges of values of this criterion that are considered satisfactory, and hence the ranges of values considered unsatisfactory. Of course these ranges may change over time. Note that those ranges of these criterion values that are considered satisfactory, and hence those ranges of values considered unsatisfactory, are determined by the analysts or planners. These ranges are subjective. They are based on human judgment or human goals, not scientific theory. In some cases they may be based on well-defined standards, but standards will not have been predefined or published for most system performance criteria.

Figure 19.2 illustrates a possible time-series plot of all simulated values of C_t along with the designated range of values considered satisfactory. In this example the satisfactory values of C_t are within some upper and lower limits. Values of C_t above the upper limit, UC_t, or below the lower limit, LC_t, are considered unsatisfactory. Each criterion will have its own unique ranges of satisfactory and unsatisfactory values. Once these data are defined, it is possible to compute associated reliability, resilience, and vulnerability statistics, as illustrated in Figure 19.3 and as defined below.

Define the

Reliability of C = Number of satisfactory C_t values/ Total number of simulated periods, T (19.5)

Resilience of C = Number of times a satisfactory C_{t+1} value follows and unsatisfactory C_t value/ Number of unsatisfactory C_t values (19.6)

Reliability is the probability that any particular C_t value will be within the satisfactory range of values. Resilience is an indicator of the speed of recovery from an unsatisfactory condition. It is the probability that a satisfactory C_{t+1} value follows an unsatisfactory C_t value.

Vulnerability is a statistical measure of severity of failure, should a failure (i.e., an unsatisfactory value) occur. The extent of a failure is the amount a value C_t exceeds the upper limit, UC_t, of the satisfactory values or the amount that value falls short of the lower limit, LC_t, of the satisfactory values, whichever is greater. It is the maximum value of the set $[0, LC_t\text{-}C_t, C_t\text{-}UC_t]$.

The extent of failure of any criterion C can be defined in a number of ways. It can be based on the extent of failure of individual unsatisfactory values or the cumulative extent of failure of a continuous series of unsatisfactory values. In the latter case, each individual extent of failure is added together for the duration of each continuous failure sequence.

Since there can be many values of these individual or cumulative extents of failure for any set of simulations, a histogram or probability distribution of these values can be defined. Thus we can define "extent-vulnerability" as

Individual Extent-Vulnerability (p) of C = Maximum extent of individual failure of criterion C occurring with probability p, or that may be exceeded with probability 1 – p (19.7)

Cumulative Extent-Vulnerability (p) of C = Maximum extent of cumulative failure of criterion C occurring with probability p, or that may be exceeded with probability 1 – p (19.8)

Extent-vulnerability can also be defined based on the expected or maximum observed individual or cumulative extent of failure. The conditional expected extent of failure indicator can be defined as

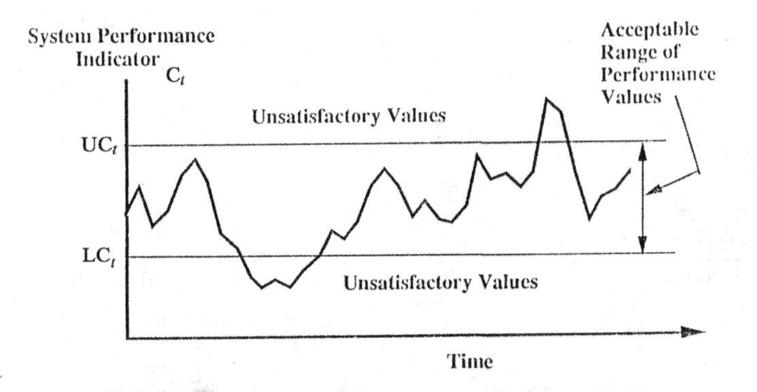

Figure 19.2. A portion of the time series of values of a system performance indicator derived from a predictive simulation model.

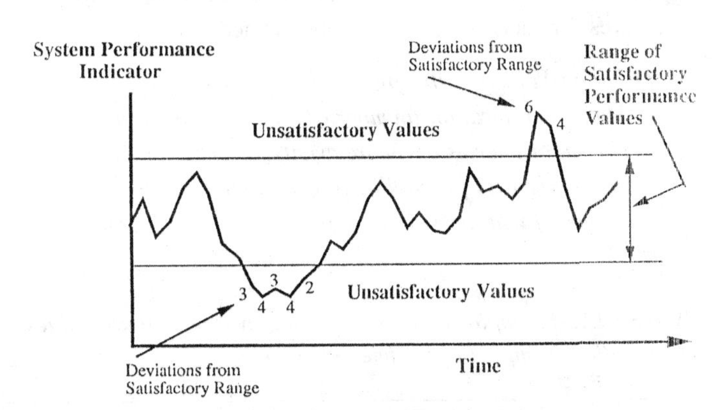

Reliability: $\frac{\text{Number of Satisfactory Values}}{\text{Total Number of Values}} = \frac{31}{38} = 0.82$

Resilience: Number of Times a Satisfactory Value Follows an Unsatisfactory Value / Number of Unsatisfactory Values

= 2 / 7 = 0.29

Vulnerability:

Extent: 6 with conditional probability of exceedance of 0 / 7 = 0.0
4 with conditional probability of exceedance of 1 / 7 = 0.14
3 with conditional probability of exceedance of 4 / 7 = 0.57
2 with conditional probability of exceedance of 6 / 7 = 0.86

Expected extent given unsatisfactory values = (3+4+3+4+2+6+4) / 7 = 3.7

Duration: 5 with conditional probability of exceedance of 0.0
2 with conditional probability of exceedance of 0.5

Expected duration given unsatisfactory values = (5 + 2) / 2 = 3.5

Figure 19.3. Deriving measures of reliability, resilience, and vulnerability from the time-series data shown in Figure 19.2.

Conditional Expected Extent-Vulnerability of C =
Σ_t *individual (or continuous cumulative) extents of failure of* C_t*/Number of individual (or continuous series of) failure events* (19.9)

The sum of equation 19.9 is over all individual or continuous sequences of failure events. Continuous sequences include those of only one period in duration. The unconditional expected extent-vulnerability of C is defined using the above equation except that the denominator is replaced by the total number of simulation time periods T.

The maximum simulated individual extent of failure can be defined as

Maximum Extent-Vulnerability of C =
$Max[0, LC_t\text{-}C_t, C_t\text{-}UC_t]$ (19.10)

For some criteria, for example droughts, the duration as well as the individual and cumulative extents of failure may be important. A histogram or probability distribution of the durations (number of time periods) of failure events can be constructed following any one or multiple simulations. From this histogram or probability distribution, duration-vulnerability measures for each criterion C can be defined as

Duration-Vulnerability (p) of C =
Maximum duration (number of time periods) of a continuous series of failure events for criterion C occurring with probability p or that may be exceeded with probability 1-p (19.11)

Expected Duration-Vulnerability of C =
Total number of periods t having failures of C_t*/ Number of continuous series of failure events* (19.12)

The number of continuous sequences of failures for criterion C includes events that last for only one period.

Once these statistical reliability, resilience, and vulnerability measures of the time-series values are defined, as appropriate for each economic, ecological, environmental, and social indicator or criterion, they can be applied to predicted criteria values over groups of years on into the future. This produces time series of reliability, resilience, and vulnerability data for each criteria. If over time these statistical measures are improving, that is, the reliabilities and resiliences are increasing, and the vulnerabilities are decreasing, the system being studied is getting increasingly sustainable. Often, however, one will find the predicted reliabilities, resiliences, and vulnerabilities of some criteria improving, and for other criteria they may be worsening. Or for any given criterion, some of these statistical measures may be improving, and others worsening. This will force one to give relative weights to each measure of each criterion.

One of the better ways to present and examine and compare the values of all these measures for all relevant criteria is through the use of a scorecard. A scorecard is a matrix that presents the values of each of these measures for each of the criteria associated with each alternative system. Table 19.1 illustrates such a scorecard that uses actual values of the crite-

Table 19.1. *Example scorecard showing average annual values of various criteria for regional development alternatives. Units of each impact are defined elsewhere*

	Policy Alternatives and Components			
Impacts	*Agricultural* irrigation pumps drainage	*Industrial* water storage groundwater use	*Environmental* water recreation treatment tax on water use	*Mixed* water storage treatment canal transptort
Annual investment costs	[300]	400	(700)	(700)
Annual economic benefits	[1,200]	700	(100)	1,000
Agricultural production	[800]	150	(50)	600
Drinking water cost	(1.40)	[0.90]	1.20	1.10
Pollution index at site A	150	(220)	[30]	70
Power production	200	[1,200]	(50)	800
Fisheries production	70	(20)	[80]	40
Flood protection %	[99]	98	(96)	[99]

[Best] (Worst)

ria that are not assumed to vary over time. Table 19.2 is a scorecard of statistical measures of these criteria that do vary over time. One can see the increasing or decreasing predicted values over each of the five 10-year periods examined in the study on which this scorecard was based.

The scorecard in Table 19.2 shows that some alternatives are predicted to be better with respect to some criteria, and other alternatives are predicted to be better with respect to other criteria. In these situations, which are common, the multiobjective decision-making process will involve making trade-offs among incommensurate objectives. In some cases this negotiation or decision-making process can be facilitated by attempting to rank each of the alternatives, taking into account each of the criteria. One such approach is through the use of sustainability indices.

To rank each of the alternatives that are nondominated (i.e., ones that are not inferior to others with respect to all criteria) the information contained in the scorecards can be combined into a single sustainability index. When defining an index using these statistical measures of reliability, resilience, and vulnerability, it is convenient to convert each vulnerability measure into a measure that, like reliability and resilience, ranges from 0 to 1 and in which higher values are preferred over lower values. This can be done in two steps. The first involves identifying the largest vulnerability value for each criterion C among all the alternative systems being compared and then dividing each system's vulnerability measure for the criterion by this maximum value. The result is a relative vulnerability measure, ranging from 0 to 1, for each criterion C. One of these relative vulnerability values for each criterion C will equal 1, namely that associated with the system having the largest vulnerability measure.

$$\textit{Relative Vulnerability(C)} = \textit{Vulnerability(C)}/ \textit{Max Vulnerability(C) among all alternatives} \qquad (19.13)$$

This definition of relative vulnerability will apply to each type of vulnerability identified above, and for any specified level of probability, when applicable.

The second step in converting the vulnerability measure to one that is similar to reliability and resilience in that higher values are preferred over lower values is to subtract each relative vulnerability measure from 1.

Once this scaling and conversion have been performed, each statistical measure ranges from 0 to 1 and higher values are preferred over lower values. They can now be combined into a single index for each criterion C. One way of doing this is to form the product of all these statistical measures.

$$\textit{Sustainability(C)} = [\textit{Reliability(C)}]\,[\textit{Resilience(C)}]\,[\Pi_v \{1\text{-}\textit{Relative Vulnerability } v(C)\}] \qquad (19.14)$$

where *Relative Vulnerability* $v(C)$ is the v-th type of relative vulnerability measure being considered for criterion C. The use of multiplication rather than addition in the above index gives added weight to the statistical measure having the lowest value. For example, if any of these measures is 0, it is unlikely any of the other measures are very relevant. A high value of the index can result only if all statistical measures have high values.

The resulting product, the *Sustainability*(C) index, ranges from 0, for its lowest and worst possible value, to 1, at its highest and best possible value. This sustainability index applies to each criterion C for any constant level of probability p, and can be calculated for each alternative system or decision being considered.

To obtain a combined weighted relative sustainability index that considers all criteria, relative weights, W_C, ranging from 0 to 1 and summing to 1, can be defined to reflect the relative importance of each criterion. These relative weights may indeed be dependent on the values of each *Sustainability*(C) index. Once defined, the relative sustainability of each alternative system being compared is

$$\textit{Relative Sustainability} = \Sigma_C W_C\, \textit{Sustainability(C)} \qquad (19.15)$$

Since each sustainability index and each relative weight W_C ranges from 0 to 1, and the relative weights sum to 1, these relative sustainability indices will also range from 0 to 1. The alternative having the highest value can be considered the most sustainable *with respect to the criteria considered, the values of each criterion that are considered satisfactory, and the relative weights.* Every one of these assumptions involves the subjective judgments of those participating in the evaluation process.

Note what has been done here. A particular proposed or actual water resources system design and management policy is simulated over time. The simulated data include assumptions regarding system design, operation, and hydrologic and other inputs and demands that represent a scenario representative of what could occur in the future. Incorporated into that simulation are the variables individual stakeholders consider important and relevant to sustainability. These are called criteria. These criteria could include physical, economic, environmental, ecological, and social variables. A time series of each of these criterion values at different locations is produced from the simulation. These time series are divided into subperiods and statistical measures of reliability, resilience, and vulnerability are used to summarize each subperiod's time series of each criterion.

The results can be presented on scorecards, or can be combined into a sustainability index for each of the simulated

Table 19.2. *Example scorecard showing average annual values and trends in reliability, resilience and relative invulnerability of various criteria values for regional development alternatives. Units of all criteria are defined elsewhere. Relative weights of Sustainability Index were assumed to be equal and sum to 1*

	Policy alternatives and components			
Impacts	*Agricultural* irrigation pumps drainage	*Industrial* water storage groundwater use	*Environmental* water recreation treatment tax on water use	*Mixed* water storage treatment canal transport
Annual investment cost	300	400	700	700
Reliability trends	0.89–0.93	0.85–0.89	0.82–0.84	0.88–0.89
Resilience trends	0.77–0.76	0.78–0.80	0.67–0.75	0.74–0.78
1-Rel.vulnerability				
Extent trends	0.84–0.91	0.75–0.77	0.66–0.72	0.78–0.83
Duration trends	0.91–0.96	0.92–0.90	0.89–0.87	0.90–0.92
Overall product	0.52–0.62	0.46–0.49	0.32–0.39	0.46–0.53
Annual economic benefits	1,200	700	100	1,000
Reliability trends	0.83–0.88	0.82–0.85	0.82–0.88	0.78–0.89
Resilience trends	0.71–0.81	0.76–0.83	0.77–0.79	0.84–0.88
1-Rel.vulnerability				
Extent trends	0.94–0.91	0.79–0.76	0.76–0.77	0.88–0.89
Duration trends	0.91–0.86	0.90–0.93	0.85–0.88	0.93–0.96
Overall product	0.50–0.56	0.44–0.50	0.41–0.47	0.54–0.67
Agricultural production	800	150	50	600
Reliability trends	0.92–0.94	0.65–0.60	0.72–0.77	0.78–0.79
Resilience trends	0.71–0.75	0.72–0.76	0.85–0.90	0.76–0.77
1-Rel.vulnerability				
Extent trends	0.88–0.93	0.62–0.67	0.86–0.92	0.88–0.84
Duration trends	0.93–0.86	0.71–0.75	0.89–0.85	0.90–0.95
Overall product	0.53–0.56	0.21–0.23	0.47–0.54	0.47–0.49
Drinking water cost	1.40	0.90	1.20	1.10
Reliability trends	0.83–0.85	0.88–0.89	0.82–0.86	0.85–0.91
Resilience trends	0.81–0.80	0.88–0.85	0.77–0.78	0.78–0.88
1-Rel.vulnerability				
Extent trends	0.84–0.93	0.85–0.87	0.86–0.92	0.88–0.93
Duration trends	0.92–0.95	0.82–0.90	0.89–0.88	0.90–0.91
Overall product	0.52–0.60	0.54–0.59	0.48–0.54	0.53–0.68
Pollution index at site A	150	220	30	70
Reliability trends	0.79–0.83	0.75–0.79	0.92–0.94	0.78–0.83
Resilience trends	0.73–0.77	0.76–0.80	0.97–0.95	0.90–0.92
1-Rel.vulnerability				
Extent trends	0.74–0.81	0.71–0.73	0.86–0.92	0.88–0.89
Duration trends	0.81–0.86	0.82–0.90	0.89–0.94	0.84–0.88
Overall product	0.35–0.45	0.33–0.42	0.68–0.77	0.52–0.60
Power production	200	1,200	50	800
Reliability trends	0.78–0.80	0.77–0.76	0.67–0.75	0.74–0.83
Resilience trends	0.85–0.89	0.89–0.93	0.82–0.84	0.88–0.92
1-Rel.vulnerability				
Extent trends	0.92–0.90	0.91–0.96	0.89–0.87	0.90–0.89
Duration trends	0.75–0.77	0.84–0.91	0.66–0.72	0.78–0.78
Overall product	0.46–0.49	0.52–0.62	0.32–0.39	0.46–0.53

	Policy alternatives and components			
Impacts	*Agricultural* irrigation pumps drainage	*Industrial* water storage groundwater use	*Environmental* water recreation treatment tax on water use	*Mixed* water storage treatment canal transport
Fisheries production	70	20	80	40
Reliability trends	0.82–0.84	0.78–0.83	0.84–0.91	0.92–0.90
Resilience trends	0.68–0.75	0.90–0.92	0.91–0.96	0.75–0.77
1-Rel.vulnerability				
Extent trends	0.65–0.72	0.88–0.89	0.89–0.93	0.78–0.80
Duration trends	0.89–0.87	0.74–0.78	0.77–0.76	0.85–0.89
Overall product	0.32–0.39	0.46–0.53	0.52–0.62	0.46–0.49
Flood protection %	99	98	96	99
Reliability trends	0.89–0.98	0.95–0.99	0.90–0.92	0.89–0.87
Resilience trends	0.97–0.96	0.78–0.80	0.74–0.83	0.67–0.72
1-Rel.vulnerability				
Extent trends	0.94–0.91	0.85–0.87	0.78–0.78	0.66–0.75
Duration trends	0.91–0.92	0.92–0.95	0.88–0.89	0.82–0.84
Overall product	0.71–0.81	0.58–0.65	0.46–0.53	0.32–0.39
Overall sustainability index	0.55–0.56	0.44–0.50	0.46–0.53	0.47–0.55

Best Worst

criterion variables considered important with respect to system sustainability. These individual indices for succeeding time periods can be compared to judge whether sustainability is increasing or decreasing over time.

Now, consider some complications. These criterion values are in all likelihood spatially dependent. It is very likely that these relative sustainability indices will vary depending on the site where the time-series values are observed and computed. If so, these relative sustainability indices can be computed for various sites for each alternative water resources system being evaluated and compared. Each of these site-specific relative sustainability indices can then be considered using scorecards and other multiobjective analyses methods. Alternatively, one could develop an overall system indicator of reliability, resilience, and vulnerability through some averaging scheme.

Another complication occurs if the time-series data for any criterion show trends. In such situations some partitioning of time may be appropriate. Any worsening (or improving) situation in the future should not be hidden by including the poorer future (or present) values with the better present (or future) values when calculating these statistical measures. In these cases the relative sustainability indices will be time dependent as well as spatially dependent.

19.5 FINAL REMARKS

It is important to remember that these relative sustainability indices are all based on subjective assumptions concerning future hydrology, costs, benefits, technology, ecological responses, and the like. They are also based on subjectively determined ranges of satisfactory or unsatisfactory values, and on subjectively determined relative weights. Nevertheless, if the range of criteria used are comprehensive and identify the concerns and goals of everyone now and, we hope, on into the future, as best we can guess, this relative sustainability index can be used by itself to identify the preferred design or policy alternative. Otherwise it can be used along with other criteria in a multiobjective analysis.

There is no guarantee that analyses such as these, performed by different groups and/or at different times, will end up with the same conclusions.

Once again, change is with us, and until someone convincingly provides us with a scientific definition of sustainability, one that does not involve human judgments, it is very possible that different groups and different generations will have different views of just what is sustainable. Using methods such as the ones just proposed, however, does force us to look into the future as best we can, to evaluate the multiple physical, economic, environmental, ecological, and social impacts of what we may wish to do now on individuals living in our own generation and those living in future generations. It also ensures that we have that information in some summarized form in front of everyone involved during the entire decision-making process.

REFERENCES

Baan, J. A. (1994) Evaluation of Water Resources Projects on Sustainable Development. *Proceedings*, International UNESCO Symposium, Water Resources Planning in a Changing World, Karlsruhe, Germany, June 28–30, IV 63–72.

Falkenmark, M. (1988) Sustainable development as seen from a water perspective. In *Perspectives of Sustainable Development*, Stockholm Studies in Natural Resources Management, Stockholm 1: 71–84.

Goodland, R., Daly, H., and El Serafy, S. (1991) Environmentally Sustainable Economic Development: Building on Brundtland. World Bank Environmental Working Paper No. 46, World Bank, Washington, DC, 85 pp.

Norgaard, R. B. and Howarth, R. B. (1991) Sustainability and discounting the future. In *Ecological Economics: The Science and Management of Sustainability*, edited by R. Costanza, Columbia University Press, New York, 88–101.

Pezzey, J. (1992) *Sustainable Development Concepts, An Economic Analysis*. World Bank Environment Paper Number 2, World Bank, Washington, DC. 71 pp.

Toman, M. A. and Crosson, O. (1991) Economics and "Sustainability," Balancing Trade-offs and Imperatives. ENR 91-05, Resources for the Future, mimeo, Washington, DC.

UN (United Nations) (1991) *A Strategy for Water Resources Capacity-building: The Delft Declaration*. UNDP Symposium, Delft, June 3–5.

WCED (World Commission on Environment and Development) (1987) *Our Common Future*. (The Brundtland Report), Oxford University Press, New York, 383 pp.

20 Irreversibility and sustainability in water resources systems

H. P. NACHTNEBEL*

ABSTRACT

One of the main characteristics of the sustainability concept is in the long-term evaluation of the possible set of outputs from any decision. Due to the fact that water resources projects have an extremely long physical lifetime and quite broad and diverse impacts, ranging from social, to environmental and economic outputs, the impact evaluation procedure is subjected to a substantial degree of uncertainty. Another approach is seen in the identification of actions that are as far as possible reversible to be able to cope with unexpected and disadvantageous outputs. It is the objective of this chapter to analyze the usefulness of measures such as reversibility to characterize sustainability. Two examples are investigated from which one is based on utilities that are time dependent, while in the other example a physically based approach is emphasized. Both examples refer to water and environmental management.

20.1 INTRODUCTION

Water management structures are designed for a long life time. Several reservoirs in the Middle East have been continuously operated for centuries and irrigation schemes date back over millennia (Garbrecht 1985; Garbrecht and Vogel 1991; Hartung and Kuros 1991; Glick 1970; Schnitter 1994; Petts, Möller, and Roux 1989). Similarly, navigation channels in Europe are being utilized since the medieval age, first for shipping purposes and now for recreation and tourism. On the other hand, many examples are known where reservoir capacity has been quickly decreased due to sedimentation processes, and large irrigation schemes are referenced which quickly lead to salinization of soils to such an extent that the irrigated area had to be abandoned (Goldsmith and Hildyard 1984).

These few examples clearly demonstrate that the decision-making process in water management has to consider beneficial and adverse impacts over long operation periods. In many examples, an increase in economic benefits is accompanied with a loss in environmental quality or socioeconomic inequity (Goldsmith and Hildyard 1984). This was one of the reasons that an extended framework for water-related decision making was already developed in the 1970s including economic, social, and environmental objectives (U.S. Water Resources Council, 1973, 1979, 1980). And about a decade later, the concept of sustainability entered the political and scientific discussion (WCED 1987; Jordaan et al. 1993; IHP National Committee 1994). This WCED report, a political document, gained broad interest by the public and in the scientific community. The concept that sustainable development attempts to meet the needs and aspirations of the present without compromising the ability to meet future requirements is directly related to the statement that radical changes are called for in the way the world economy is run and that environmental constraints on growth have to be faced. Similar statements were already made in "The Limits to Growth" (Meadows et al. 1972) over ten years before. Although both address similar topics and reach similar conclusions (Common 1995), the reactions were quite different. While Meadows et al. (1972) demand to keep the world total output at a certain level to be able to maintain it in the long term, and which would also require an intensive redistribution of products among rich and poor nations, the Bruntlandt report (WCED 1987) clearly states that growth must be revived in developing countries and that there must be also a growth rate of about 3–4 percent in developed countries (WCED 1987, p. 51). Such growth rates could be environmentally sustainable when more efficient technology is implemented to achieve less material and energy-intensive consumption.

In both documents large periods have to be considered in evaluating the full range of impacts related to any decision. The longer the lifetime of a scheme, the more analysis of the possible set of adverse outcomes it would require. Simultaneously, it would lead to increasing uncertainty and imprecision. In some cases, the uncertainty could be too large to allow any rational decision and therefore the degree of reversibility of an

* Department for Water Management, Hydrology and Hydraulic Engineering, University for Agricultural Sciences, Vienna, Austria.

action would gain importance in decision making. In such cases the objective would be in the identification of alternatives with an acceptable level of reversibility, low risk, and high social equity (Nachtnebel, Eder, and Bogardi 1994). This should express that if adverse impacts of an action are observed, the system should be at least partially reversed to reduce disadvantageous outcomes. Complete reversibility is in most cases impossible, but this criterion could be measured by the degree to which an engineered natural resource system such as a contaminated groundwater can be remedied to its original, unengineered state. The time horizon and the required energy to achieve this state are seen as possible and stable indicators for reversibility, while an economic indicator might dramatically change within such a long time period.

In this chapter, two examples are analyzed where environmental impacts are evaluated by different indicators. The objective is to develop measures that evaluate the state of a system under consideration of several sources of uncertainties, like changing preference structures and the long-term performance of a system. Risk and reversibility related characteristics are addressed. The first example refers to fully irreversible actions such as utilizing exhaustively a nonrenewable resource. To reduce the risk, the long-term outcomes are evaluated in an extended framework considering various uncertainties in the decision-making process. In the second example, any preference structure or monetary loss function has been dropped. Rather, the question is posed what could be done if disadvantageous states of the system are observed in the future and which effort would be necessary to re-establish an acceptable state. The methodology tries to follow physical principles to assess the degree of reversibility.

20.2 ECONOMY-BASED IRREVERSIBILITY CONSIDERATIONS

In economic sciences many decisions, especially related to the utilization of environmental resources, are analyzed in the context of irreversibility. This means that a decision has to be taken to utilize completely a resource and to destroy it or to preserve it for further uses (Krutilla 1967; Fisher, Krutilla, and Ciccetti 1972; Arrow and Fisher 1974; Conrad 1980; Mitchell and Carson 1989; Pindyck 1991). This rigorously simplified decision making where at each stage in time a yes/no decision is considered has been modified by Reed (1993), and Beltratti, Chichilnisky and Heal (1994) by taking only partial destruction of the resource into account.

The nondestructive decision to preserve the resource under consideration will result in a certain yield by utilizing some part of the resource dependent on its growth rate and simultaneously, it keeps the option open for further utilization if the preference structure of the decision maker would change. The possibility in keeping options open is credited with an option value due to not taking an irreversible action. The option value is defined here as the amount that people will pay for a contract which guarantees them the right to obtain a good for a specific price at a specific time in the future (Mitchell and Carson 1989). This problem is directly related to a stopping problem (Kim 1992; Conrad 1992), defining the time when an irreversible action becomes more favorable than to continue the preserving practice. Obviously, considering that future generations might have quite different preferences about environmental values, the problem of irreversibility is directly linked to sustainability concepts.

Although water resources cannot be destroyed like a forest by clear cutting, there are some examples that can be described within a similar framework based on irreversible decision making. Examples are in groundwater mining, in abstraction of water from its fluvial system for agricultural irrigation, and polluting a freshwater body by industrial wastes instead of preserving it for any other uses like water-based recreation or for ecological needs.

Here, the "optimal decision procedure" will be analyzed briefly if and when such an irreversible action should be taken. The crude approach evaluating the expected present value of benefits and the expected present value of costs associated with any action and then identifying the point when the expected value (EV) of benefits exceed the EV of costs will only be an appropriate tool if further development is completely known together with all value-related changes. Therefore, an optimal decision should be based on all possible future values expressed in monetary terms and on future amenity services taking into account the associated probabilities.

Consider a resource R which if completely utilized and subsequently excluded from any other use at time t results in a net revenue $NR(t)$ which is subjected to a trend (drift), a random component (Reed 1993; Nachtnebel, Konecny, and Fürst 1995) and another random component correlated to a non-monetary value function $A(t)$ which represents the amenities from the resource at time t (Figure 20.1).

$$\frac{dNR(t)}{NR(t)} = adt + \alpha dw_1 + \rho dw_2 \qquad (20.1)$$

where

$NR(t)$ = net revenue
t = time
a = drift in net revenue
α = standard deviation of fluctuation
ρ = correlation coefficient

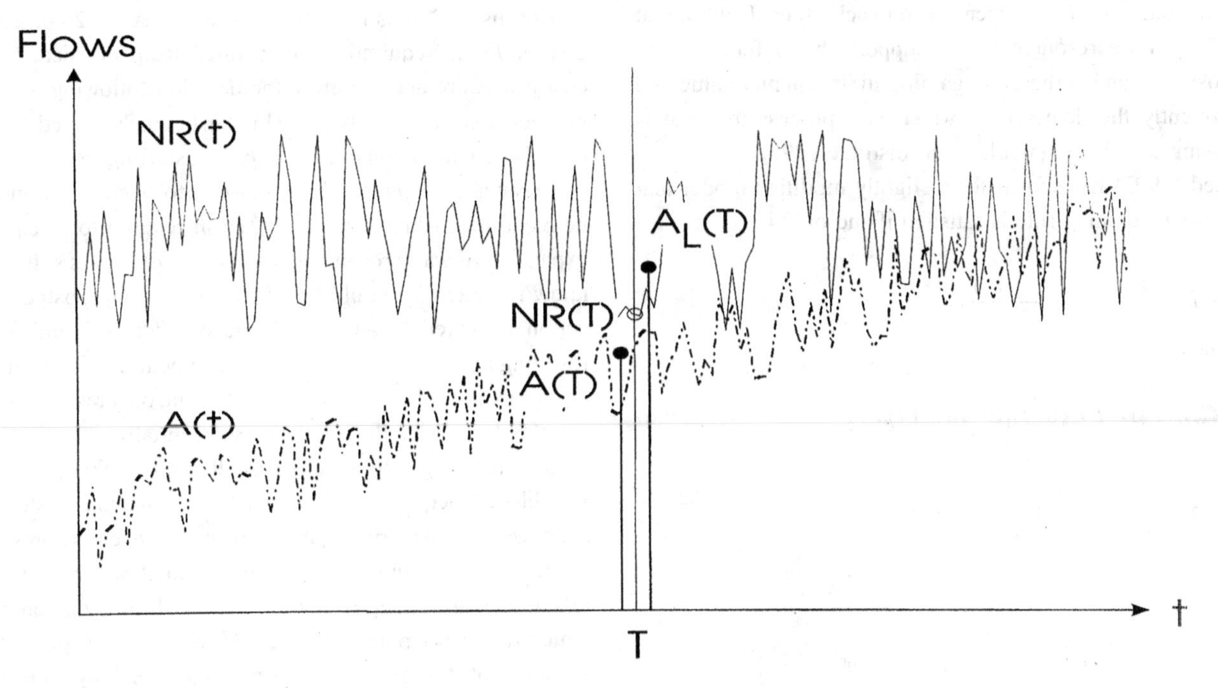

Figure 20.1. Example for one set of development paths for $NR(t)$ and $A(t)$.

$$A(T) = \int_0^T c_t \cdot A(t)dt$$

$$A_L(T) = \int_T^\infty c_t \cdot A(t)dt \qquad (20.2)$$

where

$c_t = e^{-\delta t}$ = discounting factor
t = time
$A(t)$ = amenity function
$A_L(T)$ = discounted losses in amenities after time T
T = stopping time

$w_1(t)$ = Wiener process
$w_2(t)$ = Wiener process

As long as the decision is not made there is a revenue of external benefits $A(t)$ stemming from preserving the ecosystem or from public access for recreation. Obviously this value, however it is expressed, is also subjected to trends and fluctuations. Similar to equation 20.1, the amenity flows $A(t)$ are defined by

$$\frac{dA(t)}{A(t)} = bdt + \beta d\,w_2 + \rho d\,w_1 \qquad (20.3)$$

where

$A(t)$ = amenity value
t = time
b = drift in amenity value
β = standard deviation of fluctuation
ρ = correlation coefficient

This set of coupled stochastic differential equations yields solutions of the type

$$A(t) = A_0 \cdot e^{bt+\beta w_2(t)+\rho w_1(t)} \qquad (20.4)$$

The total return flow (monetary and amenity values), $J(T)$, is expressed by the expected values, EV, given by

$$J(T) = EV\left\{\int_0^T e^{-\delta t} A(t)dt + e^{-\delta T} \cdot EV(NR(T)) - \int_T^\infty e^{-\delta t}(t)dt\right\}$$

with

$$EV(NR(T)) = EV\left\{\int_0^\infty e^{-\delta t} \cdot NR(T+t)dt | T\right\} \qquad (20.5)$$

and the optimal decision is obtained by maximizing $J(T)$ which only depends on the time T when the irreversible decision is taken. Then the resource is completely utilized and no longer accessible for any other use.

This approach considers an infinite number of development paths under different realizations of $NR(t)$ and $A(t)$ reflecting future appreciation of environmental utilization by both monetary and non-monetary terms. The correlation introduced in equations 20.1 and 20.3 assumes that the future realizations of these random terms are dependent on each other. Think about a surface water resource which supports basic functions of an ecosystem and is therefore gaining in its amenity value and subsequently the demand for water as a process material is increasing worldwide (Worldwatch Institute 1996).

Reed (1993) has shown for a slightly modified model that the optimal stopping rule is satisfied if and only if:

$$NR(T) \geq \frac{1+x}{x} \cdot \frac{EV(A_L(T))}{e^{-\delta T}}$$

with

$$EV(A_L(T)) = EV\left\{e^{-\delta T}\int_0^{\infty} e^{-\delta t} A(T+t)dt \Big/ T\right\} \quad (20.6)$$

$$\frac{1}{2}\sigma^2 x^2 + \left(a - b + \frac{1}{2}\sigma^2\right)x + (a - \delta) = 0 \quad (20.7)$$

where

σ^2 = the variance of stochastic process $\left\{ln\left(\frac{NR(t)}{A(t)}\right)\right\}$ given as

$$\sigma^2 = \alpha^2 - 2\sigma\beta\rho + \beta^2 \quad (20.8)$$

Accordingly, the optimal decision is obtained when the benefits $NR(T)$ from utilization at time T exceed the discounted accumulated foregone amenities, $EV(A_L(T))$. The coefficient $x \geq 0$ is obtained from the solution of equations 20.7 and 20.8.

This decision rule is still not satisfying because it is based on expectation values obtained by integrating over many (infinite) development paths and this implicitly assumes that many similar decision problems under the same degree of information and environmental settings have to be solved, which is rather unlikely. This approach is also based on the full information about the process, neglecting uncertainty in model parameters and limited observation. Therefore, the obtained "optimal rule" should be replaced by a distribution function considering the parametric uncertainty in a, b, α, β, δ, ρ. Still using an expected value as a stopping criterion would imply that about roughly 50 percent of the decisions have been taken at the wrong time. To reduce risk it would be recommended to estimate instead of the expectation value as given in equation 20.5, the value $J(x, T)$ where x refers to a given quantile (Nachtnebel et al. 1995) or to x in equation 20.7. This coefficient $C = \frac{1+x}{x}$ has some useful characteristics (Reed 1993). It is greater than one, it increases with a, α, β, (drift in $NR(t)$ standard deviations in the random components), while it is inversely dependent on b, δ, and ρ (drift in amenity, discount rate, and correlation among random components).

Because $C \geq 1$ is incorporated into equation 20.6, a higher $EV(NR(T))$ is required than obtained from classical benefit-cost procedure and therefore the decision following the above procedure is more risk averse. This can be easily proved by trying different combination of a, b, α, β, δ, ρ, starting from zero.

In water management the question as to when an action would be recommendable is mainly asked in relation to a sequenced expansion of water resources structures. Often, it is sufficient to know if an action should be taken now or later. Postponing an action is linked to an assemblage of additional information resulting in improved estimation. This updating approach could also be included in the process, but it would only result in another x without changing the procedure substantially.

Summarizing, the problem of decision making with irreversible consequences is treated in an extended stochastic framework considering outputs from a system expressed in monetary and nonmonetary terms. Both flows (revenues and amenities) are modeled by stochastic differential equations which describe potential future revenues and preferences. Because of the limited information, especially concerning amenity values of the next generation, this task is classified as extremely difficult or impossible for long periods such as the lifetime of water-related structures and utilization.

20.3 PHYSICALLY-BASED REVERSIBILITY CONSIDERATIONS

In 20.2, already a rather complex model was obtained in considering several uncertainties in a socio-economic framework and still it is felt that the model is a crude abstraction of reality and the decision will be strongly dependent on future preferences which might be substantially different from ours. Now, these assumptions are being dropped and only physical principles that are independent of any economic level or preferences are applied to water management practices, such as to evaluate the effort in cleaning a polluted groundwater system. First, some principles of thermodynamics are briefly described and then some generalized applications are discussed.

Many physical laws like classical mechanics are independent of the time arrow. The movement of large bodies could be completely described under reversed trajectories in time and constitutes a perfect example of reversibility. With the formalization of thermodynamics, entropy defined in the second law of thermodynamics was introduced and provided useful information about the direction of processes in time. Its formulation by Clausius in 1850 states that it is impossible to bring heat from a lower to a higher temperature without any changes in the environment. According to Thomson's definition in 1851, it is impossible to construct a periodically working machine by cooling only one reservoir (Becker 1966; Landau and Lifschitz 1970; Prigogine and DeFay 1954). A summary, including an interesting account of the history of thermodynamic development, has been given also by Partington (1949). Here, only a short selection of the basics of thermodynamics is given to provide the background for applied case study. The entropy concept will be applied to estimate the minimum energy necessary to reverse a system to a certain degree. The required energy can be seen as a useful indicator for describing and quantifying reversibility.

The Gibbs formula relates any change in energy of a system to a linear combination of all other energy changes which have to be independent from each other:

$$dU = \sum_{i=1}^{N} dE_i = \sum_{i=1}^{N} \xi_i dX_i \qquad (20.9)$$

where

E_i = energy
ξ_i = intensive variables
X_i = extensive variables

An intensive variable will not change its numerical value when two subsystems with the same level in ξ_i are combined, while the X_i is will add up. For example, someone can think of two boxes under the same pressure. When they are connected to each other, the pressure remains constant while the volumes will add up. Dependent on the form of energy E_i, there are different sets of intensive and extensive variables. In this context, the first thermodynamics law states that energy can neither be generated nor annihilated, corresponding to

$$dU = dQ + dA \qquad (20.10)$$

where

Q = heat flow
A = mechanical energy

Applying this equation to the mixing of two gases which are initially separated into two boxes with volume V and which remains constant and utilizing the universal gas law (Figure 20.2) it follows:

$$dU = 0 = TdS - pdV = TdS - \frac{NRT}{V} \cdot dV$$

$$dS = \frac{NR}{V} dV$$

$$\Delta S = \Delta S_1 + \Delta S_2$$

$$= \Delta(S_1(V_1 + V_2) - S_1(V_1)) + \Delta(S_2(V_1 + V_2) - S_2(V_2))$$

$$= 2N \cdot R \cdot ln2$$

$$V_1 = V_2 = V$$

$$N_1 = N_2 = N \qquad (20.11)$$

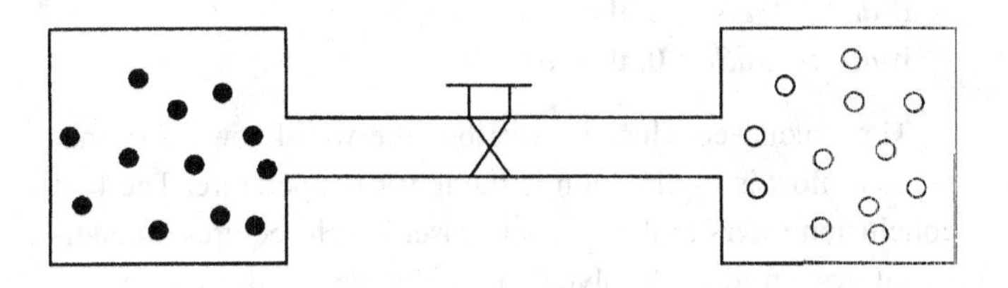

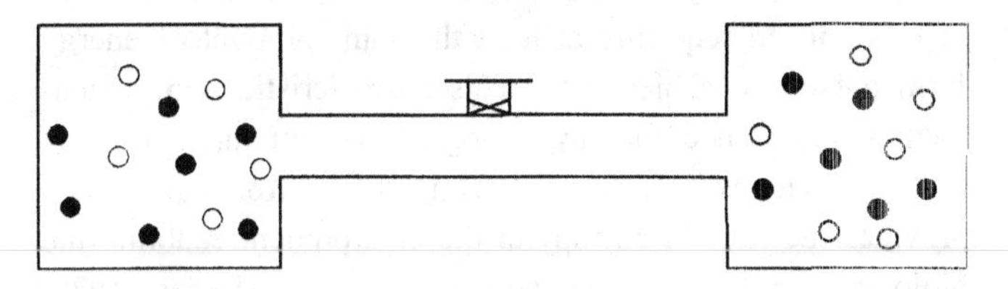

Figure 20.2. Complete mixing of two gases (initial state, final state).

where

P is the pressure; V_i is the volume; T is the temperature; S is the entropy; R is the gas constant; and N_i is the mol-number

The second law, in a simplified form, is given by

$$dU = T\,dS \qquad (20.12)$$

and due to $dU \geq 0$ and $T \geq 0$, dS is also always ≥ 0. Entropy can only be generated but not annihilated but considering more than one system entropy can be exchanged among them. In the whole system, entropy can be only increased but the entropy exchange can be utilized to decrease the entropy in a subsystem (equation 20.13).

$$dS_1 = dS_{EX} + dS_{GEN} \qquad (20.13)$$

S_{EX} = amount of exchange entropy between 1 and 2
S_{GEN} = generated entropy in the whole system ≥ 0

The entropy of system 1 might be decreased by transferring dS_{EX} to system 2, but the transferred amount must exceed the generated entropy associated with the exchange process. A simple example will demonstrate the exchange of entropy, assuming two reservoirs with T_1, S_1 and T_2, S_2 and exchanging energy only by heat flow.

$$T_1 dS_1 + T_2 dS_2 = 0$$

$$dS_2 = -\frac{T_1}{T_2} = dS_1$$

$$dS = dS_1 + dS_2 = \left[1 - \frac{T_2}{T_1}\right] \cdot dS_2 \qquad (20.14)$$

Assuming $T_1 > T_2$ the following conclusions can be drawn:

if $dS_1 > 0$, $dS_2 < 0$, then $dS < 0$
if $dS_2 > 0$, $dS_1 < 0$, then $dS > 0$.

The second conclusion describes the well-known fact that the heat flow is in direction to the lower temperature. The first conclusion refers to the opposite case, which requires an additional system to be involved compensating the decrease in S_2. Independent of any technical realization, the energy $T_1\, dS_1$ has to be removed requiring at least the same amount of energy from outside to achieve this. This characteristic is of fundamental importance for any biological system interacting by flows of energy and matter with the environment which provides the flow of entropy to the environment enabling the biological system to maintain is structure (DeFay 1929; Bertalanffy 1932; Schrödinger 1945; Prigogine 1947; Nicolis and Prigogine 1977).

For abiotic systems at least two aspects are interesting referring to the spatial and temporal scales. The spatial scale is important because decrease in entropy requires an external system that will absorb the entropy flow, while the temporal scale is important for nonequilibrium states.

20.4 REVERSIBILITY IN ENVIRONMENTAL MANAGEMENT

20.4.1 Complete cleaning of a polluted groundwater system

Assume a homogeneously polluted groundwater system that should be cleaned up. This means that the opposite procedure has to be applied. It is assumed that a substance m is available in dispersed form and the question is, what would be the effort to enrich and extract the material from the environment? To extract a certain raw material, at least two steps are necessary (Figure 20.3). First, an amount Q of material containing the pollutant with a certain concentration c_m has to be extracted from the environment. Second, assisted by technological processes such as purification, filtering, etc., the separation of pollutant has to be achieved (Faber, Niemes, and Stephan 1983).

$$c_m = \frac{N_m}{\sum_{j=1}^{e} N_j} = \frac{N_m}{N} \tag{20.15}$$

j = substance indicator $\{j = 1, I\}$
m = polluting material
N_j = number of moles of substance j in volume V
c_m = initial concentration of substance m in volume V
N = total number of moles in volume V

Assuming for the sake of simplicity that the various substances do not interact and that the chemical energy is sufficiently described by the number of particles, the necessary energy amount for the enrichment process can be estimated by using Gibbs potential (Falk and Ruppel 1976). The volumes V_i associated to the moles N_i remain thus unchanged during purification. This process is associated with increased ordering of the location of the particles and this results in a decrease of entropy. Consider the theoretical case of full extraction of material m. Then the material m which was contained in V is reduced in its pure state to V_m and according to equation 20.11 this is related to a change in entropy ΔS_1

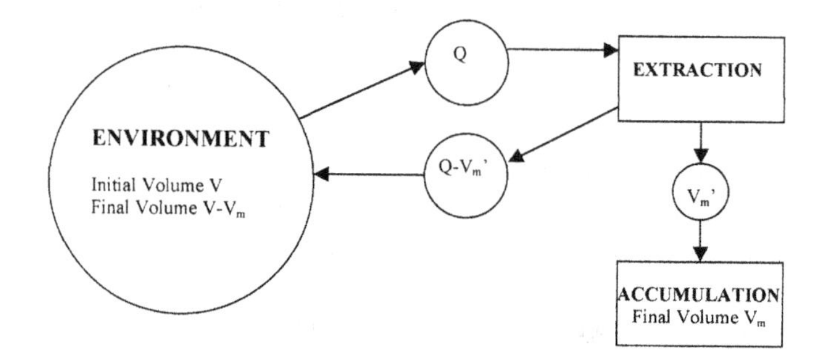

Figure 20.3. Extraction of material m from groundwater.

$$\Delta S_1 = -N_m \cdot R \cdot ln\frac{V}{V_m} \tag{20.16}$$

Also, the polluted groundwater has undergone a change of its volume from V to $(V - V_m)$ corresponding to a change in entropy of

$$\Delta S_2 = \sum_{i\neq m}^{I} \Delta S_{2i} = -\sum_{i\neq m}^{I} N_i \cdot R \cdot ln\frac{V}{V - V_m} \tag{20.17}$$

The total change (reduction) in the entropy of the system is

$$\begin{aligned} \Delta S &= \Delta S_1 + \Delta S_1 \\ &= -\left\{N_m \cdot R \cdot ln\frac{V}{V_m} + \sum_{i\neq m}^{I} N_i \cdot R \cdot ln\frac{V}{V - V_m}\right\} \end{aligned} \tag{20.18}$$

which can be transformed into a functional expression dependent on the initial concentration c_m.

$$\Delta S = N_m R\left\{lnc_m + \frac{1 - c_m}{c_m} \cdot ln(1 - c_m)\right\} \tag{20.19}$$

The specific reduction in entropy, defined per mole of substance m, is

$$\Delta S = R\left\{lnc_m + \frac{1 - c_m}{c_m} ln(1 - c_m)\right\} \tag{20.20}$$

The results for different initial concentrations are given in Figure 20.4 together with their first derivative (Faber et al. 1983).

According to equation 20.9, the minimum required energy to change entropy is $dU = TdS$ assuming that there is no inefficiency in the energy utilization to run the purification process. Thus, following equations 20.18–20.20, the energy for purification is given by

$$U = \int dU = \int T \cdot dS$$
$$= RT\left\{lnc_m + \frac{1-c_m}{c_m} \cdot ln(1-c_m)\right\} \qquad (20.21)$$

where T refers to the temperature of the groundwater system. It can be seen from Figure 20.4 that at a low grade of the pollutant a high amount of energy is required to extract it. This also implies that a nondegradable substance (conservative tracer) in a groundwater system being dispersed during its transport will pollute an increasing volume with a smaller concentration and will therefore require an energy for removal which increases in time.

Considering both the abstraction of the material and its enrichment and using approximately only the first term of equation 20.21, which constitutes a lower bound for the energy, the total amount of energy necessary to obtain finally a unit (mole) of pollutant m is

$$E(c_m) = l_A(c_m) + l_E(c_m) = \frac{\alpha}{c_m} + [-RT \cdot ln \cdot c_m] \qquad (20.22)$$

where α is the energy (pumping work per mole of groundwater); l_A is the pumping work; l_E is the extraction work; c_m is the initial concentration of pollutant; T is the temperature.

This approach implicitly assumes that the concentration is constant in time and space, which is only possible if there is no groundwater recharge, neither from outside nor from the waste water from the purification process.

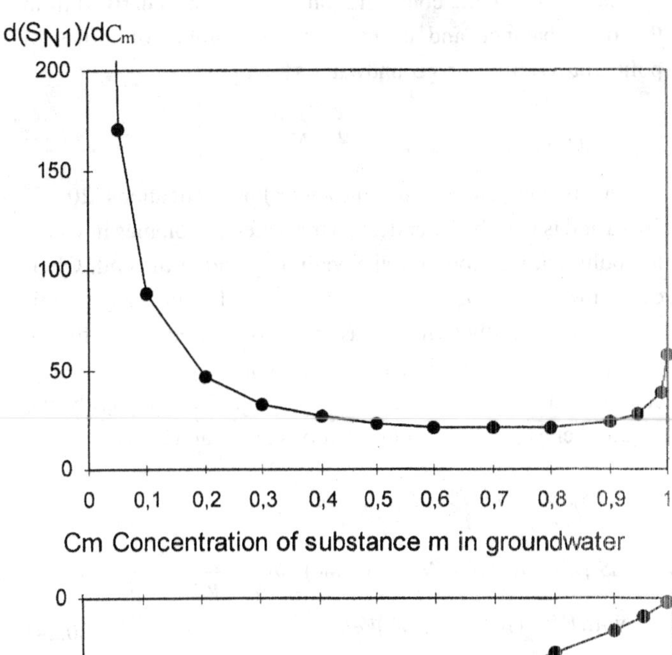

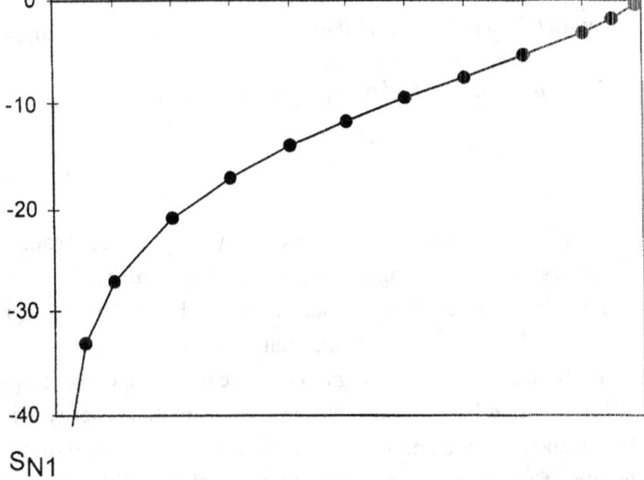

Figure 20.4. Specific change in entropy versus initial concentration and first derivative.

20.4.2 Partial cleaning of a polluted groundwater system

Next, a polluted zone with recharge Q of the process water containing a certain concentration $c_{m,MIN}$ which defines the efficiency of the purification system is considered. It is also assumed that only the pollutant will be extracted.

The initial conditions for the system are

$$N_m(t = o) = N_m^o \quad \text{(moles of pollutant)}$$
$$N(t = o) = N \quad \text{(moles of water)}$$
$$c_m^0 = \frac{N_m V_o}{N \cdot V_o} = \frac{V_m}{V}$$

with the mole volume V_o, the total volume of the polluted groundwater V, and the volume V_m of the pollutant.

It is assumed that the volume Q containing $N_m(t) \cdot \frac{Q}{V}$ moles is pumped and that the same amount of water is recharged containing $N_{m,MIN} \frac{Q}{V}$ moles per unit volume $V(1)$. The difference is extracted and stored separately. The assumption that the same volume is pumped and recharged is a simplification that holds in any practical case. Within each cycle, full mixing of the recharged water is assumed and thus the concentration of the pollutant is continuously decreasing to the level $N_{m,MIN}/N$.

The change in the concentration of pollution is derived from the mass balance and is given by the number of moles of pollutant $N_m(t)$ in the groundwater system

$$N_m(t) = \left(N_m^0 - N_{m,MIN}\right) \cdot e^{-\frac{Qt}{V-Q}} + N_{m,MIN} \quad (20.23)$$

The related change in entropy follows equations 20.16–20.18 and is mainly dependent on the different volumes in which the pollutant was found at the beginning and at the end. Obviously, the state of $N_{m,MIN}$ in groundwater will be only asymptotically reached. Although this has no major consequences for the entropy, it has severe impacts on the pumping energy term in equation 20.22. The change in entropy and subsequently the required energy are given in equations 20.24 and 20.25.

$$\Delta S_m = -R \cdot \left(N_m^0 - N_{m,MIN}\right) \cdot ln\frac{V}{V'}$$

$$\Delta S_{GW} = -R \cdot \left(N - N_m^0 + N_{m,MIN}\right) \cdot ln\frac{V}{V - V'}$$

$$\text{with } V' = \left(N_m^0 - N_{m,MIN}\right)V_0 \quad (20.24)$$

$$U = R \cdot T\left\{lnc_m^0 + \frac{1-c_m^0}{c_m^0} ln\left(1-c_m^0\right) - lnc_{m,MIN} + \frac{1-c_{m,MIN}}{c_{m,MIN}} \cdot ln\left(1-c_{m,MIN}\right)\right\} \quad (20.25)$$

The required energy $l_E(c_m^0)$ to clean up a polluted groundwater system considering recharge of partially purified water is finite. This implies that the state is reversible to the unengineered conditions. Although the change in entropy is finite and therefore the amount of energy to achieve the change in entropy is limited, one has to consider that the abstraction energy will be constant in time and is therefore linearly increasing with the length of pumping period. Therefore, it is impossible to reach the level $c_{m,MIN}$, corresponding here to the "unpolluted state," if recharge is considered.

From this simple example several conclusions can be drawn:

- Entropy is a measure to describe reversibility also for water-related problems;
- Instead of preference structure and loss functions a simple methodology based on physical principles is proposed;
- Several water-related problems can be described in a thermodynamic framework;
- Energy requirements can be estimated to reverse a system close to its initial state;
- Further, phase transitions could be also considered in the methodology. These processes could result in a large increase of energy to remove such a pollutant.

20.5 SUMMARY AND CONCLUSIONS

Sustainability concepts are based on long-term aspects of decision making. Water-related decisions often have long-term repercussions that are difficult to assess. One way to cope with adverse consequences would be in trying to reverse the system under concern into its previous state. In this chapter, two examples were investigated which addressed the degree of reversibility of a system. The first example refers to an irreversible decision resulting in the complete consumption of a resource excluding it from any other future use. Some aspects of uncertainty describing net revenue and amenity values are analyzed. These explicitly formulated uncertainties lead to a more risk-averse decision making. Still many other sources of uncertainties should be additionally considered.

To overcome the time dependence in preference functions which are extremely difficult to estimate, a physically based approach is proposed to assess the reversibility of a system. Instead of societal values, the thermodynamic laws are used to estimate the required amount of energy to reverse a system. This approach was demonstrated by the remediation of a polluted groundwater system. The energy input is seen as a useful indicator to describe some aspects of sustainable development. It was also shown that the improved state could only be achieved by the help of an external system. Thus, the adequate definition of boundaries attracts attention.

Any action in nature is related to a modification of flows of energy, water, and entropy. Therefore, the thermodynamic laws and their extension to nonequilibria and interconnected systems provide a useful tool for rational decision making. This framework could provide information about the possibility and about the minimum required energy to reach a certain state. No information about the technology is necessary because the estimated energy amount constitutes a lower bound and any improvement in technology will only smoothly reduce the additional energy losses.

Acknowledgments

This work was partially supported by a research grant of the National Bank of Austria under contract No. 5064.

REFERENCES

Arrow, K. J. and Fisher, A. C. (1974) Environmental preservation, uncertainty and irreversibility. *Quart. Journal of Economics* 88: 312–19.

Becker, W. (1966) *Theorie der Wärme.* Springer Verlag, Heidelberg.

Beltratti, A., Chichilnisky, G., and Heal, G. (1993) *Preservation, Uncertain Values, Preferences and Irreversibility. Research Paper* 59.93, Fondazione ENI Enrico Mattei, University of Siena.

Bertalanffy, L. (1932) *Theoretische Biologie.* Vol. 1, Bornträger, Berlin.

Common, M. (1995) *Sustainability and Policy*. Cambridge University Press, UK.

Conrad, J. M. (1980) Quasi option value and the expected value of information. *Quart. J. of Economics* XCIV (June): 813.

(1992) Stopping rules and the control of stock pollutants. *Natural Resource Modelling* 6(3): 315–27.

Defay, R. (1929) *Bull. Acad. Roy. Belg. Cl. Sci.* 15(1).

Faber, M., Niemes, H., and Stephan, G. (1983) *Entropie, Umweltschutz und Rohstoffverbrauch*. Lecture Notes in Economics and Mathematical Systems 214, Springer Verlag, Berlin, Heidelberg, New York.

Falk, G. and Ruppel, W. (1976) *Energie und Entropie*. Springer Verlag Berlin, Heidelberg.

Fisher, A. C., Krutilla, J. V., and Ciccetti, C. J. (1972) The economics of environmental preservation. A theoretical and empirical perspective. *American Economic Review* 57: 605–19.

Garbrecht, G. (1985) Sadd-el-Kafara, the World's Oldest Large Dam. In IWPDC (International Water Power and Dam Construction) Handbook, IWPCD Publishing, Sutton, UK, 71–76.

Garbrecht, G. and Vogel, A. (1991) Die Staumauern von Dara. In *Historische Talsperren* edited by K. Witwer, Stuttgart, 2: 263–76.

Glick, T. F. (1970) *Irrigation and Society in Medieval Valencia*. Belknap, Cambridge, MA.

Goldsmith, E. and Hildyard, N. (1984) *The Social and Environmental Effects of Large Dams*. Sierra Club Books, San Francisco.

Hartung, F. and Kuros, G. R. (1991) Historische Talsperren im Iran. In *Historische Talsperren*, edited by K. Witwer, Stuttgart, 1: 221–74.

IHP-National Committee of Germany (1994) *Proc. of International UNESCO Symposium on Water Resources Planning in a Changing World.* Karlsruhe.

Jordaan, J., Plate, E. J., Prins, E., and Veltrop, J. (1993) *Water in our Common Future*. COWAR, UNESCO, Paris.

Kim, S. H. (1992) *Statistics and Decisions.* Van Nostrand Reinhold, NY.

Krutilla, J. V. (1967) Conservation reconsidered. *American Economic Review*, 57: 777–86.

Landau, H. and Lifschitz, B. (1970) *Lehrbuch der Theoretischen Physik*, Vol. 5, Springer Verlag, Berlin.

Meadows, D. H., Meadows, D. L., Randers, J., and Behrens, W. W. (1972) *The Limits to Growth*. Universe Books, New York.

Mitchell, R. C. and Carson, R. (1989) *Using Surveys to Value Public Goods: the Contingent Valuation Method*. Resources for the Future. Washington, D.C.

Nachtnebel, H. P., Eder, G., and Bogardi, I. (1994) Evaluation of criteria in hydropower utilization in the context of sustainable development. *Proc. International UNESCO Symposium: Water Resources Planning in a Changing World*, p. IV-13, IV-24, Karlsruhe.

Nachtnebel, H. P., Konecny, F., and Fürst, J. (1995) Irreversibility in economic context. Working Paper, IWHW-BOKU, Vienna.

Nicolis, G. and Prigogine, I. (1977) *Self Organisation in Non-equilibrium Systems*. John Wiley Interscience, NY.

Partington, J. R. (1949) *An Advanced Treatise on Physical Chemistry*. Longmans, Green and Co., NY.

Petts, G. E., Möller, H., and Roux, A. L. (1989) *Historical Change of Large Alluvial Rivers. Western Europe*. John Wiley and Sons.

Pindyck, R. S. (1991) Irreversibility, uncertainty and investment. *J. of Economic Literature* 20: 1110–48.

Prigogine, I. (1947) *Etude Thermo-dynamique des Processus Irreversibles*. Desoer, Liège, Belge.

Prigogine, I. and DeFay, T. (1954) *Treatise on Thermodynamics*. Vol. I–III, Longmans, Green and Co., NY.

Reed, W. J. (1993) Uncertainty and the conservation or destruction of natural forests and wilderness. In *Statistics for the Environment*, edited by V. Barnett and K. F. Turkmann, John Wiley and Sons, NY.

Schnitter, N. J. (1994) *A History of Dams*. A. A. Balkema, Rotterdam.

Schrödinger, E. (1945) *What Is Life*. Cambridge University Press, London.

U.S. Water Resources Council (1973) Water and Related Resources: Establishment of Principles and Standards for Planning, Federal Register, Vol. 38, 174(24): 778–24: 869.

(1979) Procedures for Evaluation of National Economic Development (NED) Benefits and Costs in Water Resources Planning (Level C); Final Rule, Federal Register, Vol. 44, 242.

(1980) Environmental Quality Evaluation Procedures for Level C Water Resources Planning; Final Rule.

WCED (World Commission on Environment and Development) (1987) *Our Common Future*. Oxford University Press, Oxford.

Worldwatch Institute (1996) *Dividing the Waters. Food Sensitivity, Ecosystem Health, and the New Politics of Scarcity*. Edited by Sandra Postel, Washington, D.C.

21 Future of reservoirs and their management criteria

KUNIYOSHI TAKEUCHI*

ABSTRACT

There are nearly 40,000 large dams in the world, increasing around 250 a year. More reservoirs need to be built, in developing countries, for anticipated population growth, upgrading standard of living, urbanization, flood control, hydroelectric energy, and so on. In developed countries, however, instead of reservoir construction, more emphasis will be placed on demand management and efficient use and reuse of water to meet higher environmental quality needs. Climate change would increase the importance of reservoirs but societal adaptation measures should precede the physical counteractions. The average sedimentation rate to fill reservoirs in the world may not be very high, but the rate is much higher in East and Southeast Asia where many reservoirs would suffer from sedimentation problems in the latter part of the twenty-first century. Reservoirs are the most important component for risk and uncertainty management of water resources systems. But for a reliable and robust water resources system, an integrated management and the administrative structure that makes it possible are most important. The integrated management is also the key strategy for sustainable reservoirs and water resources management.

21.1 INTRODUCTION

Reservoirs are indispensable to utilize running water for sustaining life and civilized activities. They must have been built ever since the beginning of human history and clearly since the emergence of irrigated agriculture. The oldest ruin of a dam may be the Sad el-Kafara Dam, 30 km south of Cairo, Egypt (length 104 m, height 11 m and storage capacity $0.57 \cdot 10^6 m^3$), that was built around 2800 B.C. for water supply, and believed to be destroyed by the first hit of a flood (Biswas 1970). One of the oldest existing reservoirs in the world is "Mannoh-Ike" in Japan (height 32 m, area 1.4 km^2 and storage capacity $15.4 \cdot 10^6 m^3$) built prior to the ninth century A.D. and still used for paddy field irrigation.

According to World Register of Dams (ICOLD 1988), there were 36,235 large dams defined as higher than 15 m in 1986 in seventy-nine ICOLD (International Commission on Large Dams) member countries and fifty-four non-member countries. There were 427 dams in 1900, and about 29,900 dams were built during 1951–82. The rate of dam construction is decreasing in the long run but in recent years, it is not necessarily decreasing. The average number of dams built during 1983–86 was 267 annually. The total number of large dams now must be approaching 39,000. The number of large dams under construction in the world was about 1,200 during 1991–94. Among the 1,242 dams under construction in 1994, the dams higher than 100 m constitute 6.36 percent as compared to 1.1 percent of the total 36,235 dams (Veltrop 1995).

As far as the total amount of reservoir capacities and the total inundated area in the world are concerned, Avakian (1987) estimated the 6,000 km^3 of storage and the 400,000 km^2 of area. These figures correspond to around 5 percent of the total precipitation on the land (119,000 km^3) or 13 percent of the total runoff from the land (46,000 km^3) (UNESCO 1978) and the approximate area of Sweden, respectively. Thus, quite a large part of the world's fresh water is stored in man-made reservoirs and this portion is still increasing.

Table 21.1 shows the gross capacity, inundated area, and purposes of the world's ten largest reservoirs by storage volume and two additional reservoirs with large inundated areas. It also shows the sum of 2,575 Japanese reservoirs in 1994. Although climate, topography, and other conditions are different, it is quite striking that a single dam-reservoir, Akosombo (Volta Lake), for instance, stores 150 km^3 of fresh water and inundates 8,500 km^2 area, which are, respectively, 7.8 times and 6.8 times more than the aggregated impact of 2,575 reservoirs of Japan. Most of the large reservoirs are primarily for the purpose of hydroelectric power generation.

Table 21.2 compares the hydroelectric power generated in the eight countries where the reservoirs listed in Table 21.1

* Department of Civil and Environmental Engineering, Yamanashi University, Kofu 400, Japan.

Table 21.1. *Some large dams in the world and Japanese total*

	Name of dam (river, country)	Capacity (km^3)	Inundated area (km^2)	Purposes
1	Bratsk (Angara, Russia)	169.0	5,470	HNS
2	High Aswan (Nile, Egypt)	162.0	6,500	IHC
3	Kariba (Zambezi, Zambia)	160.4	5,100	H
4	Akosombo (Volta, Ghana)	148.0	8,482	H
5	Daniel Johnson (St. Lawrence, Canada)	141.9	2,000	H
6	Guri (Orinoko, Venezuela)	135.0	4,250	H
7	Krasnoyarsk (Lena, Russia)	73.3	2,000	HN
8	WAC Bennet (Mackenzie, Canada)	70.3	1,650	H
9	Zeya (Amur, Russia)	68.4	2,420	HNC
10	Cahora Bassa (Zambezi, Mozambique)	63.0	2,580	IHC
11	Kuibyshev (Volga, Russia)	58.0	6,150	HNIS
12	Rybinsk (Volga, Russia)	25.4	4,550	HNS
Total of 2,575 Japanese reservoirs in 1994		18.7	1,250	

H: hydropower, I: Irrigation, C: Flood control, N: Navigation, S: Water supply.
1–10 are the largest 10 reservoirs of the world in gross capacity. 11 and 12 are two reservoirs with large inundated area.
Sources: ICOLD (1988), Japan Dam Association (1996).

Table 21.2. *Production of hydroelectric energy in 1990*

Country	Hydroelectric energy generated (GWh)
Brazil	207,230
Canada	296,685
Ghana	5,235
Japan	95,836
USA	290,964
USSR	233,000
Venezuela	37,245
Zambia	7,731

Source: UN (1992) *UN Energy Statistics Yearbook* 1990.

belong. The Japanese figure includes not only those at dam sites but also of run-of-the-river type and pumped peak-hour generation. Especially, its extensive development of run-of-the-river type hydroelectric power generation makes Japanese dams highly efficient in hydropower with respect to inundated area.

Figure 21.1 shows the relation between the gross capacity V ($10^6 m^3$) and the inundated area A (km^2) of 7,936 reservoirs (7,602 dams from ICOLD World Dam Register 1988 and 334 dams from Yearbook of Dams of Japan 1990, selected from those that have both capacity and inundated area data). Although some data are not necessarily reliable as the average water depth becomes greater than 1,000 m or less than 10 cm, it clearly demonstrates the general tendency of the V-A relation of reservoirs in the world. The average relation is

$$V = 9.2A^{1.1} \tag{21.1}$$

If the land surface is assumed to be covered by a common topographical shape, that is, the cross section of a valley is in a unique power function $h = ax^b$, where h is the elevation from the river bed at the distance x normal from the river, then the equation 21.1 implies that $A \propto h_0^{8.8}$ and $V \propto h_0^{9.8}$, that is, the inundated area and the capacity of a reservoir are a very high power function of the dam height h_0. This indicates that a large reservoir inundates exponentially more land with its dam height and that there is no scale merit of having large reservoirs as long as environmental effects due to inundated area are concerned. There is rather an environmental scale demerit in large dams (see Takeuchi 1997a).

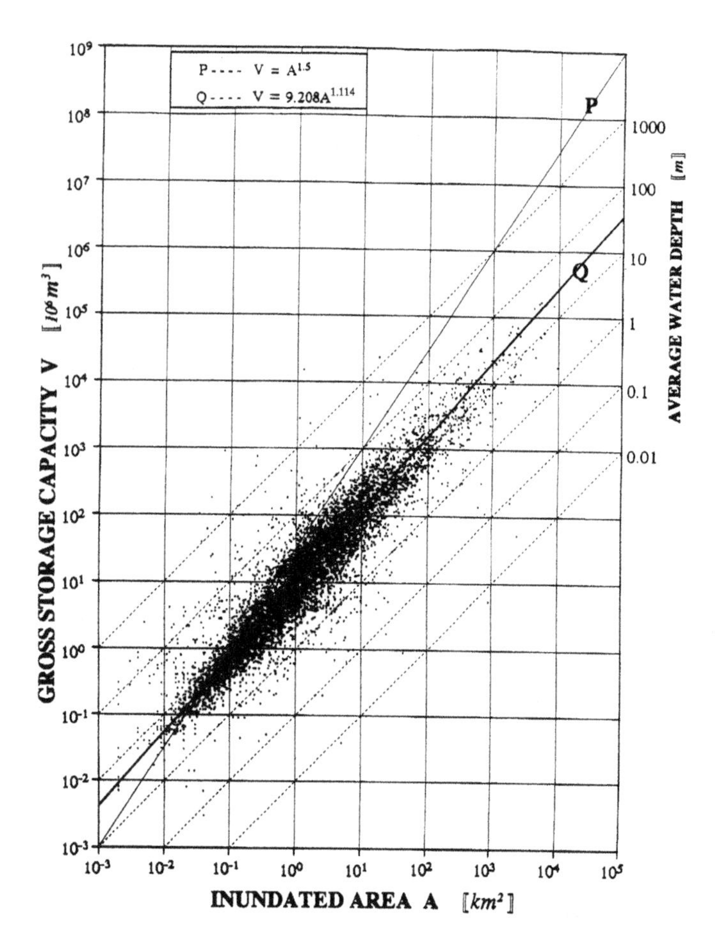

Figure 21.1. The relation between the inundated area (A km^2) and the gross storage capacity (V 10^6m^3) of selected 7,936 reservoirs in the world. After: ICOLD World Dam Register (1988) and Yearbook of Dams of Japan (1990).

21.2 CONTROLLING FACTORS OF FUTURE NEEDS OF RESERVOIRS

Main controlling factors of the future needs of reservoirs would be the growth of developing countries, environmental conservation movement especially in developed countries, needs of clean hydroelectric energy, and the anticipated climate change, of which implications will be briefly reviewed below.

21.2.1 Developing countries

Developing countries are expected to have nearly or more than double their present population in the next half century or so, and their extra needs of water resources are tremendous. It is necessary for agricultural expansion, both for people and cattle feeding, for municipal and industrial water supply especially for megalopolis, for flood control, for safer and more efficient land use, and for hydroelectric power generation for industrialization. Most of the extra water supply will be directed to surface water development, because substantial groundwater depletion causes serious land subsidence and if it is nonrenewable fossil water in arid regions, it is too costly as well. Since surface water can never be stored and transferred without reservoirs, reservoir development will be indispensable regardless of the environmental concern. The only remaining question will be how to construct and manage them in a sustainable manner.

21.2.2 Developed countries

The situation is quite different in developed countries where the population is getting stabilized and the most of economical dam sites have already been developed. The municipal water demand is still increasing due to increasing standards of living, but agricultural and industrial water demands are getting stabilized because the irrigated land is no longer increasing and the rate of recycling of industrial water has been advancing (77 percent in Japan).

Water shortage is, however, fairly common in developed countries, too. Some Japanese cities are suffering from water shortage almost every year, due to a frequent visit of dry spells on a large regional scale. A large part of Japan, especially the southwestern part, had only 50–60 percent of the average annual precipitation in 1978, 1984, and 1994, and the connection with the global climate change is speculated. Such a tendency naturally creates the demand of further water resources development for large urbanized areas.

However, it seems that the public are not appealing for more dams. On the contrary, more and more people tend to appeal for better management of water and for safe and efficient water use. Efficient allocation of water rights among users, a priority use for urgent needs such as temporary allocation of agricultural or hydropower water rights for urban water use during drought, emergency import of water from outer regions, as well as various water-saving operations are now popular options in public opinion. The municipal water use in Northern Kyushu, which experienced the worst droughts that have ever occurred in Japan a few times in the last twenty years, is at less than 300 liter per day per capita, the lowest in Japan excluding a cool region, Hokkaido (National Land Agency 1995). This clearly reflects the recent global environmental movement. Recycling of goods, care for life cycle of artificial goods, preservation of nature and the like are becoming common virtues in people's consciousness. Water is not an exception.

Nonetheless, there are about 600 new dam construction plans endorsed by various levels of governments in Japan.

Many of them are in conjunction with flood control, which is the responsibility of the central government and the local government has incentives to get subsidies. Its complete realization may, however, be far in the future if indeed realized. A large part of rainfall-induced disasters in Japan are localized landslides and mudflows for which large reservoirs are of no help.

The Bureau of Reclamation Commissioner stated at the 18th ICOLD Congress that "the dam building era in the United States is now over" (Beard 1994). "We will emphasize water conservation, demand management and efficient use, including reuse, whenever possible. Every problem we must address has a common theme. That is: there isn't enough water in the river. This sounds elementary, but it isn't. Most western streams are over-allocated and under stress. Excessive use has been condoned – even encouraged – and legitimate in-stream uses have been ignored or prohibited. To solve these problems, we cannot build new reservoirs. Instead, we will have to encourage the movement of water from one use to another. We believe conservation, demand management, efficiency improvements, and reuse offer the best opportunities for doing this, if structured to provide real economic benefits to all participants."

Thus in developed countries, for one reason or another, reservoir construction for water supply, flood control, hydropower is no longer an eminent thrust; rather, an environmental concern is prevailing. Other symbolic examples are the two dams in the Elwha River, Olympic, Washington. Elwha Dam, 132 m high built in 1913, and Glines Dam, 64 m high in 1927, were decided by Elwha River Ecosystem and Fisheries Restoration Act of October 24, 1992 to end their three-fourths of a century of service for hydroelectric power generation and to be removed to restore the original flow regime of the river for natural fish habitat such as salmon, on which the Klallam Indians used to live (Sumi 1996). Thus social preference for human adjustment over control of nature reflects the wider movement of modal shift of the society toward the new paradigm of human existence.

21.2.3 Hydroelectric power

Hydroelectric power is another major potential source of reservoir needs. Especially in developing countries, hydroelectric power is one of the key elements for economic development and an important means of getting foreign currency. It is also expected to be a clean energy source together with biomass, geothermal, wind, wave, solar and others. According to the *Water Power and Dam Construction Handbook 1991* (International Water Power and Dam Construction 1991), the technically exploitable hydroelectric power is estimated as $15 \cdot 10^6$ GWh/y and the exploited one is $2.1 \cdot 10^6$ GWh/y, 14% of the potential. Potential is especially large in many developing countries in South and Central America, South and Middle East Asia, the former USSR, China, and Africa, where the total potential is estimated as $13 \cdot 10^6$ GWh of which $0.9 \cdot 10^6$ GWh or only about 7 percent have been developed in contrast to 55–70 percent in Europe, Japan, and North America.

As shown in Table 21.4, however, it should be noted that the energy consumption of the world is so large that even a full development of hydro-potential can not solve the energy problem. The world energy consumption is estimated as 330,000 petajoule (10^{15} joule) in 1991 (World Resources Institute 1994), of which hydro-energy is 8,000 or only 2.4 percent and nuclear energy is 6.8 percent. Hydroelectric power generation CO_2 increase and air pollution and is considered more advantageous than thermal power with respect to global warming and acid rain. This is often used as a strong support of hydropower development.

This does not mean that hydropower generation is of no use. Since the availability of hydropower potential is highly localized, its development, where it is available, would benefit the local economy a great deal. Especially in developing countries where the proportion of hydro-energy in their total energy consumption is even less, it can be a significant contribution if their hydropotential is developed. Thus, the new reservoir

Table 21.4. *Commercial energy production in 1991 (peta (10^{15}) joules)*

Total	Liquid	Gas	Solid	Primary electricity		
				Hydro	Nuclear	Geothermal and Wind
334,890	93,689	132,992	76,275	8,049	22,669	1,261

Note: The primary electricity indicated are the electric energy produced from the gross nuclear and geothermal energy with efficiencies 0.33 and 0.1, whereas for hydro and wind, it is 1.0.
Source: World Resources Institute (1994), p. 332.

construction needs would continue to be high for hydropower generation in developing countries where large potential is available.

The methods of development would, however, become more carefully selected to enhance local economy and potentials of sociocultural activities and to avoid the degradation of biodiversity. It is said that reservoir impact is not a global issue, while fossil fuels and nuclear energy have direct impact on the global climate and environment. But it is true that many reservoirs in Africa, South America, and other humid regions inundated huge ecologically rich rain forests. Extinction of some species or decrease of numbers of large mammals does not affect the global energy and water circulation, but it seriously impacts the condition of future generations, violating the principle of intergenerational equity.

21.2.4 Climate change

Climate change due to global warming could be the most decisive factor in the future needs and potentials of water resource development in any region. Although its timing, magnitudes, seasonal patterns, and regional characteristics are yet unknown, some climate change seems to have started already with frequent occurrence of extreme events all over the world. It would have a major impact on the water resources in the twenty-first century and later. But the countermeasures to the significant climate change would not be just reservoirs or the mere extension of existing technologies. It would necessitate the increase of social adaptability to a large variation of climate and eventually the gradual but drastic shift of land-use pattern and population distribution. For unknown climate change, we cannot start constructing large reservoirs, large dikes, etc. but should prepare for any future by improving our ability to cope with the current problems.

21.3 RESERVOIR SEDIMENTATION

Reservoirs continuously receive sediments from upstream. However small the sedimentation rates are, they accumulate and eventually fill up the whole reservoir in the long run if not deliberately released. Table 21.5 shows some statistics calculated from the available estimates of reservoir sedimentation. According to Milliman and Meade (World Resources Institute 1988, p. 193), the world average annual sediment discharge to the sea is 13 billion metric ton, and according to Meybeck (World Resources Institute 1988, p. 192), 2–5 billion metricton is trapped behind dams each year. If this figure is correct, the 6,000 km^3 of world reservoir storage will be filled on the average in 1,400–3,600 years assuming that 1.2 metric ton of sediments occupy 1 m^3. It sounds very far in the future.

However, the Mead Lake is losing its storage at the rate of 0.33 percent every year, implying three hundred years to be completely filled (Gottschalk 1964). In the case of the Aswan High Dam, according to Shahin (World Resources Institute 1988, p. 190), the mean annual suspended solid flow through the Aswan site was, prior to the Aswan High Dam, 125 million metric ton, therein 98 percent during floods, wheras the present

Table 21.5. *Reservoir sedimentation rates*

	Total capacity ($10^6 m^3$)	Sediments ($10^6 m^3$, %)	Annual sediments ($10^6 m^3/y$)	Life time (years)
729 Japanese reservoirs[1] (>$10^6 m^3$)	17,322	1,188 (6.9) (in '94)	40.8	395
Lake Nasser (Aswan High Dam)	162,000		109.0[2]	1,600
Lake Mead (Hoover Dam)	34,852		126.8[3]	300
World reservoirs	6,000,000	2,000–5,000[4]	1,400–3,600	

(1) Kobayashi (1996).
(2) Said (1993).
(3) Gottschalk (1964).
(4) Meybeck in World Resources Institute (1988).

flow downstream of the dam is 2.5 million metric ton. It implies that $30 km^3$ dead storage saved for sediments out of the total $164 km^3$ will be filled in some three hundred years and the whole dam will be filled in 1,300 years. The estuary sand bank of the Nile retreated 4 km in thirty years since the completion of the Aswan High Dam in 1964 (Biswas 1992). This, together with the seawater intrusion, is one of the major reasons that Vietnamese oppose the development of dams on the main channel of the Mekong.

Sediment yields, however, differ greatly region by region. In East and Southeast Asia, where about two thirds of the world sediments discharge to the sea come from, there are many reservoirs filled in much shorter time. In northeastern Thailand, a number of irrigation purpose reservoirs have been filled soon after their completion with sediments from deforested areas. There are a number of dams in Japan, mostly smaller than $10 \cdot 10^6 m^3$, built in very steep mountains for hydroelectric power generation that often experience a quick filling of sedimentation, such as within ten years after their completion. Although the head for hydropower generation does not change, small effective storage results in little temporal adjustability of power generation.

Various sediment flashing techniques such as sediment outlet gates and bypass channels are developed. Most methods are still in an experimental stage (Investigation Committee on Sediment Release Impact of Dashi-Daira Dam of the Kurobe River 1995). The Dashi-Daira dam of the Kurobe River, Japan gives a precious lesson. It is a concrete gravity dam of 76.7 m height, $9.01 \cdot 10^6 m^3$ storage for hydropower generation completed in 1985 with two built-in 5 m × 5.5 m sediment discharge gates at 30 m from the bottom. Prompted by large landslides that have occurred upstream, it tried to release sediments from the gates in 1991. It had an unexpected impact. The sediment contained a lot of humus and broken logs in an anaerobic condition which produced sulfur dioxide brought 28 km down to the sea. Even fishermen near the estuary could not stand the odor. Besides, the sediments destroyed many aquatic weeds in the sea and put local fisheries out of work. It is not necessarily an extraordinary case, as similar experiences are reported elsewhere in Japan.

21.4 OUTLOOK OF RESERVOIRS IN THE FUTURE

It is impossible to keep constructing reservoirs forever if they are not destroyed and rebuilt. Large dams and reservoirs are practically irreversible. Once dam sites are exhausted, no more dams can be constructed. What then are the dam site resources? As is shown in Table 21.3, only 14 percent of the technically exploitable hydroelectric energy potential has been developed so far in the world, occupying $6{,}000 km^3$ of space and $400{,}000 km^2$ of land. If two thirds of the potential energy is to be developed as in Japan, still nearly five times more reservoirs have to be constructed. Then, by simple extrapolation, the total storage of reservoirs would become $30{,}000 km^3$ and the inundated area, $2{,}000{,}000 km^2$. They are more than three quarters of average annual river runoff in the world and two thirds of the area of India, respectively.

Will such a level of development be realized? It is possible in many parts of the world if the world economy continues to

Table 21.3. *Technically exploitable hydropotential and hydrogeneration in 1989*

Region	Technically exploitable hydropotential (GWh/y)	Hydrogeneration in 1989 (GWh)	Hydropotential used (%)
World	15,000,000	2,100,000	14.0
Europe	872,156	480,544	55.1
Canada, US	968,982	552,710	57.0
South/Central America	3,400,000	375,725	11.1
USSR	3,831,000	219,800	5.7
Southern Asia, Middle East	2,300,000	166,314	7.2
China	1,943,304	123,294	6.3
Japan	130,524	87,968	67.4
Africa	1,283,101	43,958	3.4
Australia	205,000	37,280	18.2

Source: *International Water Power & Dam Construction* (1991), 39.

grow. If the world population doubles and the economic growth brings all countries to the level of the current industrialized countries, it would not be surprising that what happened in Japan will happen in many other countries. The Japanese population doubled in seventy years since 1925 and it was in this period that 99 percent of storage capacities of reservoirs were constructed.

It is very difficult to estimate how the sediment refill process proceeds in the world reservoirs. It must be true, however, that many small reservoirs, especially in Asia, will be filled by sediments to a considerable extent during the next century. The proportion of reservoirs lower than 30 m is more than 78 percent and, of those lower than 60 m, is more than 95 percent of all large dams higher than 15 m in the world (ICOLD 1988). They may be the ones subject to relatively early fillings. With many reservoirs being filled, it is foreseeable that various water resource problems including water supply and flood control will arise due to the lack of reservoirs. In the latter part of the twenty-first century, the reservoir dredging will be a major task in reservoir management.

Thus, the picture of the twenty-first century may include more construction of more new reservoirs in developing countries and serious maintenance works in relatively heavy sediment recipient reservoirs. Both will have strong environmental impacts on river ecology as well as riparian communities. In the thirtieth century, most of the potential dam sites may have been exhausted and even the largest reservoirs will be to a large extent filled up by sediments. Many reservoir sites would have been turned to flood-prone flatlands with nice waterfalls at the dam sites. We do not know whether people will be living or forests and animals will be resettled on those man-made flatlands. We do not know yet alternatives of reservoirs for water supply and flood control nor the technology of reversing the sediment-filled reservoirs to the original river valleys. The degree of uncertainty in technological innovation may be equivalent to that of climate change and societal change.

21.5 RELIABLE AND ROBUST MANAGEMENT OF RESERVOIRS

Reservoirs are the key component of the water resources systems that cope with risk and uncertainty. As even ants in the Aisopos' story know, storage is the most certain way of preparing for the expected and unexpected variations in the future. Under the increasing uncertainty of society and climate, the needs of reservoirs are even larger. However, reservoirs are still only one component of the total water resources system. Any system excessively relying on one component is always vulnerable to surprise. In order to increase reliability and robustness of water resources systems, therefore, the integrated water resources management is necessary, where not only reservoirs but also all other physical as well as nonphysical measures are utilized in a combined way.

It should be noted that as Chapter 18 of Agenda 21 (UNCED 1992) indicated, the integrated water resources management is the key strategy for sustainable freshwater resources management, too. Furthermore, it is the key to the least marginal environmental impact rule and for sustainable reservoir development and management (Takeuchi 1997b). It broadens the feasible solution space, the set of qualified engineering and managerial options that can serve for the given objectives and can substitute each other. The diversity of system components and the flexible combination of them make the system robust as well as environmentally less stressed and sustainable.

In order to make an integrated water resources management possible, however, the following are prerequisite. One is that there should be new concepts for integration, that is, ideas on what and how to integrate the available means for a better coordinated total system. Another is the techniques that can make a large number of components and knowledge manageable to decision makers. How do we move on to a computer-aided decision support system maintaining human ability in emergency cases? Probably the most important is the institutional domain where the various combinations of integrated management take place. Is the administrative system flexible enough to introduce and manage various ideas and combination of measures?

21.5.1 Integrated urban water management

An example of integrated water management is a comprehensive urban water management widely considered in Japanese cities. This is basically made up of combinations of two physical systems: rainfall storage facilities (cisterns) in houses, buildings, parks, playgrounds, underground tunnels, etc., and rainfall infiltration facilities in house gardens, public open spaces, roads, etc. A large portion of rainfall on the urban area is, since it is impermeable, immediately drained out to the river or the sea without infiltrating to the ground nor being used as a city water supply. Thus urban river discharge becomes highly concentrated during floods, very low during dry conditions, and the cities become more reliant on external water supply. By promoting infiltration and small-scale storage of rainwater in houses and buildings, flood peaks become low, low flows are augmented, and water supply is partially met internally (Musiake et al. 1987). This management is sometimes called an urban oasis plan since old springs and small streams once dried up came back flowing and a number of scattered small water storages in urban areas serve as emergency water reserves during crises such as earthquakes and, at the same time, create popular water-oriented landscapes. Thus, by recovering the natural hydrological circulation back to urban areas, it increases the reliability and robustness of urban water systems.

21.5.2 Computer-aided decision support system

As for technological media for managing ever-increasing experiences, knowledge, available physical and institutional components, etc., computer-aided technologies are in progress. Although its introduction is still cautiously received in the field operation offices, a flood of artificial intelligence systems seems inevitable in the near future, reflecting the increased system complexity, a large number of facilities to be operated, and necessary but limited manpower. From the reliability and robustness points, dependence on computer knowledge base and decision supports coupled with automated remote controls is quite risky. If electricity goes off, if a machine malfunctions, if surprises appear and so forth, human ability to control automated systems is very limited, as was seen in the cases of aircraft crashes, nuclear power stations' overrunning, and others. There are many efforts to get around such problems and to preprogram for such emergency cases in the system. But we have to take a cautious approach to artificial intelligence systems until truly intelligent robots with self-learning, self-thinking, and risk-conscious "minds" become available. Human ability to control the emergency cases should always be reserved and trained. The integrated water resources management should therefore be decently subdivided into manageable sizes.

21.5.3 Institutional modification for integrated water resources management

In order to make integrated water resources management possible, it is most important to establish an institutional setting that can allow various policies to be implemented. The integrated water resources management includes basinwide multiobjective management, optimal allocation of water with economic and environmental evaluation, use of demand management with pricing mechanism and regulatory measures, development of alternative sources of water supply such as waste-water reuse and water recycling, water conservation through improved water-use efficiency, water-saving and effluent control. Many of them may be possible, implemented within the current administrative and institutional structure. They would all require the drastic modification and rearrangement of current administrative sectoring, regulations, water rights, etc. In particular in reservoir operation, water rights allocation set separately to each reservoir is an obstacle to utilize available water in efficient ways and to meet the needs for basin-wide integrated, multiobjective water management. Even during severe droughts, a temporary transfer from agricultural water use to municipal water use is not easy, still less for introducing new demands like environmental use. The reliability and robustness can never be realized with stiff administrative sectionalism and regulations. The opportunity cost of administrative stiffness is very high, creating inefficient water use and difficulty for risk management.

21.6 CONCLUSIONS

Reservoirs are a necessary evil, good for the majority of people but often not invited by the affected people. They are good for human beings but destructive to existing nonhuman lives. The development plan of reservoirs should be evaluated under the least marginal environmental impact rule and managed in the most environment-friendly manner. Potential as well as existing reservoirs are scarce resources to be allocated in accordance with sustainable development principle. Reservoirs should contribute to both the current and future generations. In order to prolong the lifetime of reservoirs, sedimentation control is very important. It will be one of the most serious issues in the latter part of the next century.

A large number of reservoirs were built in the world, inundating a huge area of precious water-rich lands. Some of the giant dams store water and inundate land much more than the sum of thousands of other reservoirs. Their environmental as well as social effects are great and become subject to strong criticisms and opposition. Nevertheless, the needs of reservoirs seem to be increasing in developing countries, reflecting anticipated population growth, upgrading standard of living, urban concentration, and clean energy demands.

Reservoirs are the key components to the risk and uncertainty management of water resources systems. Yet in order to increase the reliability and robustness, an integrated water resources management is absolutely necessary. For the integrated water resources management to be realized, various ideas of managing water resources systems in a holistic way should be developed and the computer-aided artificial intelligence systems should be wisely utilized. Above all, the flexible administrative and institutional systems should be installed on which the integrated management may be implemented. The integrated management serves for robustness as well as for sustainability of water resources systems.

REFERENCES

Avakian, A. B. (1987) Reservoirs of the world and their environmental impact. In UNEP and UNESCO (1990) *The Impact of Large Water Projects on the Environment*, Proc. of UNESCO/UNEP Intern. Symposium in Paris, 29–36.

Beard, D. (1994) note in USCOLD *Newsletter* 105, November, 12–15.
Biswas, A. K. (1970) *A History of Hydrology*. North Holland Publishing, Amsterdam.
(1992) The Aswan High Dam revisited. *Ecodecision*, 67–69.
Gottschalk, L. C. (1964) Reservoir sedimentation. Chapter 17-I. In *Handbook of Applied Hydrology*, edited by Ven Te Chow, McGraw-Hill, New York.
ICOLD (1988) World Dam Register 1988.
International Water Power and Dam Construction (1991) The world's hydro resources. *Water Power & Dam Construction Handbook*, Reed Business Publishing, Surrey, 35–41.
Investigation Committee on Sediment Release Impact of Dashi-Daira Dam of the Kurobe River (1995) unpublished document (in Japanese).
Japanese Dam Association (1990) *Yearbook of Dams* 1990.
(1996) *Yearbook of Dams* 1996.
Kobayashi, E. (1996) Current measures against sediments in dam reservoirs, In Japanese Dam Association, *Yearbook of Dams*, 1996, 21–27.
Musiake, K., Ishizaki, K., Yoshino, F., and Yamaguchi, T. (1987) *Conservation and Revitalization of Water Environment*. Sankaido, Tokyo (in Japanese).
National Land Agency (1995) *Water Resources in Japan*. Ministry of Finance Press, Tokyo.
Said, R. (1993) *The River Nile: Geology, Hydrology and Utilization*. Pergamon Press, Oxford.
Sumi, K. (1996) History of dam construction in USA. In: Parliament Members Association for Realization of Check Mechanism of Public Works, *Why did USA stop building dams*?, Tsukiji-Shokan, Tokyo, 22–54 (in Japanese).
Takeuchi, K. (1997a) On the scale diseconomy of large reservoirs in land occupation. In *Sustainability of Water Resources Under Increasing Uncertainty*, edited by D. Rosbjerg et al. (Proc. IAHS Rabat Symposium, April 1997), IAHS Publ.
(1997b) Least marginal environmental impact rule for reservoir development, *Hydrol. Sci. J.*, 42(4): 583–98. Special Issue edited by S. Simonovic.
UN (1992) *United Nations Energy Statistics Yearbook* 1990.
UNCED (1992) *Agenda 21: Programme of Action for Sustainable Development*. United Nations Conference on Environment and Development (UNCED), June 3–14, Rio de Janeiro, UN Publications New York.
UNESCO (1978) *World Water Balance and Water Resources of the Earth*. Studies and Reports in Hydrology 25, USSR Nat. Comm. for IHD.
Veltrop, J. (1995) Benefits of dams to society, USCOLD *Newsletter*, March 1995.
World Resources Institute (1988) *World Resources* 1988–89. Basic Books, New York.
(1994) *World Resources* 1994–95. Oxford University Press, New York.

22 Performance criteria for multiunit reservoir operation and water allocation problems

DARKO MILUTIN* AND JANOS J. BOGARDI**

ABSTRACT

A genetic algorithm model was developed to derive the best water allocation distribution within a multiple-reservoir water supply system. Three different objective functions were used to test the applicability of the model on a real-world seven-reservoir system. The appraisal of obtained solutions was carried out through the respective system's performance evaluation using a number of performance indicators. Due to the difference in the objective functions, the use of performance indicators proved crucial in the comparison of the solutions proposed by the three models. In addition, in all of the three cases the resulting release distributions produced in repeated runs of the same model showed substantial variability. The variability, however, was not reflected in the respective objective function achievement, indicating that there might be a number of potential solutions to the problem. In this respect, the comparison of the related performance indicator estimates was found to be a valuable means to provide a better insight into the essential difference between different solutions.

22.1 INTRODUCTION

Genetic algorithms (GA) fall into a group of search strategies that are based on the Darwinian concept of biological evolution. They apply the principles of natural genetics and selection to solve optimization problems related to artificial systems (Holland 1975). By using the objective function as a fitness measure, GAs emulate the Darwinian concept of "survival of the fittest" on a population of artificial beings to search the solution space of the optimization problem. The artificial "creatures" that the search is based on represent a specific coding of potential solutions to the problem.

To name just a few, the reported applications of GAs to water-related problems deal with model calibration (Babovic, Larsen, and Wu 1994), pipe network optimization (Dandy, Simpson, and Murphy 1996), reservoir operation (Esat and Hall 1994), and groundwater monitoring (Cieniawski, Eheart, and Ranjithan, 1995). This chapter presents an application of genetic algorithms to derive the optimal release distribution within a multiple-reservoir water supply system. Three different objective (fitness) functions were used within the developed GA model. Fitness evaluation of a potential solution was based on simulation of the system's operation according to the standard reservoir operating rule. The simulation itself was carried out on a monthly scale over a forty-four-year-long period of available monthly flow records.

The choice of genetic algorithms for solving this particular water allocation problem is due to their ability to search effectively complex, multidimensional solution spaces without imposing high computation space and time requirements. The applicability of the methodology has been extensively tested and some aspects of the obtained results and experience gained have been reported in Milutin (1996) and Milutin and Bogardi (1996a, 1996b). The cited reports include the application of the GA model alone as well as its use in conjunction with a decomposition, stochastic dynamic programming-based methodology for the optimization of a multiple reservoir water supply system operation. It has been found that the use of the GA-based water allocation model can significantly improve the performance of the latter decomposition algorithm. The use of the GA model in this contribution is, however, restricted to providing a sufficiently broad basis for the assessment of the role of performance criteria in the evaluation of a multiple-reservoir system operation.

Therefore, particular emphasis was placed on the comparison of solutions identified by the three models. Obviously, the variation of a single value of the objective function achievement was neither sufficient nor adequate a characterization of the system's performance for this task. In that respect, the basis for comparison was extended by a number of performance indicators (PI). Several criteria have been suggested so far for performance assessment of water supply systems (Duckstein and Plate 1987; Bogardi and Verhoef 1995). Seven PI's were

* AHT International GmbH, Essen, Germany.
** UNESCO, Division of Water Sciences, Paris, France.

selected in this particular study: quantity-based reliability, time-based reliability, average duration of full supply, average duration of failure, mean monthly deficit, maximum duration of failure, and maximum observed deficit in a single month. In addition, the PIs were also used to compare solutions obtained in repeated runs of the same GA model. Namely, a batch of ten initial population samples was generated at random. Subsequently, a GA model was run ten times, each time starting with a different initial population. Although the objective function values associated with the solutions obtained in different runs of the same GA model did not vary much, the resulting release distribution patterns exhibited a substantial variability. In this respect, the PIs proved crucial in assessing the impact of the derived release distribution patterns on different aspects of the system's performance.

22.2 CASE STUDY SYSTEM

The analyses were carried out on a real-world seven-reservoir water supply system (Figure 22.1). The reservoirs interact by means of both serial and parallel interconnections. Available water in a reservoir may be distributed both to local users within the basin as well as to remote users and/or reservoirs situated in other basins. The complexity of feasible water allocation patterns is reflected in the fact that one reservoir may provide water for a number of demand centers while, at the same time, a single demand may be supplied by more than one reservoir. In addition to water supply, all of the reservoirs serve for flood protection while some of them have hydropower generation facilities. However, these purposes were not taken under consideration in this study. The salient features of the seven reservoirs are given in Table 22.1.

The average annual inflow to the system is estimated at $963.832 \cdot 10^6 m^3$ and the total active storage of the seven reservoirs amounts to $1000.7 \cdot 10^6 m^3$. However, the great variability of inflows under the prevailing semi-arid climatic conditions tends to constrain the utilization of the available resources. The efficient exploitation of the available water is further limited by water losses, mainly due to evaporation. The total monthly elevation losses due to evaporation were estimated to vary between 0.171 m in January and 1.201 m in July for the entire system.

The total estimated demand imposed on the system is $469.504 \cdot 10^6 m^3$. As Table 22.1 reveals, a considerable number of the demand centers gets water from more than one reservoir. In four cases there are two reservoirs supplying a demand center (irrigation water demands labeled IBH, IAEA, IBV, and IMSC), whereas five demand centers (drinking water demands labelled NA, MO, SO, SF, and TO) receive water from three reservoirs,

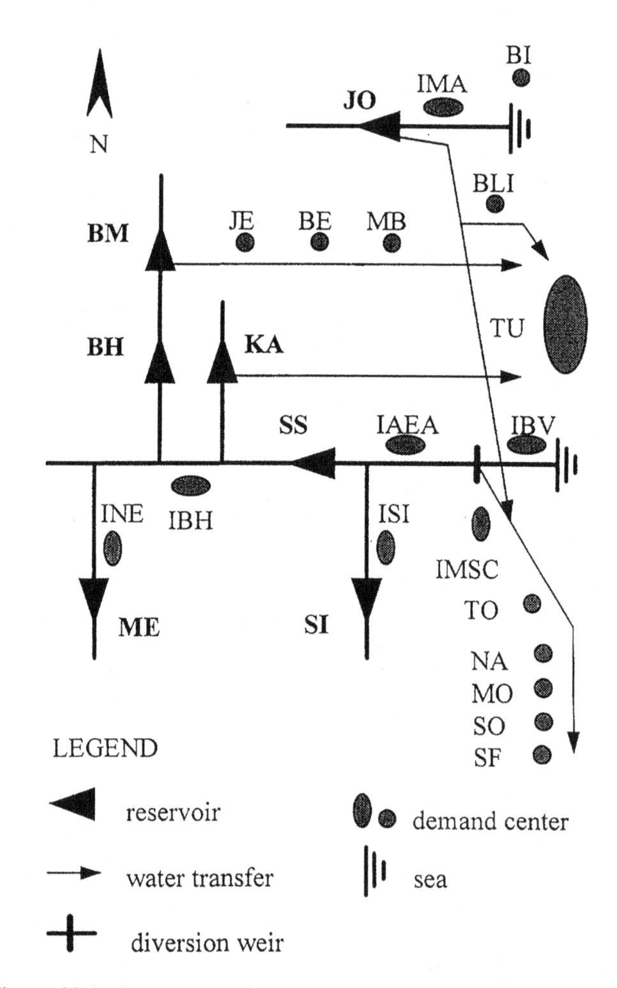

Figure 22.1. Seven-reservoir system.

and only one user (demand TU) gets water from five reservoirs. The remaining eight demand centers depict reservoirs' local users: irrigation schemes named IMA, INE, and ISI; municipal users BI, JE, BE, and MB; and water requirements for the recharge of a natural lake (BLI).

The backbone of the system is the SS reservoir. This reservoir's capacity amounts to 51 percent of the total system active storage while, on an annual average scale, its incremental inflow reaches 45 percent of the total inflow to the system. SS also regulates and utilizes any excess flow that may originate from the three reservoirs situated directly upstream: KA, BH, and ME.

On the other hand, SI is a small reservoir with a relatively poor inflow. Although the list of potential SI's users is quite long, its main purpose is to provide water for the local irrigation scheme (ISI). Any additional release from this reservoir may be directed toward the remaining users from the list.

The JO reservoir, however, can contribute significantly toward the supply of all the associated demand centers. In addi-

Table 22.1. *Salient features of the seven reservoirs [$10^6 m^3$]*

Reservoir	Active storage	Mean annual inflow	Demand targets
JO	121.3	132.959	BI, IMA, BLI, TU, TO, NA, MO, SO, SF
BM	44.2	42.325	TU, BE, JE, MB
KA	72.2	48.389	TU
BH	102.5	92.015	IBH
ME	89.0	175.859	INE, IBH
SS	510.0	429.188	IAEA, TU, TO, NA, MO, SO, SF, IBV, IMSC
SI	61.5	43.099	ISI, IAEA, TU, TO, NA, MO, SO, SF, IBV, IMSC

tion to the three local users, JO plays an important role in providing water for the remaining demand centers which it supplies jointly with other reservoirs from the system.

Reservoirs BH and ME serve primarily for irrigation water supply. They jointly supply the IBH irrigation scheme while ME also provides water for the local irrigation areas (INE). They both may contribute to the increase of inflow to SS.

The sole purpose of the KA reservoir is to provide drinking water for the TU demand center. Any excess available water may be used to cover the SS supply shortage. The BM reservoir also contributes primarily toward drinking water supply and, if any surplus of water is available, compensates for potential shortage in BH's water deliveries.

22.3 GENETIC ALGORITHM MODEL

Genetic algorithms are essentially search strategies based on the Darwinian evolutionary principles: they transform the search for the solution of an optimization problem into a simulated evolution of an artificial species. The link between the actual problem and the evolution of the representative fictitious being is provided through a specific interpretation of the problem, and a potential solution thereof, in terms of the essential evolutionary parameters and processes.

It should be noted here that the detailed description of the genetic algorithm theory is not going to be presented in this chapter. This is, on the one hand, due to the limited space available. On the other hand, the emphasis is put on the principal topic of this contribution, which is an illustration of the role and importance of the use of a broader set of performance evaluation criteria to assess a complex water resources system operation rather than relying on a single metric being the expectation of the selected objective function in deriving the best water allocation strategy. For a comprehensive introduction to the genetic algorithm theory, the interested reader can refer to Holland (1975) and Goldberg (1989). The detailed description of the genetic algorithm model used to address this particular water allocation problem can be found in Milutin (1996) and Milutin and Bogardi (1996a, 1996b). Consequently, the description of the GA model given in this chapter is only providing a general overview of the GA theory along with the principal issues related to the problem in question. The intrinsic mathematical relationships within the GA formulation are therefore omitted.

GAs require that a potential solution to the problem be represented by a specific code, which is analogous to chromosomes in biological systems. The encoded representation of a solution is further regarded as a complete genetic material (i.e., genotype) of an artificial being whose characteristics (i.e., phenotype) are reflected in the actual values of the parameters it represents (the terms: string, chromosome, and individual are used interchangeably in this chapter to refer to the encoded representation of a potential solution). The most frequently used encoding technique is the representation of a solution as a binary integer of a predetermined fixed length. The transformation is fairly straightforward if the solution itself is an integer, whereas the cases involving solutions that are real numbers from a known interval require that the solution interval be mapped over the interval of its binary representation.

A GA starts its search from a population of strings, that is, potential solutions. An initial population consisting of a predetermined number of strings is generated at random. The subsequent actions emulate the evolution of the initial population through creation of new populations by applying three principal GA operators: selection, crossover, and mutation.

Selection of an individual for reproduction is based on its fitness: the likelihood that an individual be selected for reproduction is proportional to its fitness relative to the aggregate fitness of all individuals in the generation. In terms of the actual optimization problem, fitness is actually the objective function value obtained for the potential solution represented by that particular chromosome. Therefore, prior to selection itself, each individual from a population is decoded into its respective parameter values and its fitness is evaluated. Once selected, pairs of strings are crossed over to create two new chromosomes which enter the subsequent population. It is necessary to stress that the application of the crossover operator is performed with a certain probability. Therefore, a fraction of the new generation is created by crossing over pairs of selected strings while the rest of the new population members is provided by simply copying the selected chromosomes from the previous generation. Each bit position of a newly created individual is subsequently subjected to potential mutation. Being regarded as a "secondary" operator in GAs, mutation is employed with a certain low probability. Common practice has been to apply mutation once over approximately one thousand bit positions.

The process of creation of new populations proceeds with the expectation that the overall improvement in individuals' fitness would lead toward the optimum solution to the problem. Throughout generations, newly created individuals inherit good traits from their parents and, by means of mutation, new qualities are introduced into chromosomes as populations emerge, bringing about the convergence of the GA procedure toward a stable population. The termination of a GA run can be decided upon a number of criteria. The convergence of the maximum or the average population fitness, the time available for execution, or the maximum number of generations are the most frequently used ones.

The problem in this particular study was to determine the best distribution of releases from groups of reservoirs toward their common demand. It was assumed that the relative contribution of a reservoir to supplying an associated demand remained constant over a year. Therefore, the ultimate solution consisted of twenty unknown variables. As Table 22.1 reveals, and after introducing a few simplifications into the presented supply-demand patterns, there are six demand centers supplied by more than one reservoir:

1. TU demand is covered by JO, BM, KA, SS, and SI.
2. TO demand is jointly supplied by JO, SS, and SI.
3. The demand created by aggregating NA, MO, SO, and SF is covered by JO, SS, and SI.
4. IBH demand is covered by BH and ME.
5. IAEA demand is supplied by SS and SI.
6. The demand created by aggregating IBV and IMSC is covered by SS and SI.

Three additional unknown variables are due to the fact that BH, ME, and KA may contribute to the increase of the natural inflow of SS. Therefore, potential supply shortage of SS was considered as a joint hypothetical demand imposed upon these three reservoirs. However, any of the three reservoirs was allowed to allocate water to compensate for SS shortage only after it had fully covered its consumptive demands.

The coding of a potential solution was performed over a binary alphabet. As the solution in this particular case was a vector in a twenty-dimensional space, the encoded string was created by concatenating the binary codings of vector coordinates into a single string, keeping track of locations of substrings that represented different components of the solution. Due to the fact that the solution coordinates were real numbers from the interval [0, 1], each coordinate was mapped over the interval of L-bit-long binary integers between 0 and $2^L - 1$.

The adopted selection procedure is known as "proportional selection" (Holland 1975) or "biased roulette wheel selection" (Goldberg 1989). It states that the likelihood that an individual be selected for reproduction is proportional to its fitness relative to the total aggregated fitness of all individuals in the generation. The GAs developed in this study apply the so-called one-point crossover: both of the selected strings are cut at a randomly chosen bit location and the parts between the crossover site and the strings' ends are swapped. The subsequent mutation of a newly created string inverts, with a substantially low probability, a bit value of 0 to 1 and vice versa.

Fitness evaluation was based on simulation of the system operation over the forty-four-year-long time period of available inflows. It was assumed that all the reservoirs were full at the beginning of the simulation period. The operating strategy used in simulation of each reservoir was the standard reservoir operating rule: release all the water available to meet the imposed demand during a time step. Three GA models were tested, each minimizing a value of a different objective function:

1. OF_1: The accumulated monthly squared supply deficiency toward each demand center over the entire simulation period.
2. OF_2: The square of the monthly supply deficit observed for the whole system, aggregated over the entire simulation period.
3. OF_3: The maximum monthly supply deficit observed for the entire system.

As GAs are essentially maximization procedures, the original problem had to be inverted into a maximization task. In the case of the first two fitness functions, this was accomplished by subtracting the obtained objective achievement from the maximum possible value it can take for the given system. As to the third objective function, the maximization problem is created by simply taking the inverse of the observed maximum deficit. In addition, the devised GAs applied linear scaling of the derived fitness. This was done to prevent premature convergence of the procedure due to high variability of fitness in the initial random population and to promote more fit individuals in later generations when fitness variability was generally falling toward rather low values (Goldberg 1989).

The particular problem to be solved by the GA is obviously a highly constrained one. Namely, for a given demand center, the sum of relative contributions from the related reservoirs must add to unity. This means that the sum of their respective L-bit-long binary representations must equal the maximum value of a binary integer of that length. However, as GAs do not take into consideration what is behind the encoded material they deal with, it is highly unlikely that a newly created individual will represent parameter subsets that meet all the constraints. Therefore, the developed GA models employed a "repair" algorithm to test the feasibility of all substrings in an individual and, if necessary, to modify them. The adjustment of an infeasible parameter subset was performed on their integer representations. The "repair" procedure selected a member of the infeasible subset at random and increased or decreased it by one, depending on whether the sum of the representative integers was smaller or larger than required. The procedure was repeated until the condition on the aggregate of parameter values was met.

22.4 PERFORMANCE EVALUATION

The comparison of the system's performance under release strategies obtained by different GA models was based on a number of performance criteria. Each of the performance measures was estimated by simulating the system's operation over the entire forty-four-year-long time period. Reservoirs' monthly operating strategies were based on the standard reservoir operating rule and the respective water allocation patterns provided by different GA models. Before introducing the selected PIs, some clarification regarding the notation used needs to be presented:

- T is the simulation time span (in months) while index t refers to a time stage (month).
- The demand imposed on the system at stage t is d_t, and the volume of water allocated by the system to cover the demand at t is given by r_t. Consequently, the supply shortage observed at stage t is s_t:

$$s_t = \begin{cases} d_t - r_t, & r_t < d_t \\ 0, & r_t = d_t \end{cases} \tag{22.1}$$

- A non-zero supply shortage detected during a time step constitutes a failure event in the operation of the system. Descriptor f_t indicates whether a failure event was observed at stage t or not:

$$f_t = \begin{cases} 1, & \textit{failure} \\ 0, & \textit{non-failure} \end{cases} \tag{22.2}$$

- The total number of stages (months) when failure occurred is given by T_f.
- Index i depicts a continuous sequence of months with either steady failure or steady nonfailure events.
- Failure interval is a continuous sequence of stages during which the system fails to provide the full amount of demanded water. The duration of such a period is depicted by $t_{f,i}$. The number of failure intervals is given by N_f.
- The length of a continuous sequence of stages during which the system manages to sustain full supply is given by $t_{s,i}$. The number of such intervals is N_s.

In addition to the respective objective function achievement, each of the derived alternative release distribution plans has been appraised using a set of seven performance indicators. The selected PIs include:

- *Quantity-based reliability* (PI_1) is the ratio between the total amount of utilizable system release and the total demand over the entire simulation period:

$$PI_1 = \frac{\sum_{t=1}^{T} r_t}{\sum_{t=1}^{T} d_t} \tag{22.3}$$

- *Time-based reliability* (PI_2) is the long-term probability that the system will be able to meet the full targeted demand:

$$PI_2 = 1 - \frac{1}{T}\sum_{t=1}^{T} f_t \tag{22.4}$$

- *Average interarrival time* (PI_3) is the average duration of periods the system is continuously in failure mode. It commonly measures the average recovery time from the occurrence of failure and can also be called *repairability*:

$$PI_3 = \frac{1}{N_f}\sum_{i=1}^{N_f} t_{f,i} \tag{22.5}$$

- *Average interevent time* (PI_4) is the average duration of periods the system sustains full supply. This type of performance indicator can also be named *average recurrence time*:

$$PI_4 = \frac{1}{N_s}\sum_{i=1}^{N_s} t_{s,i} \tag{22.6}$$

- *Mean monthly deficit* (PI_5) measures the average magnitude of failures detected when a particular operating strategy is followed:

$$PI_5 = \frac{1}{T_f}\sum_{t=1}^{T} s_t \tag{22.7}$$

- *Maximum duration of failure* (PI_6) is the longest interval of consecutive failure events:

$$PI_6 = \max_i(t_{f,i}), \quad i\in\{1, 2, \ldots, N_f\} \tag{22.8}$$

- *Maximum vulnerability* (PI_7) describes the magnitude of the most severe failure event:

$$PI_7 = \max_i(s_t), \quad t\in\{1, 2, \ldots, T\} \tag{22.9}$$

22.5 ANALYSES AND RESULTS

Three groups of ten experiments were set up to assess the impact of different fitness (objective) functions on the performance of the developed GA search procedure. The three resulting GA models, each utilizing a different objective function in fitness evaluation, were executed with the same set of model parameters. The preliminary calibration experiments had resulted in the following parameter values:

1. Each of the twenty solution coordinates was represented by an eight-bit-long binary number, thus creating a 160-bit-long string.
2. The number of individuals in a generation was thirty, whereas the maximum number of generations was set to one hundred.
3. Crossover and mutation probabilities were 0.75 and 0.005 respectively.
4. Linear fitness scaling factor, which represents the expected number of offspring the best individual in a generation will get, was set to 2.0.

As stated earlier GAs require a starting population of potential solutions for their search. Therefore, it is essential to repeat the search of a GA several times starting from different initial solutions to test whether the final solutions reached in different runs concentrate around a single point or dissipate over a broader region of the solution space. In that respect, the basis for the comparison of the three models was obtained by creating at random a set of ten initial populations. Each model was subsequently run ten times starting from the predefined initial solutions. The maximum number of generations per GA run was used as a termination criterion for all three models. Table 22.2 presents the ranges of relative contribution of reservoirs toward their respective demands obtained in ten runs of each of the models. Subsequently, the best solution in terms of the respective objective function achievement from each run was appraised by estimating the related performance indicator values. Tables 22.3 through 22.5 display the derived

Table 22.2. *Release distribution variability obtained with different objective functions [%]*

Demand	Reservoir	OF_1	OF_2	OF_3
TU	JO	0.0–27.8	0.0–33.3	0.4–44.3
	BM	0.0–25.9	0.0–48.2	0.0–26.3
	KA	0.0–51.0	0.0–40.8	15.7–55.7
	SS	8.2–73.3	3.1–77.6	16.1–61.2
	SI	0.0	0.0–0.4	0.0–0.4
TO	JO	14.1–76.9	20.4–74.9	13.3–63.1
	SS	20.4–85.9	15.7–71.8	33.3–75.7
	SI	0.0–2.7	0.0–11.0	0.0–11.0
NA, MO, SO, SF	JO	23.1–83.1	18.0–70.2	0.0–54.5
	SS	12.9–76.9	29.8–81.2	40.0–82.7
	SI	0.0–12.5	0.0–16.5	0.0–22.4
IBH	BH	26.7–43.5	26.3–42.7	32.2–43.9
	ME	56.5–73.3	57.3–73.7	56.1–67.8
IAEA	SS	94.9–99.2	87.5–98.4	84.7–98.8
	SI	0.8–5.1	1.6–12.5	1.2–15.3
IBV, IMSC	SS	98.4–100.0	96.9–100.0	98.4–100.0
	SI	0.0–1.6	0.0–3.1	0.0–1.6

Table 22.3. *Performance indicators estimates for the 10 GA runs (Objective function 1)*

No	OF_1	PI_1	PI_2	PI_3	PI_4	PI_5	PI_6	PI_7
1	436,910.3	0.996	0.943	3.3	55.3	2.970	8	6.345
2	436,816.3	0.995	0.934	3.9	54.8	3.178	8	6.552
3	436,785.6	0.994	0.932	4.0	54.7	3.287	8	6.658
4	436,739.8	0.994	0.924	3.6	44.4	3.277	8	6.809
5	436,737.7	0.994	0.921	3.5	40.5	3.057	9	6.598
6	436,731.7	0.994	0.924	3.6	44.4	3.303	8	6.774
7	436,731.4	0.993	0.922	3.4	40.6	3.382	8	10.212
8	436,727.1	0.994	0.921	3.5	40.5	3.196	8	6.787
9	436,709.9	0.993	0.919	3.6	40.4	3.267	8	6.806
10	436,703.1	0.993	0.917	3.4	37.2	3.278	8	6.884

Table 22.4. *Performance indicators estimates for the 10 GA runs (Objective function 2)*

No	OF_2	PI_1	PI_2	PI_3	PI_4	PI_5	PI_6	PI_7
4	1,931,597.3	0.994	0.932	4.0	54.7	3.389	8	6.698
9	1,931,285.0	0.992	0.909	3.2	32.0	3.650	8	7.168
10	1,931,216.1	0.991	0.903	3.4	31.8	3.636	8	7.277
1	1,931,194.1	0.991	0.900	3.5	31.7	3.592	8	7.237
2	1,931,150.9	0.990	0.883	3.6	27.4	3.373	9	9.631
3	1,930,960.8	0.989	0.869	4.3	28.7	3.344	9	7.431
8	1,930,908.0	0.989	0.877	3.6	25.7	3.560	8	7.630
7	1,930,604.1	0.987	0.858	4.4	26.6	3.671	9	7.903
6	1,930,549.5	0.986	0.858	4.4	26.6	3.797	9	7.919
5	1,930,392.9	0.985	0.850	4.6	26.4	3.893	10	8.042

Table 22.5. *Performance indicators estimates for the 10 GA runs (Objective function 3)*

No	OF_3	PI_1	PI_2	PI_3	PI_4	PI_5	PI_6	PI_7
10	0.155	0.995	0.934	3.9	54.8	3.083	8	6.467
3	0.152	0.995	0.936	3.8	54.9	3.251	8	6.578
1	0.146	0.993	0.913	3.5	37.1	3.255	9	6.836
9	0.146	0.993	0.917	3.4	37.2	3.327	8	6.868
2	0.145	0.993	0.917	3.4	37.2	3.241	8	6.881
6	0.145	0.993	0.915	3.5	37.2	3.338	8	6.920
4	0.139	0.990	0.875	4.4	30.8	3.243	10	7.181
8	0.139	0.991	0.903	3.4	31.8	3.621	8	7.207
5	0.138	0.990	0.888	3.5	27.6	3.380	9	7.243
7	0.119	0.984	0.841	4.2	22.2	4.055	9	8.398

*PI*s. Each table is sorted in descending order according to the objective function achievement of the presented alternative solutions.

22.6 DISCUSSION

As Table 22.2 reveals, the common feature of the resulting water allocation distributions derived by the three models is that the obtained solutions vary within rather broad intervals. This indicates that there are multiple water allocation strategies that result in similar objective function achievements.

Clearly, the supply of the demands TU, TO and the aggregate of NA, MO, SO, and SF could be achieved by distributing the water from the associated reservoirs in many different ways. However, one common recommendation emerges out of all of the three models. That is, to maximize its contribution to other of its demand centers the SI reservoir should not allocate any water for the TU demand. It could further be concluded that the BM, KA and especially JO and SS reservoirs possess ample resources to meet the imposed demands.

As to the IBH demand which is jointly supplied by the BH and ME reservoirs, the variability of the derived solutions is of a lower magnitude than in the former case. As expected, due to considerably larger inflows to ME (Table 22.1), all of the alternatives suggest a significant contribution of this reservoir toward the IBH demand.

Further inspection of Table 22.2 shows even less variability of the release distribution between SS and SI toward the IAEA and the aggregate of the IBV and IMSC demands. It should be noted here that the upper limits of the SI reservoir contribution toward the two demands are isolated points associated with the lowest objective function achievements in the respective groups of experiments (i.e., OF_2 and OF_3). In fact, based on the objective function value, the top eight out of ten GA runs using the objective OF_2 derived the relative SI supply toward IAEA at no more than 7.8 percent, whereas the top nine solutions involving OF_3 recommended that this value should not exceed 7.9 percent.

The above general conclusions drawn from the objective function achievements and the ranges of solutions obtained in the three batches of GA experiments cannot, however, shed any light on the issues that are necessary for the appropriate assessment of the operation of water supply systems. In this respect, a more comprehensive comparison of different operating strategies, or water allocation patterns for that matter, can be achieved by appraising the respective simulation results on the expected reliability of the system's operation (both quantity and time related), the expected frequency and magnitude of failures, as well as the magnitude of the most severe failure that may be expected. The selected set of seven performance indicators was therefore estimated for each of the thirty alternative solutions. Tables 22.3 through 22.5 display the PI estimates for the solutions provided by the three models. Each table presents the solutions arranged in descending order according to the respective objective function achievements.

With only few exceptions, the behavior of the PIs estimated in each ten-experiment batch generally meets the expectations. Namely, the PI_1, PI_2, and PI_4 values tend to drop with the decrease of the objective function achievement while, at the same time, the rest of the PIs show a general tendency to increase. The most significant exception to this rule can be seen in fluctuations of the average length of failure intervals (PI_3) observed in results of models that utilized objective functions OF_2 and OF_3. Contrary to the expectations, the lowest values of PI_3 in these two models are not associated with the maximum objective achievement. However, in both cases the best solution exhibits the lowest value of the maximum vulnerability (PI_7) which directly influences the value of OF_2 and is, in fact, the actual value of OF_3.

The overall comparison of the solutions obtained by the three models suggests that, in terms of the respective PI values, the model that utilizes the first objective (OF_1) results almost invariably in a superior system performance than the other two. Except in one case (PI_3 of experiment No. 9 with OF_2) the first model produces the best value of each of the PIs. This may not be too unusual an outcome due to two reasons:

1. According to the definition, the objective OF_1 indirectly pursues the minimization of supply deficit of each of the demands throughout the entire time span, whereas the OF_2 objective concentrates on the deficit on the system level (also over the whole simulation period) and the OF_3 strives toward minimizing the maximum monthly supply deficit of the system completely disregarding both the number and the magnitude of all the other observed shortages.
2. Virtually the entire observed deficit originates from the inability of the SI reservoir to meet the imposed demands (Table 22.6).

Therefore, the objective OF_1 can definitely find the most incentive to reduce the contribution of the SI reservoir toward all of its demand centers that could be supplied from other sources, thus reducing the number and the magnitude of failures in the operation of the system.

As to the best alternatives proposed by each of the models (in terms of the objective function achievement) the use of objective OF_1 led to the overall best performance of the system. The figures in Tables 22.3 through 22.5 obviously reveal that the model which utilizes the OF_1 objective outperforms the other two over all PIs, excluding PI_6, which is identical for all

Table 22.6. *The frequency (count) and the range of the average annual supply deficit of each of the reservoirs observed in the solutions proposed by the three GA models*

res.	OF_1		OF_2		OF_3	
	count	range [$10^6 m^3$]	count	range [$10^6 m^3$]	count	range [$10^6 m^3$]
JO	2	0.015–0.188	0	—	0	—
BM	0	—	1	1.059	0	—
KA	0	—	0	—	0	—
BH	1	0.055	1	0.044	0	—
ME	0	—	0	—	0	—
SS	0	—	0	—	0	—
SI	10	2.025–3.223	10	2.773–6.989	10	2.452–7.741

Table 22.7. *Performance evaluation (on annual scale) of the three best solutions*

	OF_1	OF_2	OF_3
Number of years with observed deficit	9	9	9
Min. number of failure months in a year	1	2	2
Max. number of failure months in a year	7	7	7
Average number of failure months in a year	3.3	4.0	3.9
Total number of failure months	30	36	35
Average annual deficit [$10^6 m^3$]	9.899	13.556	11.988
Max. annual deficit [$10^6 m^3$]	23.571	25.268	24.567

three of them. Similarly, the use of the objective OF_3 resulted in preferred PI values over those of the OF_2 objective function. This conclusion is further supported by a number of performance descriptors estimated on an annual scale for the three best alternatives (Table 22.7). However, it is clear that the three best solutions do not differ significantly. Nevertheless, the PI estimates of these three as well as those of the other alternatives did show that the use of the first objective function resulted in considerably better and more stable PI values throughout all the experiments.

All of the three sets of alternative solutions derived by the three different models exhibited a common feature that a fairly narrow interval of the related objective function achievements was exposed against a broad variability of the respective solution coordinates. Consequently, unless resorting to some additional performance metric(s), a potential decision maker would have been confronted with an extremely difficult task of expressing his/her preference toward a single solution within either of the sets. Therefore, it is believed that the use of the selected set of performance indicators has contributed significantly to a more comprehensive analysis of various aspects of the performance of the proposed alternative water allocation strategies, which was clearly not possible with the help of the respective objective function achievements alone. Furthermore, given the fact that the three objective functions are totally incommensurate, the comparison among solutions from different sets, as well as a preliminary conclusion on the sensitivity of the genetic algorithm model toward different objective criteria, were not possible.

REFERENCES

Babovic, V., Larsen, L. C., asnd Wu, Z. (1994) Calibrating hydrodynamic models by means of simulated evolution. In *Proc. of the First International Conference on Hydroinformatics*, edited by A. Verwey, A. W. Minns, V. Babovic, and C. Maksimovic, Balkema, Rotterdam, 193–200.

Bogardi, J. J. and Verhoef, A. (1995) Reliability analysis of reservoir operation. In *New Uncertainty Concepts in Hydrology and Water Resources*, edited by Z. W. Kundzewicz, UNESCO Water Science Series, No. 3. Cambridge University Press, 306–15.

Cieniawski, S. E., Eheart, J. W., and Ranjithan, S. (1995) Using genetic algorithms to solve a multiobjective groundwater monitoring problem. *Water Resources Research* 31(2): 399–409.

Dandy, G. C., Simpson, A. R., and Murphy, L. J. (1996) An improved genetic algorithm for pipe network optimization. *Water Resources Research* 32(2): 449–58.

Duckstein, L. and Plate, E. J. (1987) *Engineering Reliability and Risk in Water Resources*, NATO ASI Series E: Applied Sciences No. 124. Martinus Nijhoff Publishers.

Esat, V. and Hall, M. J. (1994) Water resources system optimization using genetic algorithms. In *Proc. of the First International Conference on Hydroinformatics*, edited by A. Verwey, A. W. Minns, V. Babovic, and C. Maksimovic, Balkema, Rotterdam, 225–31.

Goldberg, D. E. (1989) *Genetic Algorithms in Search, Optimization, and Machine Learning*. Addison-Wesley, Reading, MA.

Holland, J. H. (1975) *Adaptation in Natural and Artificial Systems*. Ann Arbor: University of Michigan Press.

Milutin, D. (1996) Application of genetic algorithms to water management problems: Deriving reservoir release distribution. In *Water Quality and Environmental Management*, Proceedings of the First International ICER TEMPUS Ph.D. Seminar, edited by G. Lakatos, *Acta Biologica Debrecina, Oecologica Hungarica* 7: 157–70.

Milutin, D. and Bogardi, J. J. (1996a) Application of genetic algorithms to derive the release distribution within a complex reservoir system. In *Hydroinformatics '96*, Proceedings of the Second International Conference on Hydroinformatics, edited by A. Muller, Rotterdam, Balkema, 109–16.

(1996b) Hierarchical versus distributed release allocation within optimization of a multiple-reservoir system operation. In *Proceedings of the International Conference on Aspects of Conflicts in Reservoir Development and Management*, edited by K. Rao, City University, London, UK, 485–94.

23 Risk management for hydraulic systems under hydrological loads

ERICH J. PLATE*

ABSTRACT

The process of risk management conceptualized as a sequence of actions is illustrated by a number of examples of hydraulic and hydrological systems. Starting from hazard identification, one goes on to risk assessment, risk mitigation, and finally to risk management and control. In calculation of risk, use is made of the notion of exceedance probability, failure probability, and of vulnerability assessment. Examples presented range from excessive river bed erosion, through river pollution, to dam safety study. Discussion of the notion of acceptable risk is also offered.

23.1 INTRODUCTION

The design process for hydraulic structures has not changed much in the last decades. The type of structure: canals, and conduits for transport, weirs, dams, and barrages for storing water, and the control elements, such as gates and valves, have been designed with increasing sophistication, taking account of modern technology and a rich store of ideas and experiences that have been gathered over the centuries. But the basic design process has not changed much: One assumes a static or dynamic stationary design load s, and then determines the strength and dimensions of the structures to meet the design, as expressed through a structural resistance r. In terms of decision theory, the design is a vector of decisions $\vec{D} \rightarrow r$. Traditionally, resistance and load are average values $\bar{r}$ and $\bar{s}$ (where $\bar{s}$ might be an "average" extreme load). To account for unforeseen aspects and for uncertainties of the loads, a factor of safety is assigned both to the loads and to the structure. The design dimensions then are set to yield the safety condition:

$$\gamma_s \cdot \bar{s} \leq \frac{1}{\gamma_r} \cdot \bar{r} \tag{23.1}$$

* Professor, University of Karlsruhe, Germany.

where γ_s is the safety factor for the load, and γ_r is the safety factor for the resistance. Typical is a factor of safety of 1.3 for the load, and a factor of safety of 1.3 for the structure, so that the total safety factor is 1.69, or rounded to 1.7.

A major advance in the design theory has been the introduction of the theory of reliability, in which (i) both resistance and loads are considered random variables, and (ii) the design life of the structure is taken into account.

The theory of reliability has the advantage of overcoming a weakness of the conventional practice, where structures are being designed for assumed loads, by means of modern technology and the calculating power of large computers, to an accuracy unheard of in the past, whereas load information is available only very crudely and loads have usually unknown large error bands. The safety factors are supposed to compensate for the uncertainty, but they usually depend only on experience, and not on carefully evaluated measurements of statistical samples of designs. The theory of reliability gives a theoretical foundation to the factor of safety, and it provides a means of assessing the safety of structures on a common basis, through a logical chain of design steps that are outlined in Figure 23.1. It starts with the definition of the failure state, which could be either the failure of the structure so that it has to be replaced or repaired, or it could be the failure of the function of the structure, such as a gate of an emergency outlet that does not open when it is needed. The condition of failure within any one year is associated with a probability of failure $P_F = P\{s \geq r\}$ of the structure or system, which is defined as the region $s \geq r$ in the plain formed by plotting the joint probability density $f_{r,s}(r, s)$ of resistance and load. The design concept consists of designing for a lower P_F than an accepted failure probability P_{Fac}, that is:

$$P_F = P\{s \geq r\} \leq P_{Fac} \tag{23.2}$$

where P_{Fac} is set according to standards. Since loads s and resistances r are random variables, the probability of failure also is a random variable with expected value $E\{P_F\}$ and a probability distribution, which when known provides a means of integrating uncertainty evaluation into the design. The theoretical back-

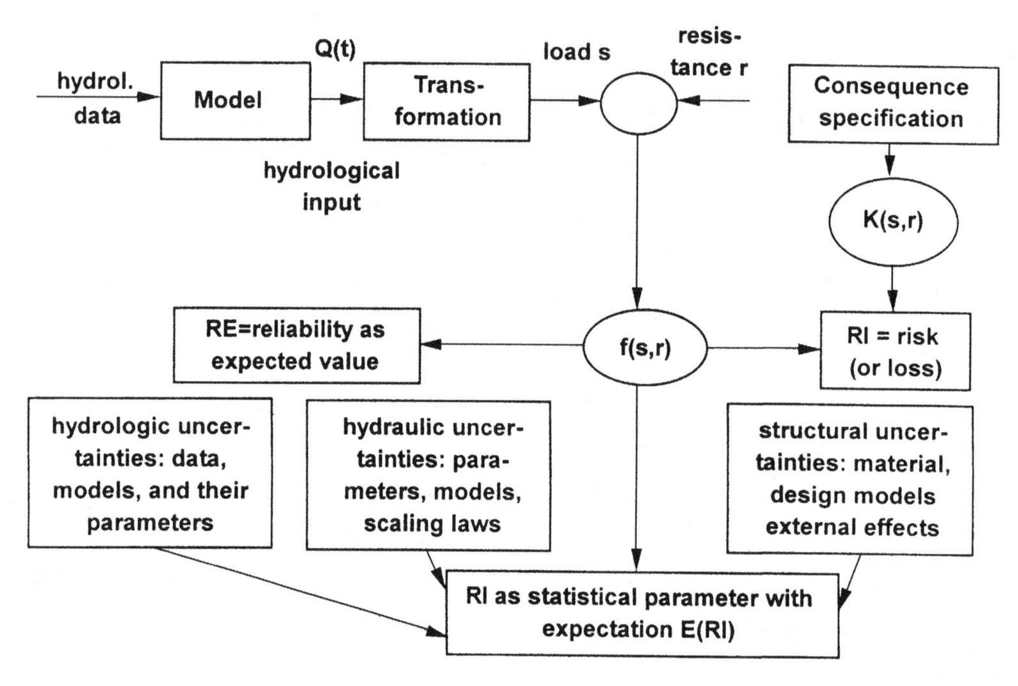

Figure 23.1. Schematics of stochastic design, based on a general concept of reliability. The functions $f_x(x)$ are probability densities, the quantity RI is the generalized risk based on a consequence function K.

ground, and the methods for calculating P_F ranging from second moment analysis to very sophisticated multidimensional probability calculations, have been well developed (see, for example, Ang and Tang 1984; Plate 1993). More elaborate models for calculating P_F use a combination of simulation and deterministic transformations, as shall be illustrated below by two examples taken from the research of the institute of the author.

The stated theory is unsatisfactory because the calculated probability of failure has to be compared with a number P_{Fac} which is also not an objective quantity, but which is set according to a feeling of the consequences of a failure – although it is often simpler to use than a more elaborate risk determination, as shall be discussed in section 23.4. A theoretically better definition of risk is based on a consequence function $K(x|\vec{D})$, where x is the magnitude of the event causing the load s, and $\vec{D}$ is the vector of decisions that influence the (usually adverse) consequences K of any event x. For example, x could be the stage of a river, which causes different kinds of loads on the dike, with consequences in terms of damages, or damage costs, which are incorporated into the consequence function K, and which obviously depend on the decisions $\vec{D}$. In terms of the pdf of the (usually annual) occurrence of x, the expected value of the consequences is:

$$RI(\vec{D}) = \int_{-\infty}^{\infty} K(x|\vec{D}) \cdot f_x(x) \cdot dx \tag{23.3}$$

which in the theory of risk analysis is called the risk.[1] Including the adverse consequences into the design can take the form

[1] Risk as defined in this manner is very different from what is usually called the risk by hydrologists. The connection which exists between these terms is through the concept of reliability (see for example Plate and Duckstein 1988).

The theory of reliability seeks to include the life expectancy of a structure into the design procedure. The failure probability usually is defined as a probability of failing within any one year, which for short lived structures, such as temporary structures designed for preventing a construction site from being flooded, is to be different than for long-lived structures with a design life of many decades. This gives rise (a) to the concept of a failure rate, or hazard function $h(t)$, and (b) to the concept of reliability RE, which is defined as the probability of a structure to not fail during its lifetime, i.e.:

$$RE = 1 - \int_0^{T_D} h(t) \cdot dt$$

The failure probability for a time interval Δt = 1 year, is expressed through $h(t)$ is:

$$P_F = \frac{1}{T} = \int_0^{\Delta t} h(t) \cdot dt$$

with T the recurrence interval, leads for $T_D = n \cdot \Delta t = n$ years to the quantity:

$$RE = 1 - \left(1 - \frac{1}{T}\right)^n$$

which is the hydrological risk.

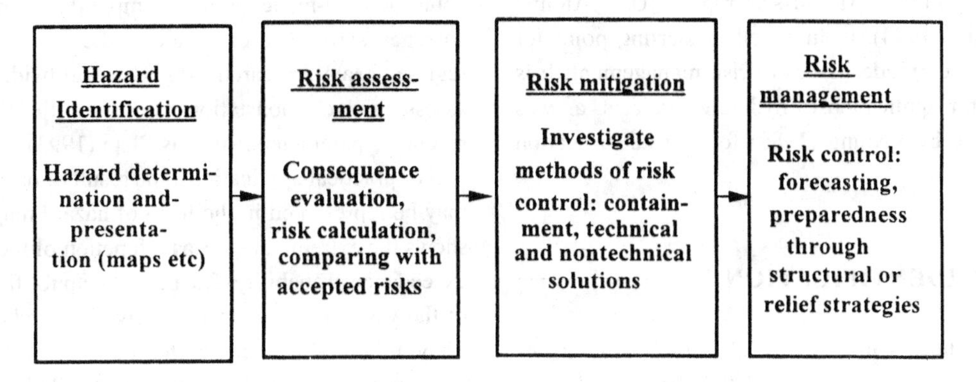

Figure 23.2. The risk management process.

of a decision process, where the decision vector has to be selected according to the condition:

$$RI^* = \min_{\vec{D} \to \vec{D}^*} \left\{ RI\left(\vec{D}\right) \right\} \tag{23.4}$$

that is, the decision is made that minimizes the adverse consequences of the process. This design condition is the basis for risk management. Unlike conventional structures, systems with large hydraulic structures usually include operation rules, and failures are likely not only failures of the structures, but also failures in the purpose of the structure, and minimization of the risk involves not only decisions on the structures and their dimensions to be selected during the design process, but also includes optimization of the operation to avoid failures. An example is the case of a reservoir, which fails if the purpose of the reservoir cannot be satisfied – a failure which is remedied by natural causes, and which therefore has a high resilience: where high resilience is defined as the capability of resuming operation soon after a failure. This type of failure is termed operational failure. The other failure mode is structural failure, in this case failure of a dam which destroys the reservoir. Therefore, the design that minimizes the risk must include both these failure modes, which should be considered as mutually dependent.

It is evident that the calculation of the failure probabilities or of failure consequences is not the end of the process. An even broader approach includes also the operation of the system in the operational failure mode, as well as emergency plans for the case of structural failure. The approach that considers all these aspects is risk management, and it is the purpose of this chapter to give an introduction to this concept as it applies to hydraulic systems which are subjected to hydrological or other random loads.

23.2 THE RISK MANAGEMENT PROCESS

Risk management is a methodology for giving rational consideration to all factors affecting the safety or the operation of large structures or systems of structures. It identifies, evaluates, and executes, in conformity with other social sectors, all aspects of the management of the system, from the identification of loads to the planning of emergency scenarios for the case of operational failure, and of relief and rehabilitation for the case of structural failure. It is a technical and social and also an economic process based on balancing costs and benefits both in monetary and in social terms. An excellent summary of risk management activities as they relate to natural disaster reduction has been given in a manual by the United Nations Disaster Relief Organisation (UNDRO 1991), to which one should also turn for many valuable references.

The process of risk management can be conceptualized as a sequence of actions shown in Figure 23.2. It is a framework for a series of methods to be employed in obtaining a solution for a system, or the operation of a system, that minimizes risk, where risk has to be defined appropriately. Figure 23.2 will be explained in the course of this presentation. Central to this framework is the method of risk analysis, outlined in the introduction, whose roots are found in traditional engineering approaches. Its integration into a management tool is a result of safety studies for technical installations, whose failing could cause large-scale disasters, such as nuclear power plants, or some large chemical plants. As the population is highly threatened by the potential failure of these installations, high-level scientific councils in many countries (for example in Great Britain, Royal Society, 1983) have been charged with evaluating all aspects of risk. They found that risk analysis, as used for example in the Rasmussen report (U.S. Atomic Energy Commission 1974), is an excellent starting point for the discussion of the broader issues of risk management. It is well suited also for application to hydraulic projects, as was shown for example by Vrijling (1989) for coastal protection works.

23.3 HAZARD IDENTIFICATION

The first action indicated in Figure 23.2 is the procedure of hazard determination. According to UNDRO (1991) the hazard is defined as the probability of occurrence associated with an extreme event that can cause a failure. Risk management starts with a procedure of identifying the events which may cause a structural failure, for example, extreme floods, or storm surges in estuaries. But in a broader perspective, a hazard is also a threat to the performance of the system, such as the case of long-duration drought for the case of a reservoir. The next steps are:

1. the determination of the hazards for the event, which usually is defined as the probability of the event x to occur within any one year, and
2. the presentation of the hazards in a useful form, for example in maps.

The events are characterized by their strength, such as the height or volume of a flood, or of a storm surge, or of a drought, and their statistics.

The strength of an event is given by a variable x. It is a variable with the property that the extreme event is rare. Engineers express the probability of occurrence of the rare event by the statistically average time T_x that elapses between consecutive occurrences of events that exceed a certain threshhold x_{max}. T_x is called the recurrence interval, and if T_x is expressed through the number of years, then the quantity $1/T_x$ is the hazard expressed as exceedance probability P_E. For example, floods with recurrence interval of 200 years are termed "200-year flood" and describe a flood that is exceeded once every 200 years on the average.

The traditional way of determining P_E is by doing an extreme value analysis on a statistical record of extreme events x taken, for example, from a time series of discharges. The time series is assumed to be stationary, and the sample of events x is considered a homogeneous sample of a statistical population. Modern approaches are broader and do not stop at assuming stationarity of the time series. One realizes that this assumption is in many cases not justified. Nonstationarities of natural events can have two important causes. They may be due to natural variability, as exemplified by climate changes, or they could be man-made, as for example due to river works, when barrages across, or dikes along the river, change the flood regime. Many research projects worldwide are directed to assessing such nonstationarities and their effect on water resources parameters, such as Plate (1993).

For practical applications, the result of hazard determination may be represented in the form of hazard maps. A hazard map shows the extreme event x as a function of location, for a given exceedance probability. The hazard map for floods gives heights of the water above ground for given exceedance probabilities. They might also include the velocity at that point, which requires the availability of a very detailed 3-D model for the hydraulic flood calculation. Similar maps can be prepared for storm surges showing the height of the surges and include additional information on the height of the maximum waves. The utility of such maps is undisputed, and the Scientific and Technological Committee (STC) for the UN International Decade for Disaster Reduction (IDNDR) therefore rightly gives the development of hazard maps a very high priority. One of the main targets set by the STC for the IDNDR is that all countries should, by the end of the decade, have identified all hazards in the country, and have available maps for them.

23.4 RISK ASSESSMENT

A refined analysis of a hazardous situation gives proper weight to all adverse and positive consequences, and to the severity of effects on property and health of the people affected. This is where the concept of a risk, used in the technical sense, comes into the picture. Risk involves both the recurrence interval and the consequences if the event leading to failure should occur. Therefore, risk assessment is the second step in a risk management procedure.

Risk assessment requires that the consequences of structural or functional failure can be quantified. A formal procedure for calculating the local risk is available (for example, Plate 1993). Analytically, the risk is determined by equation 23.3, where $K(x|\vec{D})$ quantifies the consequence occurring when the event x happens. It depends on the design, which is expressed through the conditional dependency of $K(x|\vec{D})$ on $\vec{D}$.

23.4.1 Examples of risk calculations

To give a first general example, one might consider two decision situations D_1 and D_2 corresponding to the case of (i) building a flood protection dam and (ii) not building the dam.

Obviously, the consequences that would occur during a large flood without the dam are quite different from the case with the dam. There are a number of ways by means of which the consequence function can be expressed. A refined analysis breaks the consequences down into affected units or "elements at risk," which can be objects or persons threatened if the event x occurs. Then K is expressed through the product:

$$K = \sum_{i=1}^{n_0} v_i \cdot k_i \tag{23.5}$$

where the sum is to be taken over all persons or objects affected; n_o is called the "number of elements at risk," v_i is the (specific) vulnerability, and k_i is the extreme consequence happening to the i-th element at risk affected by the event x. From this analysis we obtain the risk formula:

$$RI(\vec{D}) = \int_0^{\infty} n_o(x|\vec{D}) \cdot v(x|\vec{D}) \cdot k(x|\vec{D}) \cdot f_x(x|\vec{D}) \cdot dx \tag{23.6}$$

For the given example, x is a flood event. Two types of risks occur which are to be evaluated. The first risk is the monetary risk, which is equal to the damage costs for all the objects affected by event x. One of these consequence functions expresses the damage to houses in a flood plain for flood event x. In this case, n_o is the number of houses affected, k_i is the damage cost if the i-th house is totally destroyed, or replacement cost, and v_i is the degree of partial destruction of a building, expressed through the percentage of the repair costs to the total cost of replacing the i-th building. The risk depends on the decision that is to be taken. Consider two extreme cases; decision D_1 is to build a large dam, whereas decision D_2 is the decision not to build any dam. In this case, the risk is the expected value of replacement or damage cost, which for decision D_1 has to be seen as a part only of the total costs. The total cost consists of the sum of investment for the dam plus damage cost in the event of dam failure – either as dam storage (i.e., the spillway overflows) or in extreme cases of structural failure. Obviously, it is useful to build the dam if total costs are lower for case D_1 than for case D_2.

But there might be compelling reasons for building the dam, even if it is not cost effective for flood protection. One of these reasons is if the expected loss of human lives for case D_2 is much greater than for case D_1. This is the second type of risk to be considered. For this risk, K is the number of people killed when event x occurs with n_o people affected. If $k_i = k = 1$ happens, the person affected is killed. The vulnerability v_i in this case expresses the probability that a person affected is killed. On the average, the number of persons losing their lives is given by the expected value of K.

For assessing the regional distribution of the risk, hazard maps are useful which are overlaid by the land-use and population information for the area at risk. In the past, such overlays had to be prepared by hand, with careful drawings being superimposed. Nowadays, the use of computer-based Geographic Information Systems (GIS) is well suited for the purpose, as they not only allow the superposition of different maps of the kind required by the analysis, but in conjunction with appropriate data banks they make possible the analytical evaluation of data from different maps.

23.4.2 An example of a risk calculation for risk of life

Consider the case of a risk caused by hazard event x with pdf $f_x(x)$. For each occurrence of the hazard event x, the consequence is expressed by the formula:

$$K(x) = n_o \cdot v(x) \cdot k(n_o) \tag{23.7}$$

where $v(x) = n_x/n_{oo}$ is the vulnerability. The vulnerability is a value obtained from the experience in many different locations, involving many different populations, with a total number of n_{oo} people at risk, of which n_x would suffer the failure consequence if the event x occurs. The vulnerability is distributed over the random variable x. For many practical cases, it is quite sufficient to express the vulnerability by a linear function, as is shown in Figure 23.3.

The linear vulnerability curve expresses that, for an event x which is lower than a threshhold value x_{min}, no failure would occur, whereas for any $x > x_{max}$ failure occurs with certainty, yielding the full consequence k. The maximum consequence could be, for example, the total cost incurred to each of the n_{oo} elements by a failure, or it could be the certainty of losing one's life, in which case $k = 1$. In many practical cases such upper and lower limits can be readily determined, whereas the shape of the curve between the limits might be quite uncertain.

The size of the actual population at risk is assumed to be n_o, which is a random variable with pdf $f_{no}(n_o)$. This pdf is independent of x, and therefore the risk can be written:

$$RI = E\{n_o \cdot k \cdot v(x)\} = E\{n_o\} \cdot E\{v(x)\} \cdot k \tag{23.8}$$

where:

$$E\{v(x)\} = \int_0^{\infty} \frac{n}{n_o}(x) \cdot f_x(x) \cdot dx$$
$$= \int_{x_{min}}^{x_{max}} \frac{x - x_{min}}{x_{max} - x_{min}} \cdot f_x(x) \cdot dx + P_E(x_{max}) \tag{23.9}$$

For the simple case of $x_{min} = x_{max}$, and k = constant (such as is the case if a large arch dam fails) the risk is equal to the product of the number $n_o = \bar{n}_o = E\{n_o\}$, and the exceedance probability for the failure condition $x \geq x_{max}$. One obtains for the risk:

$$RI = \bar{n}_o \cdot k \cdot P_E(x_{max}) \tag{23.10}$$

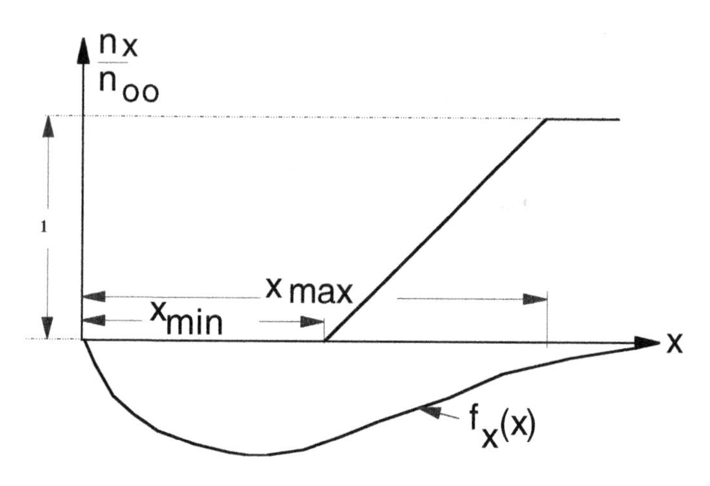

Figure 23.3. Schematic representation of a consequence function with linear vulnerability above a threshold value $x = x_{min}$.

where

$P_E(x_{max}) = \int_{x_{max}}^{\infty} f_x(x) \cdot dx$ is the exceedance probability $P\{x \geq x_{max}\}$.

If we consider the case of expected number N of fatalities of a population of size n_o under the influence of events x, that is, if we assume $k = 1$, then $RI = N$.

Usually it is assumed that n_o is constant and $K(x)$ is a deterministic function of x, that is, it is assumed that for every event x there exists one and only one value of $K(x)$. But actually K is a random variable, because both k and v cannot possibly be given as exact numbers. In the case of equation 23.10, RI is n_o times the expectation of $K(x)$, which implies that with a probability of roughly 50 percent it will be exceeded. The risk in this form may thus not be a very suitable decision criterion for important decisions with far-reaching consequences. However, in principle it presents no problem (apart from the availability of data) for $K(x)$ to be a random variable conditional on x (and of course also on $\vec{D}$), and to let n_o also be a random variable. Another possible way of describing RI is by considering K as a fuzzy number, obtained as the product of two fuzzy numbers.

23.4.3 Examples of risk calculation on the basis of exceedance probabilities

If the assumption is made that the consequences arise only if x exceeds a critical threshold x_{crit}, and that they are independent of the magnitude of x, we obtain the special case where $K(x)$ is zero for $x < x_{crit}$, and constant and equal to K for $x \geq x_{cri}$. In this case equation 23.3 or equation 23.6 reduces to:

$$RI = K \cdot \int_{x_{crit}}^{\infty} f_x(x) \cdot dx = K \cdot P_E \tag{23.11}$$

where P_E is the exceedance probability. This argument can easily be extended also to the case of resistance and load as random variables, in which case the exceedance probability P_E has to be replaced by the failure probability P_F defined according to equation 23.2.

A typical procedure of risk assessment considers risk consequences only implicitly, by defining an acceptable theoretical probability P_{Eacc} of exceedance (assumed equal to the failure case), and a design is considered safe if the condition:

$$P_E\{x \geq x_{crit}\} = \int_{x_{crit}}^{\infty} f_x(x) \cdot dx \leq P_{Eacc} \tag{23.12}$$

is met. In this case, x_{crit} is considered a fixed value.

Example: Failure of the bed of the Danube at the confluence of Danube and Isar

As a first example, we consider an interesting problem that occurred during the planning of the Rhine-Danube canal. It required that the Danube be developed into a navigable channel over a large reach, which included the confluence of the river Danube with the river Isar, as shown in the map of Figure 23.4.

Danube and Isar have quite different hydrological regimes. The Danube has its headwaters in the Black Forest regions, carrying maximum runoff during winter and early spring. The Isar has its headwaters in the Alps, with maximum discharges occur-

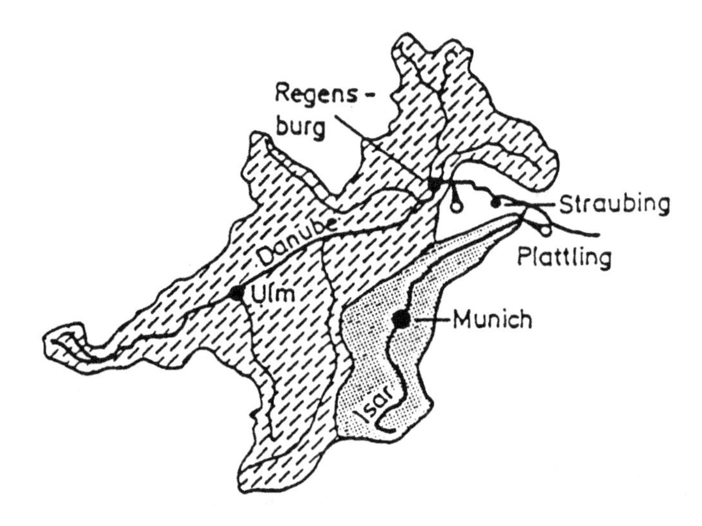

Figure 23.4. Catchments of Danube at gage Regensburg (35,400 km^2) and Isar at gage Plattling (8,839 km^2) in Southern Germany.

ring during snowmelt in summer. Since the Danube is the much larger river, flood discharges from the Isar have only a small effect on the floods in the Danube. Therefore, the Isar flood drops much of its heavy alpine sediment load into the Danube, which has a low flow. This rather coarse sediment forms a fairly thin deposit over a tertiary layer of soft and erodible sediments underneath. The Danube floods partially remove this layer in the winter time. It is important that the coarse deposit is not removed completely, because that would cause very quick and deep erosion which severely affects shipping on the Danube. The problem to be solved is to determine the failure probability for this armoring layer. Failure occurs if the thickness of the gravel layer is reduced to zero, as is shown schematically in Figure 23.5.

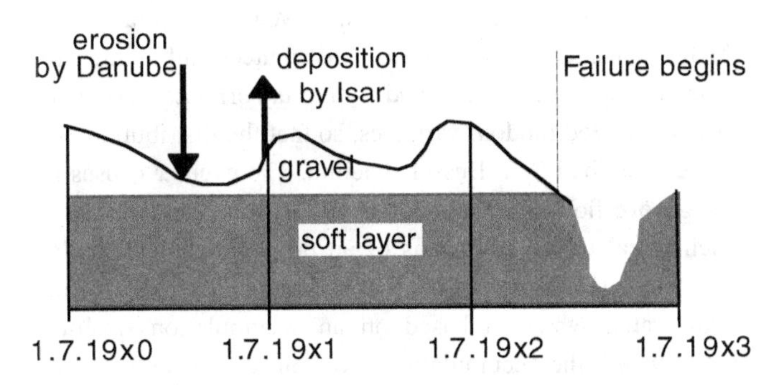

Figure 23.5. Sediment erosion and deposition at the confluence of Danube and Isar (schematic representation).

The method used for solving this problem was stochastic simulation. A stochastic model was used of correlated daily discharges in both rivers, which was based on a basic shot noise model, as is shown schematically in Figure 23.6. This model had been developed by Treiber (1975) for daily discharges on a single river. He had the ingenious idea of using the record of daily discharges itself to generate the shot noise input. This idea was extended by Kron and Plate (1992) to the case of two rivers. The interesting aspect of their model is that the correlation among the two rivers was not established from the two discharges themselves, but by a regression over the input shots of the two rivers. The shots are analogous to an effective areally averaged rainfall, and consequently they were found to be quite well correlated. By means of this model, it was possible to generate a thousand years of daily discharges for the two rivers. Combining the discharges with an appropriate sediment model then allowed the calculation of the failure probability of the river bottom, by counting the number of years, out of the total number of years generated, during which a failure occurred.

Because no accurate sediment erosion model exists that can be used in a long-term simulation, the actual application of the model was, however, not for a long-term sediment erosion and deposition case, but to identify scenarios of joint flood events, which then were analyzed by means of a hydraulic model (Söhngen, Kellerman, and Loy 1992).

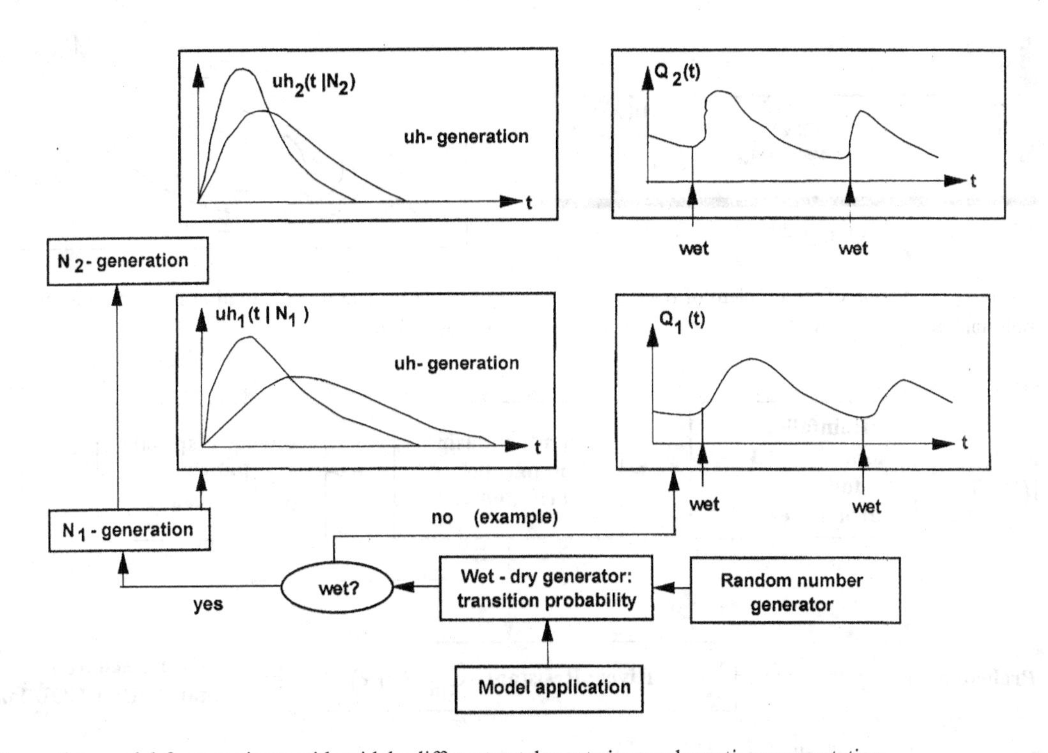

Figure 23.6. Generation model for two rivers with widely different catchment sizes: schematic representation.

Examples of risk calculations by means of failure probabilities

Very many problems cannot be formulated directly in terms of an exceedance probability, because the critical value of x is not a constant. For example, consider the case of the safety of a river against pollution, as shown schematically in Figure 23.7. The water quality of the river is expressed by the critical concentration c_{crit}, and the river is considered to be in the failure state if the actual concentration $c(x) \geq c_{crit}$. However, the pollutant carrying capacity of the river depends also on the discharge $Q(x)$, and the rate of pollutant input $M(x)$, a random variable caused by event x, has to be compared with an acceptable load which also is a random variable, defined as $M_{acc}(x) = c_{crit}(x) \cdot Q(x)$, where $M(x)$ and $Q(x)$ might even be correlated random variables. In a formal way, this problem may be presented in the plane formed by the joint probability density $f_{M,Q}(M, Q)$ of $M(x)$ and $Q(x)$, as shown in Figure 23.8. The failure region is formed by the part of the plane above the limit curve $M_{crit}(x) = c_{crit} \cdot Q(x)$, and the failure probability is the probability associated with that part of the plane, obtained by integrating the jpdf. Standard techniques exist for solving problems of this kind: through different types of second moment analysis (see Ang and Tang 1984; Plate 1993) or by means of simulation.

The problem just described can be extended to the case of the design of a retention basin in an urban system of sanitary sewers. Nonpoint pollution is generated by rainfall events which wash off pollutants that have accumulated during dry periods. This water is often collected in retention basins, whose size is dependent on limited available space inside the sewer system of a city. When the basin cannot hold the storm runoff, it overflows and discharges into receiving waters, in our case into a river. Pollution exceeding a critical level has to be avoided in the river, but since pollutant loads and discharge of receiving waters are random variables, a given probability has to be accepted that critical loads are exceeded. The purpose of our study was to determine the exceedance probability of a critical pollutant load of the river for a given size of a retention basin and a given combination of urban and rural catchment. The situation is shown in Figure 23.9.

Input into this model is a rainfall event x of duration T_D, which is considered as a uniformly distributed rainfall intensity $I(t)$ over a duration $T_D \cdot I(t)$ and T_D are uncorrelated exponentially distributed random variables, so that the distribution of x can be described by a Bessel function. The event x causes an average overflow discharge $Q_E(x, T_D)$ of the retention basin, which is calculated by means of an urban catchment model. Also calculated is the pollutant concentration $c_E(x, T_D)$ of the storm water, which is based on an accumulation wash-off model, where the accumulation takes place at a uniform rate

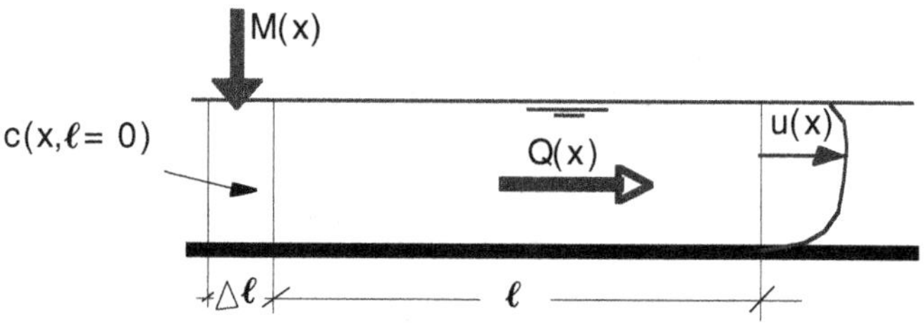

Figure 23.7. Schematic presentation of the problem of river pollution from a pollutant emission at rate $M(x)$.

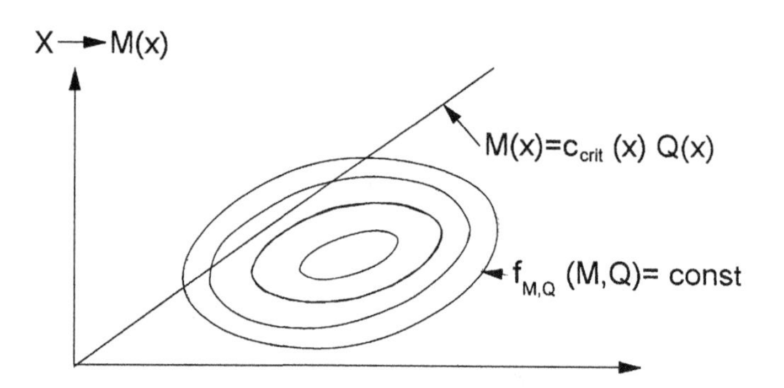

Figure 23.8. The formal solution for finding P_F for the case of river pollution.

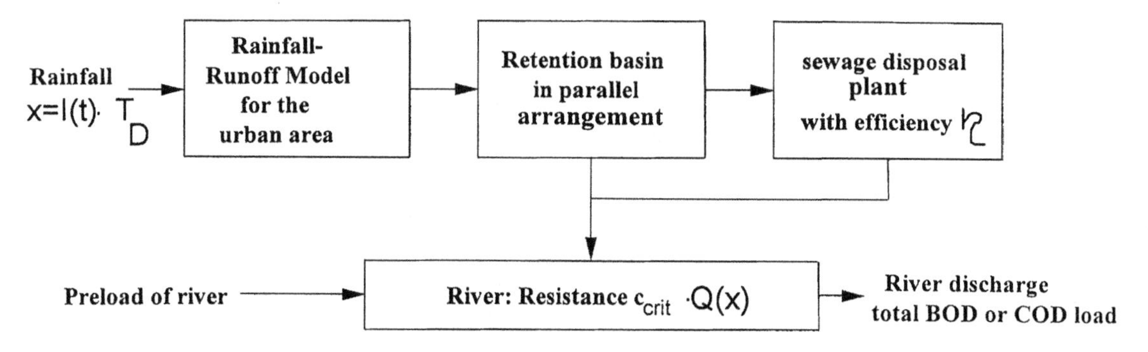

Figure 23.9. Stochastic model of river pollution from urban storm water runoff.

during the dry period, while the pollutant is washed off exponentially during the storm period. Thus, the input into the river is $M(x) = c_E(x) \cdot Q_E(x)$. Further calculations have to be as in the previous example, where the pollutant carrying capacity of the river has been defined as $c_{crit}(x) \cdot Q(x)$, with the quantities without index referring to the river.

The problem has a number of aspects that make it rather difficult to solve. There is the problem of converting rainfall to runoff for each event, and there is the problem of determining the period between rain events, during which the accumulation of pollutants on the urban surfaces takes place. Also, rainfall events and discharges in the river are correlated, because large rainfalls cause storm runoff both in the urban areas and in the river. Therefore, the problem was first solved by means of simulation and for one special case only, but from the result of the study the effect of a number of simplifications could be studied leading to a simplified model, which can be used in a more general case (Schmitt-Heiderich and Plate, 1995).

Example: Dam safety analysis by means of exceedance probabilities

Plate and Meon (1988) considered the safety of a dam containing a reservoir as a problem of determining the failure probability for the dam, according to the condition that the load s exceeds the resistance r. The resistance r was defined as the volume available for flood storage at the beginning of a flood event, and the load s is the volume required to take up the flood volume during extreme storm events, as is indicated in Figure 23.10.

Suitable expressions had to be derived for freeboard, design flood volume as a function of the season, and storage available from not using the design storage of the reservoir at the time that the flood occurs. Also, the overflow over the spillway had

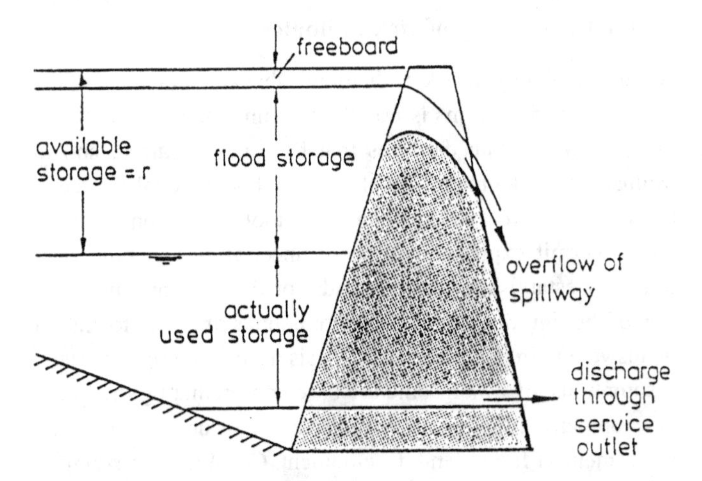

Figure 23.10. The problem of determining the safety of an existing dam against failure due to large floods.

to be quantified, involving the event based calculation of inflow flood volumes by means of suitable models. For details, reference is made to the original papers (Plate and Meon 1988), in which a method is given for calculating the failure probability on the basis of simulations, as well as by a simplified analysis based on a direct integration of the equation for failure probability (for uncorrelated random resistances and loads):

$$P_F(D) = \int_{-\infty}^{\infty} [1 - F_s(r)]_s \cdot f_r(r) \cdot dr \qquad (23.13)$$

where $F_S(r)$ is the probability function (and $1 - F_S(r)$ is the exceedance probability) of the load evaluated at the resistance r.

23.4.4 Risk evaluation

The concepts presented up to this point have been based on the comparison of a calculated risk with an accepted risk RI_{ac}. But what is an accepted risk? Bondi (1985) has pointed out that for large projects such as the storm barrier on the Thames River, the concept of a risk expressed as the product of a very small probability P_E and extremely large consequences K, such as a mathematical risk based on a failure probability of the order of, say 10^{-5} or 10^{-7}, has no meaning, because it can never be verified and defies intuition. Therefore, if money is available for it, why not use it to obtain the best protection that can be had for the available funds? Indeed, if one looks at large disaster mitigation projects, it is found that the major constraint for disaster management is imposed by the available resources.

As an objective approach, the engineer may use a rational risk, which can be determined objectively on the basis of decision theory through the optimum expressed by equation 23.4. However, the difficulties in obtaining the data required for this approach, or the problem of defining an acceptable risk, are reasons why engineers usually do not consider consequences but work on the basis of exceedance probabilities and/or safety factors. Structural engineers prefer design risks specified in terms of failure probability P_F, whereas hydraulic engineers tend to use exceedance probabilities. These figures of merit do not express any real natural conditions, but if they are determined "according to the rules of practice," by an accepted design process, they provide safety standards that are based on experience and that yield a compromise between safety and economics. The advantage of this approach is that by using the same figure of merit for all alternative solutions, it permits comparing them on a common ground.

On the other hand, risk in itself is a meaningless number if it is not compared to a standard, which is the risk accepted by a society or by a decision maker. An acceptable risk in the engineering sense is a number, which is of the same dimension as the risk calculated for the hazard under investigation. For example, if the risk calculated is an expression of the expected monetary losses caused by the disaster, then the acceptable risk also should be a numerical value expressed in dollars. Therefore, exceedance probabilities or failure probabilities become meaningful if they are compared with acceptable values P_{FD} or P_{ED}. For example, along the lower Elbe River and for the city of Hamburg, the dike height selected in the past has been according to the 200-year recurrence interval; with the growing concern over consequences of a storm surge disaster it has recently been changed to 1,000 years.

23.4.5 Definitions of acceptable risks

What may appear to be an acceptable risk in engineering may not be an acceptable risk in the public's perception. For most people, the acceptable risk is not a numerical quantity. Rather, it is the intuitive weight that they give to the impact of a natural hazard. It is a well-known fact that risk perception, for example of the threat of natural disasters, is time dependent. It is felt most strongly directly after the occurrence of a disaster, or a near-disaster. With re-establishment of the ways of life as before the disaster the memory of the impact fades, and with it the will to take action to be protected. The willingness to pay has diminished and other needs are felt more strongly and draw away the resources.

An interesting framework for defining an acceptable risk taking these facts into account has been given by Vrijling et al. (1995), who defined it as the acceptable probability of losing one's life from an action or an event, based on equation 23.13. They consider the case of a design which can protect against events x up to a value $x = x_d$. The design fails, threatening the lives of people, when $x > x_d$. The design is acceptable, at location j, if the probability $P_{Ej}(x_d) = P\{x \geq x_d\}$ is smaller than an accepted value P_{Acc}. Definitely, the acceptance is different for a person than for a nation as a whole, and therefore Vrijling, Van Hengel, and Houben (1995) propose to distinguish between the two. For the nation they propose:

$$P_{Ej}(x_d) \leq P_{Acc} = \frac{10^{-6}/year}{v_{ij}} \tag{23.14}$$

where v_{ij} is the vulnerability of an individual to the event x_i. However, if v_{ij} is assumed constant, then the risk does not reflect the feeling that the risk of large disasters involving many people should be smaller than a risk involving only a few persons. Therefore, the vulnerability should be modified. Vrijling et al. (1995) use a social definition of vulnerability which increases with the number of people affected, given by the formula:

$$v_{ij} = 10^{-3} \cdot n_j^2 \quad \text{for } n \geq 10 \text{ casualties} \tag{23.15}$$

so that:

$$P_{Ej}(x_d) \leq \frac{10^{-3}/year}{n^2} \quad \text{for } n \geq 10 \text{ casualties} \tag{23.16}$$

For the individual, the accepted risk $P_{Epj}(x_d)$ reflects the preferences of the person. A possible way of expressing $P_{Epj}(x_d)$ (according to Vrijling et al. 1995) is by writing:

$$P_{Epj}(x_d) \leq P_{PAcc} = \frac{\beta_i \cdot 10^{-4}/year}{v_{ij}} \tag{23.17}$$

where β_i can range from 0.01 for a high risk for an action that yields no benefit to the person, to 10 for a highly voluntary risk bringing high satisfaction to the person (such as mountaineering).

Obviously, a weak point of a design based on any failure probability is that this probability is a rather arbitrary quantity, one that is difficult to be found from objective criteria. It works well if used by engineers in an agreed upon framework of rules and regulations. For decision processes involving politicians and the public, as is usual for large structures or systems, the importance assigned to failure avoidance is a matter of the value perception of a society, and decision makers first of all use criteria from the locally accepted value system, which may range from a fatalistic view to let things be as they are to a demand to be protected from the remotest possible danger, depending on the attitude of individuals toward risks, which is very subjective and depends on the magnitude of the consequences and on the direct benefit that an individual might derive from taking a risk. If this is not given a priori – and it practically never is – then its value – the "acceptable risk" – has to be determined during the political decision process.

23.4.6 Uncertainty of risk estimates

Another difficulty in risk evaluation is that an assessment of the risk in numerical terms is fraught with uncertainties: it is very difficult to estimate damages that have not occurred, and to evaluate impairments for developments that are to take place in the future. There exist mathematical tools based on statistics which permit expression of these uncertainties in numerical terms by establishing error bounds for the risk, but the question of the significance of such error bounds often has to remain unanswered. In fact, a risk analysis is depending on many assumptions, of which some are based on predicting the future development of the region. A major cause of uncertainty is the assessment of future land development. Consider, for example, the case of providing flood protection through dikes along the river. The higher protection of the diked area attracts people and industry to settle in it, and when an extreme event occurs that overtops the dikes, then the disaster might be larger than if the unprotected but less utilized area had been flooded. In comparison to these uncertainties of human actions it appears that the possible effects of natural variability, such as anticipated climate changes, in most regions are minor – perhaps with the exception of the effect of sea level rise on low-lying countries.

23.5 RISK MITIGATION

The next step in the risk management chain is risk mitigation, in which alternatives for possible disaster mitigation measures are considered and the best solution for the local mitigation strategy is decided. In contrast to hazards imposed by human activities, natural hazards cannot be avoided. People who live in a disaster prone area have no choice but to live with the hazard, but they do not have to accept disasters as an act of fate and hope that a disaster will not strike. However, they must know what hazards exist, and how to cope with them, if possible. Therefore, risk mitigation is based on three parts: development of concepts for risk mitigation, evaluation of the alternatives, and the process of making a decision. Through risk analysis, hazards and the risks associated with them must be quantified first, so that on a rational basis the options available for reducing the impact of a disaster can be assessed. In fact, a risk analysis of the kind described above has its greatest merit in providing a basis for comparing alternative options.

The last part of the risk mitigation process is making the decision on the measures that are to be taken. The problem is one of optimization: to find, among the available alternatives, the one that satisfies the criteria of safety with a minimum of costs, or the one that yields the most protection for the available resources. Because other uses are often in conflict with or support disaster management actions, the problem to be solved analytically would be one of risk reduction optimization for multipurpose projects, with decisions to be made under uncertainties.

Some analytical methods exist for solving such problems, provided that the data and information basis are available, for example, by Vrijling et al. (1995). It has been lamented by a scientific community fond of optimization techniques that these methods are seldom used. The reason for this is that decision makers prefer to make decisions taking into account many criteria, among which the analytical optimum risk solution is but one. Decisions for disaster management usually are not based only on technical criteria, but on intuition and political priorities.

23.6 CONCLUDING REMARKS

The risk management chain combines all the aspects discussed in the previous sections into a chain of planning and implementation. The management of risks does not stop with the planning and the construction of protection measures. It is a process of continuous preparedness against potential failures. In this chain engineers, managers, politicians, and the public are involved. It has parts that are to be decided strictly on the level of experts: engineers, financiers, city planners. The experts can handle the more analytical aspects of their trade, and for these persons, analytical methods of risk analysis are useful. But from the previous discussions of the detriments to an analytical formulation of risk it can be concluded that designs using analytical risks are useful mainly if the frequency of the disasters is rather high, and if many structures of the same type are considered. For large risks, the engineering approach based on safety factors and worst case scenarios is the more accepted approach.

Other parts are decided by the political process, whatever that may be in a given country, and by the availability of funds and the priority given to risk reduction. The risk management chain describes the sequence of steps that are to be taken for disaster reduction measures. The steps must be done by dedicated persons who not only work on disaster mitigation during the planning stage of a project but continuously on all aspects of disaster mitigation. Risk management then is the final link of the chain. It involves the translation of the tasks identified during the risk assessment process into actions to safeguard the functioning of the mitigation measures at all times. It needs dedicated persons, but dedication is not enough: they must be well trained and supported by the necessary infrastructure. It may be said that in Europe the incidence of disasters is comparatively small – not because of the absence of large events, but because of a tradition of mitigating actions extending over many decades, and of preparedness through a well-trained staff and extensive precautionary measures, and last but not least of the willingness of the political decision makers to give risk management the priority in the national goals that it deserves.

The International Decade for Natural Disaster Reduction was established to increase, at all levels of all countries, an awareness for the contribution that disaster reduction activities can make in the development of a country, and to urge the community of nations to exchange information that can help to reach this goal. The concept of risk management can supply the conceptual framework for these activities.

REFERENCES

Ang, A. H. and Tang, W. H. (1984) *Probability Concepts in Engineering Planning and Design*. Vol. 2 John Wiley and Sons, New York.

Bárdossy, A. and Caspary, H. (1991) Detection of climate change in Europe by analysing European circulation patterns from 1881 to 1989. *Journal of Theoretical and Applied Climatology* 42: 155–67.

Bondi, H. (1985) Risk in perspective. In *Risk: Man-Made Hazards to Man*, edited by M. G. Cooper. Oxford University Press, New York.

Health and Safety Executive (1989) *Risk Criteria for Land Use Planning in the Vicinity of Major Industrial Hazards*. HM Stationery Office, London.

Kron, W. and Plate, E. J. (1992) Bed load transport at the confluence of two rivers under hydrologic uncertainty. In *Stochastic Hydraulics '92*, Proc. of the 6th IAHR International Symposium on Stochastic Hydraulics, Taipei.

Plate, E. J. (1993) *Statistik und angewandte Wahrscheinlichkeitslehre für Bauingenieure*. (Statistics and applied probability theory for civil engineers). Ernst und Sohn, Berlin.

Plate, E. J. and Duckstein, L. (1988) Reliability based design concepts in hydraulic engineering. *Water Resources Bulletin*, American Water Resources Association 24: 234–45.

Plate, E. J. and Meon, G. (1988) Stochastic aspects of dam safety analysis. *Proceedings*, *Japan Society of Civil Engineers*, No. 392/II-9 (Hydraulics and Sanitary Engineering), 1–8.

Royal Society (1983) *Risk Assessment: A Study Group Report*. Royal Society, London.

Schmitt-Heiderich, P. and Plate, E. J. (1995) River pollution from urban stormwater runoff prepared for the IHP conference in honor of J. Bernier. *Statistical and Bayesian Methods in Hydrological Sciences*, Paris.

Söhngen, B., Kellermann, J., and Loy, G. (1992) *Modelling of the Danube and Isar River's Morphological Condition – Part 1: Measurement and Formulation Proceedings,* 5th International Symposium on River Sedimentation, Karlsruhe, 1175–207.

Treiber, B. (1975) Ein stochastisches Modell zur Simulation von Tagesabflüssen, Mitteilungen Institut Wasserbau III Universität Karlsruhe, Heft 5, Karlsruhe.

UNDRO Office of the United Nations Disaster Relief Coordinator (1991) *Mitigating Natural Disasters: Phenomena, Effects, and Options*. A Manual for Policy Makers and Planners, United Nations, New York.

U.S. Atomic Energy Commission (1974) Reactor safety study: an assessment of accidental risks in US commercial nuclear power plants. Washington, DC.

Vrijling, J. K. (1989) Developments in the probabilistic design of flood defences in the Netherlands. Proceedings Seminar on the *Reliability of Hydraulic Structures*, XXIII Congress, International Association for Hydraulic Research, Ottawa, 88–138.

Vrijling, J. K., Van Hengel, W., and Houben, R. J. (1995) A framework for risk evaluation. *Journal of Hazardous Materials* 43: 245–61.